D0426767

Laboratory Techniques Series

LABORATORY TECHNIQUES IN CHEMISTRY AND BIOCHEMISTRY

LABORATORY TECHNIQUES IN CHEMISTRY AND BIOCHEMISTRY

P. S. DIAMOND, A.I.S.T., S.R.M.L.T.
Senior Chief Technician
Department of Biochemistry
Institute of Basic Medical Sciences
Royal College of Surgeons of England

R. F. DENMAN, A.I.S.T.
Senior Research Officer
Department of Biochemistry
Imperial College of Science and Technology

A HALSTED PRESS BOOK

JOHN WILEY & SONS
New York – Toronto

English edition first published in 1973 by
Butterworth & Co (Publishers) Ltd
88 Kingsway, London WC2B 6AB

Published in the U.S.A. and Canada by
Halsted Press, a Division of John Wiley & Sons Inc.,
New York

Library of Congress Cataloging in Publication Data

Diamond, Paul Sidney.
Laboratory techniques in chemistry and biochemistry.
(Laboratory techniques series)
"A Halsted Press book."
1. Chemistry—Laboratory manuals. 2. Biological chemistry—Laboratory manuals. I. Denman, Ronald Francis, joint author. II. Title. III. Series.
QD45.D49 1973 540'.28 72-14168
ISBN 0-470-21255-1

Printed in England

FOREWORD TO THE FIRST EDITION

Of our leading scientists today—those who have made substantial contributions to new knowledge or to the practical application of that knowledge—not a few would readily admit that such progress as they have made could hardly have been achieved without the skilled help of capable and expert technicians. Some at least of these scientists would go further and freely give to one or other of their knowledgeable technical assistants the major credit for providing an idea or a technique which enabled a specific discovery or invention to be made.

Till fairly recently, however, the early training of an aspirant to such first-class technical status has been haphazard—largely dependent on whether or not some competent senior person in the laboratory, who has probably acquired the necessary detailed know-how, rarely available in the textbooks, by long-continued hit-or-miss experience, is willing to devote a not inconsiderable part of his time to the practical instruction, at the laboratory bench, of the fledgeling. Such instruction from a senior technician has not infrequently been given to the young graduate, who may have had an excellent training in the more theoretical aspects of his particular discipline during his degree course, but who has yet to learn many of the facts of laboratory technique. Though this type of instruction, usually unplanned and random, has been in the past and still is, of value to the young technician, it rarely provides the background of scientific why and wherefore that is a necessary part of the equipment of the better laboratory assistant today.

In the last few years the basic training of science laboratory technicians has been put on a firmer foundation by a collaborative effort between the City and Guilds of London Institute and the Institute of Science Technology (incorporated in 1954). A systematic course and related examination have been carefully designed for those actually starting, or intending to start, careers as laboratory technicians, and a more advanced course and examination for those who wish to qualify themselves for more senior science laboratory posts. Successful completion of the first course leads to the Science Laboratory Technicians Ordinary Certificate, of the second course to the Advanced Certificate, which is now the normal prerequisite for entry to the Associate Membership of the Institute of Science Technology. Anyone who cares to scrutinize the formidable syllabi for these courses, each covering both a theoretical background and detailed laboratory work on methods and equipment (City and Guilds of London Institute, 119) will realize that a candidate who has succeeded in either, and especially in the more advanced, of these examinations has had an extraordinarily useful training. If also he has the necessary personal qualities, an interesting, varied and, today, a not unrewarding career of real value to the community, is open to him.

The flow into technical officers' posts of such young well-trained individuals, with their ability to handle not only well-established procedures on the laboratory bench, but also to use with full advantage the greatly improved and elaborate apparatus and equipment now available, is already making its very beneficial influence felt in many different types of laboratory.

The present book written, from their wide practical experience in progressive chemical and biochemical laboratories, by two outstanding technical officers, both of whom are also experienced teachers, fills a gap in the modern armamentarium of highly practical instructional manuals. It will be of special value not merely to the student technician (and his instructor!) in chemistry and biochemistry but also to many a young graduate getting down to serious laboratory work in these fields.

H. D. KAY

PREFACE TO THE SECOND EDITION

The first edition of this book was published in 1966. Since then there have been many changes both in the examination syllabus upon which it is based and in laboratory practice in general. We welcome the opportunity to take note of these changes in this second edition.

Apart from bringing many of the methods up-to-date we have introduced descriptions of several techniques which were only in limited use six years ago but which have spread very rapidly since then. Under this heading a chapter on automation in the laboratory has been inserted. In order to include this new and important material and yet to keep the size of the book to reasonable proportions, we have, rather reluctantly, cut out the introductory chapter in the first edition on the basis that a section on basic theoretical chemistry was the most expendable in a textbook on laboratory techniques. We have also omitted the details of suggested class experiments as we feel that most teachers are no longer in need of information of this kind.

The introduction of SI units has necessitated the changing of all the symbols and units in the book to conform with the new system.

We are most grateful to all those who wrote to us on the publication of the first edition either to express appreciation, suggest amendments, or point out errors. We hope that the suggestions have been acted upon and the errors expunged.

Finally we must again thank all those who have aided us so much in the preparation of this new edition, especially our publishers who have been so patient and helpful during its production.

PREFACE TO THE FIRST EDITION

Over the last quarter of a century the position of the laboratory technician has changed very radically until, today, the qualified technician is an important member of the laboratory team. Very largely this change is due to the setting up of professional bodies whose examinations have set recognized standards for qualified technical staff in hospitals, university laboratories, and industry.

Although the City and Guilds of London Institute (C.G.L.I.) and the Institute of Science Technology (I.S.T.) have been setting examinations for non-medical technicians for some years, no textbooks have been available to date which are primarily based on their syllabi and the main purpose of this book is to fill this gap as far as techniques in chemistry and biochemistry are concerned.

When we began writing this book the C.G.L.I. syllabi for chemistry and biochemistry were separate but seemed to us to overlap about 60 per cent. Since then new syllabi have been introduced in which students of these two subjects follow a common syllabus for two years and diverge to more specialized topics only in the final year of the advanced certificate course. It was for this reason that it was decided to combine the two subjects in one volume.

Although we have attempted to include most of the techniques required, some items have been omitted, either because of shortage of space, or because we feel that information is readily available in many other books. The new syllabus includes a fairly large amount of theoretical chemistry of the type that is found in the standard texts used in schools and colleges. We have included some theoretical discussion but only where this applies to the practical techniques described. There is no section on the care of laboratory animals as we feel that this is a specialized topic which has been well covered by experts in this field. No inorganic 'analysis tables' have been included, as most teachers have quite firm ideas about this section of the syllabus and we feel that it is unnecessary to add to the large number of analytical schemes already published. As far as possible we have included methods which give the student an opportunity to practise a wide variety of the manipulative skills which are in use in the working laboratory.

Although this book is based on a specific examination syllabus we hope that it will be helpful to those who enter a working laboratory after the completion of an academic training course. The apparatus and techniques of the modern working laboratory are often very different from those of the teaching laboratory, and, as instrumentation becomes more complex and more expensive, the difference is becoming even more apparent. For those whose academic knowledge is deep but whose practical experience is limited we hope that this will be a useful introduction to the techniques of the modern laboratory.

Finally we must thank the many people who have helped us so much in the writing of this book, our publishers for their help and great patience, our colleagues at Twyford Laboratories Ltd. and St. Mary's Hospital Medical School for their interest and encouragement, our colleagues and students at Paddington Technical College who allowed us to try out much of the manuscript, to our wives for their patience during a long and difficult gestation and to Messrs. F. Grover and J. H. Glen for their thorough and critical reading of the proofs.

Finally our special thanks to Professor H. Kay for his kindness in reading the proofs and writing the Foreword.

CONTENTS

CONTENTS

CONTENTS

CHAPTER 1

BASIC MATERIALS AND METHODS

The casual visitor to a chemistry laboratory would soon notice that although chemical apparatus comes in an enormous variety of shapes and sizes, it is, on the whole, constructed from a very small number of basic materials. The most common of these materials is, of course, glass. Some apparatus will be seen made of glazed porcelain. In analytical laboratories fused silica crucibles and tubing are used. Metal vessels are useful in some techniques. Plastics have been found suitable for some purposes and have become more common, although plastic apparatus has some limitations. Rubber is, of course, used almost universally for connecting tubing and stoppers.

Before describing specific types of apparatus it would be useful to discuss the properties of these materials so that the purpose of any piece of equipment can be related to the substance from which it is made.

GLASS

Glass is an eminently suitable material for scientific apparatus. Its advantages are many:

(*a*) It is extremely inert and resists attack from almost all chemicals.
(*b*) It is transparent, and changes of colour, consistency and volume are easily observed.
(*c*) It withstands considerable variations in temperature without breakage or distortion.
(*d*) It is easily worked with simple tools.
(*e*) Because it is inert and smooth it is easy to clean, and because it is transparent it can be seen to be clean.
(*f*) Glass has a low density and apparatus is comparatively light in weight.
(*g*) It is relatively cheap.

Against these advantages two main disadvantages can be set:

(*a*) Glass is brittle and must be handled with care, although it must be emphasized that breakage is more often the fault of the worker than of the glass. Scratches and abrasions on the polished surface of glass greatly lower the resistance of the apparatus to the mechanical shock of knocks and bounces. Careful handling is thus doubly important.

(*b*) Glass is attacked quite vigorously by hydrofluoric and glacial phosphoric acids and salts of the latter. Caustic alkalis etch glass noticeably, and even water can be shown to attack glass-ware very slowly.

Laboratory glass-ware is mostly made from two distinct types of glass which have quite noticeably different characteristics. These are soda ('soft') glass and borosilicate ('hard') glass (also often called by the manufacturers' trade name, e.g. 'Pyrex', 'Hysil', 'Davisil', 'Phoenix', etc.). Soda glass, called 'soft' because it has a lower melting point than 'hard' glass, is used less and less in working laboratories nowadays, although it is still quite common in teaching laboratories. Some moulded glass-ware which does not have to be heated and which is not graduated (e.g. filter funnels, reagent bottles) is still usually made of soda glass, but as the difference in price between the two types of glass narrows, soda glass is gradually disappearing from the laboratory.

Soda glass is of rather indeterminate composition and is less resistant to attack, more difficult to work, and more liable to crack when heated or cooled than borosilicate glass.

The manufacturers have given exhaustive information about borosilicate glass. Perhaps the widest range of apparatus in borosilicate is manufactured by James A. Jobling & Co. Ltd., who make 'Pyrex'. Some of the properties of this popular material are given in a published leaflet (Pyrex Glass No. 6., James A. Jobling and Co.).

Distilled water at 100°C for 6h results in a loss in weight of 0·01 mg/100 cm^2 of glass surface.

Boiling normal sodium carbonate/sodium hydroxide for 3h causes a loss in weight of 100/120 mg/100 cm^2 glass surface.

Between 0–300°C the coefficient of linear expansion of Pyrex is 33×10^{-7}/°C. A 'Pyrex' 25 cm^3 beaker may be heated in an oven to 180°C and plunged into water at 20°C with negligible risk of breakage from thermal shock.

It would be difficult to envisage any material (except perhaps platinum) which could be so successful in providing the vast range of apparatus used in the laboratory.

GLAZED PORCELAIN

Glazed porcelain is mainly used for funnels and crucibles. Its use for funnels is becoming less common as modern techniques have produced glass and sintered glass funnels which are just as efficient and, because they are transparent, much easier to clean. This will be discussed more fully hereunder.

Porcelain can be heated to a much higher temperature than glass, and porcelain crucibles can be heated in a furnace to red heat without distorting and retain a constant weight. The glaze tends to be rather susceptible to attack at high temperatures, however, and porcelain crucibles are not used in the most accurate analytical work. They have the advantage of being comparatively cheap, and suitable for class work and where the highest accuracy is not required.

FUSED SILICA

Fused silica is found in either translucent or transparent forms, the the latter being the more expensive. It is employed where very high temperatures are used and is exceptionally resistant to thermal shock. It may be cooled under the cold water tap when red hot with no risk of fracture. Combustion tubes and crucibles in analytical laboratories are frequently made of fused silica. A fairly wide range of flasks and beakers is also obtainable in this material ('Vitreosil') although they are only used when absolutely necessary because they are very expensive. Fused silica cuvettes are used for colorimetric work where transparency to light of low wavelength is necessary (see Chapter 6).

METALS

Besides the retort stands, clamps, bosses, etc. which are used for setting up chemical apparatus, metals are sometimes used for chemical vessels. Iron and nickel crucibles are used for high temperature oxidation and ashing where acids are not used. Stainless steel beakers are inert to alkalis and are used for some purposes. Perhaps the most important metal-ware is made of platinum. This is the most inert material available and is attacked only by lead and mercury at high temperatures. Before the first World War platinum was relatively inexpensive and apparatus in this metal was quite common. Today practically the only use for platinum is in platinum crucibles. These are employed for the most accurate analytical work, for instance, in industrial laboratories where the analytical method to be employed is laid down by some official body (e.g. the British Standards Institute) and where the use of platinum crucibles is specified.

Platinum wire is used in analysis for flame tests and in electrochemistry to form electrodes.

PLASTICS AND RUBBER

The chief properties of the plastics used to construct chemical apparatus are shown in Table 1.1. Methacrylate resin ('Perspex') is easily joined using a chloroform solution as an adhesive. Very little apparatus is made commercially in this material but it is used in the laboratory for the construction of tanks to hold aqueous solutions especially for electrophoresis (see Chapter 7) and for museum jars. It may be purchased in sheets of thickness from 1 mm and may be cut and polished without difficulty. New material is glass-clear but the surface is easily scratched and may become cloudy after a time through surface abrasion. It is soluble in many organic solvents especially chloroform and carbon tetrachloride.

Polyvinyl chloride (P.V.C.) is used almost exclusively as plastic tubing ('Portex', 'Tygon'). It is excellent for many laboratory purposes being almost completely inert. However, it softens at a fairly low temperature and is of little use in the transfer of hot solutions or steam. P.V.C. hardens on ageing.

Polystyrene was one of the earlier plastics used in the laboratory. It is transparent but very brittle and attacked by many solvents and has never been much used except for small boxes and pots for storage.

Polyethylene ('Polythene', 'Alkathene') is common and useful laboratory plastic. It is almost completely inert although some workers claim that some common solvents can be very slightly contaminated by prolonged storage in polythene containers. A very wide range of laboratory-ware is available in this plastic including volumetric apparatus, although this is of low accuracy compared with glass apparatus. For very large containers it is invaluable being much less dense than glass and mechanically almost indestructible. Its only disadvantages are its low melting point making it of little use for hot solutions or reactions where heat may be evolved, and its translucency which makes it less useful than glass for many purposes.

Polypropylene ('Polyprene') is of more recent origin. It has gradually taken the place of polyethylene in many manufacturers' catalogues having very similar properties to polythene, but has a somewhat higher melting point (standing boiling water without undue softening) and is slightly more transparent.

Nylon is used in the form of tubing and tube connections. Nylon

TABLE 1.1. Properties of Laboratory Plastics

	POLYTHENE *Low Density*	*POLYTHENE* *High Density*	*POLY-PROPYLENE*	*P.V.C.* *Polyvinyl chloride*	*P.T.F.E.* *Polytetrafluoro-ethylene*	*NYLON* *Polyamides*	*ACRYLIC* *Polymethyl methacrylate* *(Perspex)*	*Polycarbonate*
HEAT RESISTANCE								
Short periods	90°C	115°C	145°C	75°C	290°C	145°C	85°C	140°C
Continuous	80°C	100°C	120°C	65°C	260°C	120°C	75°C	125°C
Effect of:								
Weak acids	Resistant	Very resistant	Very resistant	None	None	Resistant	Very resistant	Resistant
Strong acids	Good but attacked slowly by oxidizing acids	Good but attacked slowly by oxidizing acids	Good but attacked slowly by oxidizing acids	Resistant	None	Attacked	Attacked slowly by oxidizing acids	Fairly resistant
Weak alkalies	Very resistant	Very resistant	None	Resistant	None	Resistant	Very resistant	Fairly resistant
Strong alkalies	Resistant	Very resistant	Very resistant	Resistant	None	Fairly resistant	Very resistant	Attacked
Organic solvents	Fairly resistant below 60°C	Fairly resistant below 80°C	Very resistant below 80°C	Soluble in ketones, esters and aromatic hydrocarbons	None	Very resistant	Soluble in ketones, esters and aromatic hydrocarbons	Attacked
Sunlight	Some discoloration and crazing	Some discoloration and crazing	Some discoloration and crazing	Slight	None	Slight	Slight	None
Burning rate	Slow	Slow	Slow	Self extinguishing	None	Self extinguishing	Slow	Self-extinguishing
Clarity	Translucent	Translucent to opaque	Translucent to transparent	Transparent or opaque	Opaque	Translucent to opaque	Transparent	Transparent
Usual colour	Cloudy white	Cloudy white	Cloudy white	Clear or coloured	Dense white	Cloudy white	Clear or coloured	Slight yellow tinge

test-tubes and centrifuge tubes are useful for some purposes. Nylon is the only one of the cheaper plastics which can be reliably sterilized with pressurized steam and is in common use in hospital laboratories where sterility is important.

Polytetrafluoroethylene (P.T.F.E., teflon) could be the most useful of laboratory plastics. It is extremely inert and withstands high temperatures. For many purposes it replaces platinum in apparatus for standard analytical techniques. It is, however, much too expensive for everyday use. A recently introduced material, similar to P.T.F.E. is Fluorinated Ethylene Propylene (F.E.P.). This has similar properties to P.T.F.E. except that it has a lower heat resistance (200–250°). It is transparent, however, which is a major advantage.

Rubber

Rubber tubing and rubber stoppers are used in every laboratory. Rubber is available in many qualities and good quality red rubber with high elasticity is suitable for most purposes. Steam and high temperatures affect rubber, although the effect lessens after frequent use. Neoprene is a synthetic rubber less elastic but more inert than the natural material. Silicone 'rubber' is very inert, withstands high temperatures, and can be obtained with high elasticity. It is rather expensive.

APPARATUS

Most of the apparatus supplied by the large laboratory furnishers is now standardized to conform to designs drawn up by the British Standards Institute (B.S.I.). These designs are vetted before publication by working chemists. This excellent system ensures that practically any apparatus ordered from laboratory supply houses in this country can be guaranteed to conform to the *British Standard Specification* and is suitable for the purpose for which it is intended. In the U.S.A. similar organizations, The National Bureau of Standards and the A.S.T.M., have a parallel function, although these cover an even wider field than their British counterpart.

Much of the apparatus of chemistry will be described in later chapters where it will be discussed in relation to the processes in which it is used. There are, however, some very basic pieces of apparatus which are of such universal use that they may be considered before specific techniques are discussed. These are the general purposes containers used in all laboratory processes.

Test-tubes

These require little description. They are simply cylindrical tubes closed at one end. The open end usually has a rim. Rimless tubes are often labelled as 'bacteriological test-tubes'. The smallest sizes are called ignition tubes, the larger sizes boiling tubes. When buying test-tubes the purchaser should ensure that the closed end is a smooth hemisphere and the wall thickness constant throughout. Tubes which do not have these characteristics will tend to crack when heated. It must be emphasized that the end of a test-tube should never be heated in a flame. The flame should be applied to the straight portion just above the round end. A test-tube which is to be heated should not be more than one third full of liquid. It should be held in a test-tube holder at an angle of 45 degrees and gently and constantly shaken while the flame is applied. The open end should not point at the operator—or anybody else.

Beakers

Various types of beakers are shown in *Figure 1.1*. In hard glass they are available in sizes from 5 cm^3 to 5 dm^3. In polythene, sizes

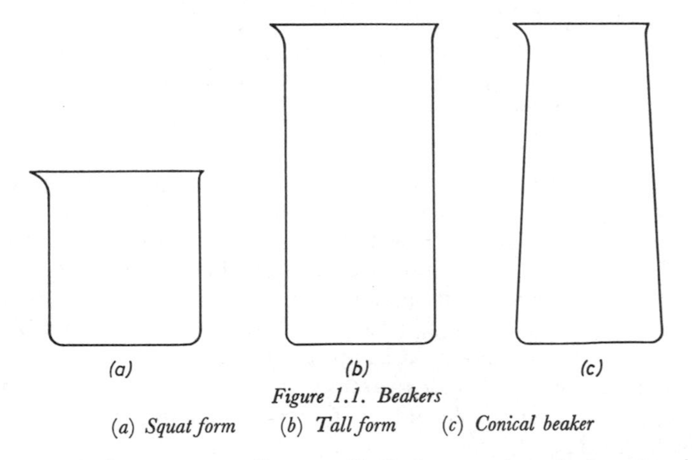

Figure 1.1. Beakers

(*a*) *Squat form* (*b*) *Tall form* (*c*) *Conical beaker*

up to 2 dm^3 are normally supplied, larger sizes being specially constructed to order. Porcelain, metal, enamelled metal and silica beakers are used in some laboratories for special purposes. The spout is not only used for pouring but allows a stirring rod or thermometer to remain immersed in the contents of a beaker when it is covered by a clock glass.

The beaker is a universal container in the laboratory used for all the routine mixing and dissolving jobs which the working chemist performs in his day-to-day work.

Flasks

There are three main types of flask (*Figure 1.2*) all made in borosilicate glass. The round-bottomed flask is used for much routine heating of liquids. The long neck acts as a condenser, or

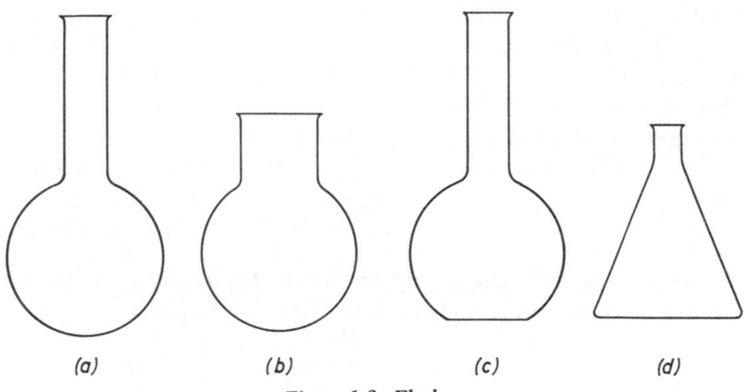

Figure 1.2. Flasks

(*a*) *Round bottomed* (*c*) *Flat bottomed*
(*b*) *Bolthead round bottomed* (*d*) *Conical* (*Erlenmeyer*)

a water-cooled condenser may be fitted to the neck. (See Chapter 2, Extraction Techniques.) The round-bottomed flask with a short wide neck is called a bolthead flask, and is used when several attachments have to be inserted through the stopper (e.g. a condenser, stirrer and thermometer all in one flask). Alternatively, multinecked flasks are available. Deep cork rings are available for standing round-bottomed flasks on the bench.

Flat-bottomed flasks (extraction flasks) are used for simple extraction of solids by liquids in the cold, and in the construction of wash bottles.

Erlenmeyer (conical) flasks are useful because the lower part holds most of the volume of the vessel. Because of this they are used as general purpose flasks in the laboratory. When full the conical flask has a low centre of gravity and with its broad base it is stable and difficult to tip over accidentally. It has 'corners' and is more likely to crack when heated.

Distillation and filtration flasks will be described under the appropriate headings in Chapter 2.

Measuring Apparatus

In 1970 the International Union of Pure and Applied Chemistry introduced a new and definitive ruling on the symbols and terminology to be used for describing quantities and units in chemistry. This is based upon the International System of Units (SI units) which has been adopted by the Conférence Générale des Poids et Mesures.

In consequence much of the familiar terminology of chemistry has been changed and many experienced workers are finding it difficult to adapt to the new expressions. At a time of flux there is bound to be some confusion and I.U.P.A.C. have been realistic in not insisting on the summary banishment of familiar but obsolescent terms. The definitive publication on the new units (Manual of Symbols and Terminology for Physicochemical Quantities and Units, Butterworths 1970) says of the older terms: 'These units do not belong to the International System of Units and their use is to be progressively discouraged. The time-scale implied by the word "progressively" need not, however, be the same for all these units, nor for any unit need it be the same in all fields of science.' It is clear that where, for practical purposes, the amount involved in the unit has remained the same, only the symbol having been changed, the obsolescent term will disappear fairly quickly. For example the units of volume have not changed, merely their description. Where, however, the new unit involves a completely different amount, as in the measurement of pressure, it will take some time for all the related devices to be rescaled and the old units will be in use for several years.

In this edition the authors have given quantities in SI units but where it is felt that these are not yet sufficiently common or where there may be confusion when using established apparatus the quantity is restated in brackets using the old units.

The unit of volume to be employed in the future is the cubic metre (m^3). This would be equivalent to 1000 litres on the old scale and is not, therefore, a practical laboratory measure. The millilitre is now referred to as the *cubic centimetre* or *centimetre cubed* (cm^3) and the litre as the *cubic decimetre* or decimetre cubed (dm^3). The old relationship of

$$1\,000 \text{ millilitres (ml)} \equiv 1 \text{ litre (l) now becomes}$$
$$1\,000 \text{ cm}^3 \equiv 1 \text{ dm}^3$$

Although the old units are not exactly equivalent to the new, they may be regarded as interchangeable for all practical purposes.

Most volumetric apparatus is still being supplied by the manufacturers bearing the old symbols.

All measuring apparatus is calibrated to contain or to deliver a stated volume of liquid in cubic centimetres at a stated temperature. Volumetric apparatus is calibrated for use at 20°C although the error at temperatures between 15°C and 25°C is less than 0·1 per cent which is well within the error due to other factors.

Besides the B.S.I. standards for volumetric glass-ware the National Physical Laboratory has laid down standards of accuracy. 'Grade B' apparatus is of sufficient accuracy for most laboratory work, but for the most accurate work 'Grade A' apparatus is used. For very special analytical procedures each piece of 'Grade A' apparatus can be provided with a certificate of calibration by the N.P.L. This is a very expensive but often very necessary procedure. Somewhat cheaper, but still quite expensive, is the 'Works Calibration Certificate' which some manufacturers are willing to supply with 'Grade A' apparatus.

The main types of volumetric apparatus together with methods of calibration and techniques for their use are discussed in Chapter 4.

Bungs and Tubing

The setting up of apparatus often requires that it should be fitted together in some way. Traditionally the various pieces of equipment are joined by means of rubber bungs or stoppers. These are available in a large number of standard sizes from 5 to 120 mm in diameter. As an example of fitting apparatus together with stoppers, the joining together of two pieces of glass tubing of different diameters will be considered.

A bung is chosen which fits snugly into the wider tube for about two-thirds of its length. A cork borer of the largest diameter which will not slip over the narrower tube is sharpened inside and out with a cork borer sharpener. Blunt cork borers are dangerous and should *never* be used. The end of the sharpened borer is moistened with a little glycerol and pushed into the centre of the bung with a gentle rotatory movement. The bung is held firmly against the side of the bench or some other convenient vertical surface. When the borer has penetrated about half way it is removed and a hole bored from the other end until the two cuts meet. This ensures that both ends of the hole are smooth and clean. The narrower tube is carefully inserted in the hole and the bung pushed into the wider tube. Glycerol is a good lubricant as it can easily be rinsed off with plain water.

Rubber of good quality is an excellent material for many purposes. New rubber may suffer surface attack from heat or reagents, either of which may cause contamination. Before use new rubber tubing must be washed free of the powdered talc which is inserted in manufacture to prevent sticking. The surface of rubber ages in use and becomes more resistant. Many experienced workers prefer to use old well-used rubber bungs, and carefully guard private stocks which have been aged by their colleagues. For some purposes it is necessary to artificially age rubber bungs and special procedures have been described for this. (See carbon and hydrogen estimations, Chapter 5.)

Standard Glass Joints

During the last 20 years or so the use of rubber bungs and tubing for joining pieces of glass apparatus has gradually been superseded by the use of apparatus having standard glass joints, and now practically all the standard laboratory set-ups can be constructed entirely of standard pieces of glass apparatus.

The most common type of standard glass joint is the cone and socket. A standard glass cone externally ground fits snugly into a standard socket which is ground on the inner surface to form a leakproof and rigid joint. Cones and sockets are fitted to all types of apparatus from flasks to flowmeters, from condensers to crucibles, so that a vast number of assemblies can be constructed entirely in glass. The most famous trade name in interchangeable laboratory glass-ware is 'Quickfit'. The 'Quickfit' catalogue lists over one thousand individual items. Typical assemblies are shown in *Figure 1.3*.

The dimensions of conical ground joints were the subject of a British Standard Specification (B.S.S. 572, 1950) which stated that the taper on the cone should be 1 in 10 on the diameter, i.e. the semi-angle of the cone is $2°51'45''$. Three series of joints are available distinguished by the prefix letters B, C or D, representing the length of engagement of the ground zone in descending order.

The International Organization for Standardization has recommended that the label on a ground joint should state:

(*a*) The series prefix letter.
(*b*) The diameter of the ground section at the wide end in millimetres.
(*c*) The length of engagement of the ground section in millimetres.

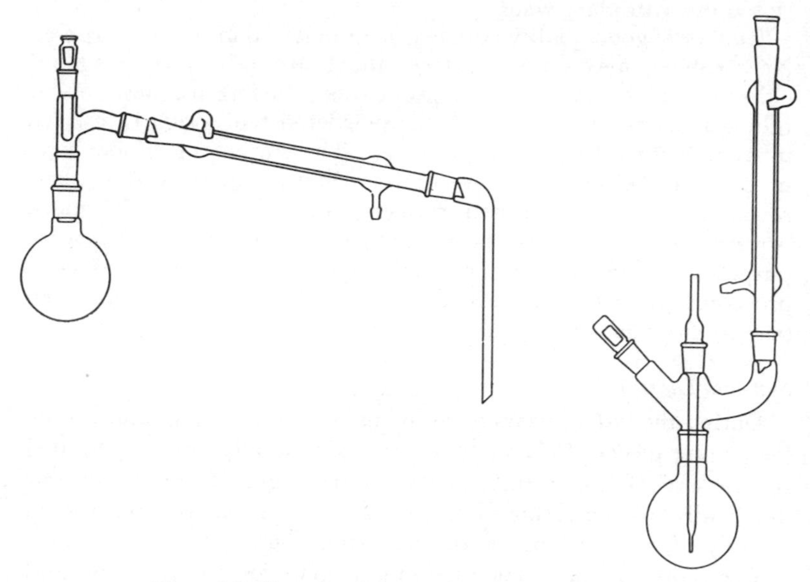

Figure 1.3. Examples of the use of standard jointed equipment

A joint labelled B29/32 is of the B series, has a nominal diameter at the wide end of 29 mm and a nominal length of engagement of 32 mm. To illustrate the significance of the series letters the three joints of 24 mm diameter are B24/29, C24/20, D24/10.

All interchangeable cones and sockets are now made in borosilicate glass and range in size from B5/13 to B55/44.

In 1967, Quickfit announced a further improvement in standard jointed equipment with the introduction of their 'Clearfit' range. The company now produces a range of equipment in which the cones and sockets are not ground but whose precision is such that satisfactory joints can be made with unground transparent glass shapes.

The advantages of Clearfit are said to be:

(i) It is usually unnecessary to lubricate the joints.
(ii) There is less danger of joints seizing.
(iii) The smooth clear glass of the joints is easier to clean and to protect from contamination than the rough ground surface.
(iv) Clearfit joints give a better vacuum seal.

(v) There is less tendency for vapours to 'creep' through the joints.
(vi) The weakening of the glass by grinding is avoided and Clearfit joints are mechanically stronger and less liable to crack.
(vii) The transparency of the joints is an advantage when thermometers are used as the whole of the scale is visible.

All Clearfit joints are compatable with standard ground joints. They are appreciably more expensive than the latter.

Another type of ground glass joint is the spherical (ball and cup) joint (*Figure 1.4a*). This has the advantage of introducing some degree of flexibility into an assembly. Ball and cup joints are not so liable to stick or 'freeze' as sometimes happens with conical joints. Special clips are needed to hold the two halves of the joint together. Conical-spherical adapters are available to add this type to the conical system. The more common sizes of ball and cup joints are shown in Table 1.2.

For some purposes flat flange joints are very useful. Besides their use in glass pipelines they are most often found in reaction vessels where they can form a standard flask which can be opened and thoroughly cleaned or emptied of solid material (*Figure 1.4b*).

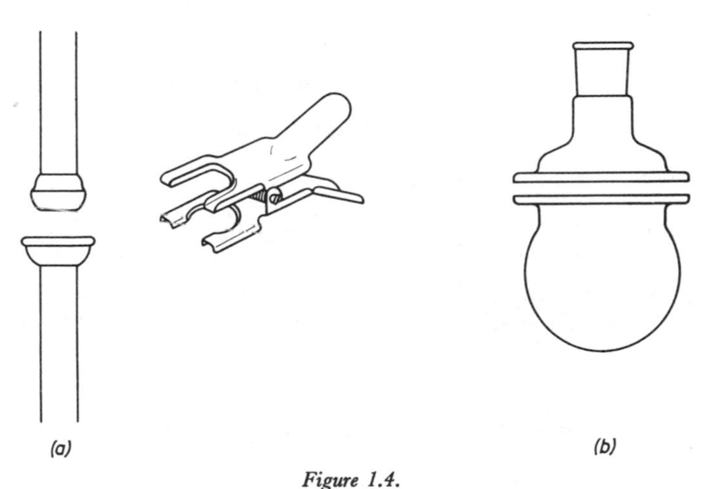

Figure 1.4.
(*a*) *Ball-and-socket joint with clip* (*b*) *Flask with flat flange joint*

TABLE 1.2. Spherical (Ball and Cup) Joints

Designation	*Bore of shank* mm	*Nominal spherical diameter* mm
S 13	2	12·700
S 13	5	12·700
S 19	9	19·050
S 29	15	28·575
S 35	21	34·925
S 41	27	41·275
S 51	35	50·800

The flat-flange lids carry one or more sockets for other apparatus to be attached. Sizes from 50 cm^3 to 20 dm^3 are available. Screw-on clips are used to hold the two flanges firmly together.

Screw Thread Joints

A recent addition to the Quickfit range is a series of screw-thread joints which hold a range of tubing diameters from 6–19 mm. The tubes have standard glass screw threads fitted with heat and chemical resistant screw caps. *Figure 1.5* shows the component parts of one of these devices.

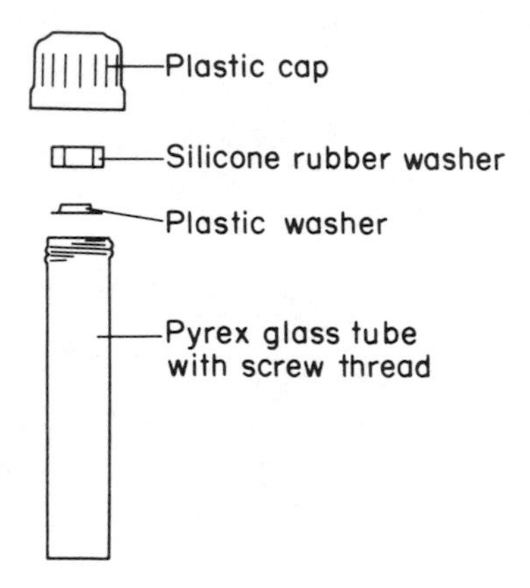

Figure 1.5. The component parts of a screw thread joint

An aperture at the top of the cap allows a narrower tube to pass through. Tightening the cap holds the insert firmly in position while unscrewing the cap by half-a-turn allows the insert to be slid to a new position or removed entirely.

A range of components is available in this form including Y and T pieces. It is claimed that the caps will hold a vacuum of the order of 10^{-3}–10^{-4} torr and a pressure of 75 lbf/in^2 for the larger sizes to 195 lbf/in^2 for the smaller sizes. Temperatures up to 300°C may be used. This type of connector is very useful for the insertion of such things as thermometers, steam tubes, sampling tubes, etc.

One extension of the screw-thread principal is the production of a range of stop-cocks or flow controllers produced by Quickfit under the trade name Rotaflo. A modified plastic cap fits on to a threaded glass tube. A P.T.F.E. plug is connected to the plastic cap and can be raised or lowered relative to a glass seating so that a glass side arm is either open or closed. The aperture which forms the pathway through the open valve can be quite finely adjusted so that liquid or gas flow-rates can be controlled within narrow limits.

The Quickfit range of screw-cap joints are intended only as an ancillary system to their standard glass-to-glass systems. Messrs Sovirel of Levallois-Perret, France have recently announced their S.V.L. system in which all the joints are of the screw cap type and complex structures of laboratory glassware can be built up from standard screw-ended parts. Further information can be obtained from Sovirel's agents, Messrs V. A. Howe.

The Lubrication of Ground Glass Surfaces

Ground glass surfaces are often lubricated to prevent sticking and freezing or where a vacuum is to be applied. For many purposes petroleum jelly ('Vaseline') is an adequate and economical grease. In high-vacuum work Shell 'Apiezon Grease' is usually used. Where high temperatures are involved silicone based greases are employed.

All these greases are formulated for specific purposes and the common practice of haphazardly using the tube of grease nearest to hand for every purpose is to be deprecated.

The layer of grease on the ground glass surface should be as thin as possible, the merest trace of grease on both surfaces being sufficient for adequate lubrication. Especial care should be taken with stopcocks (taps) through which fluids must pass. The grease should not be allowed to enter the lumen of the stopcock plug. Where liquids pass through glass taps many workers recommend that the glass should not be greased but lubricated with a little of the liquid. This is said to be especially necessary where organic grease solvents are used (see Separating funnels, Chapter 2) for obvious reasons.

Greaseless stopcocks made of teflon are very useful and the risk of contamination of liquids by greases is cancelled, but teflon stopcocks are very expensive.

An innovation which seems to give the advantages of teflon without the expense is the aerosol spray containing a solution of what is described as 'a teflon-like material'. The solution is sprayed on the ground glass surface. Within a few minutes the solvent evaporates leaving a thin layer of plastic on the ground glass. The plastic used appears to be unaffected by most solvents and has the excellent lubrication properties of teflon. The plastic coating clings tenaciously to the glass and may be washed in the normal way with no special precautions. If this material lives up to its early promise it will prove invaluable for many of the lubrication needs of the laboratory.

To lubricate standard cones and sockets thin-walled teflon sleeves are available which are shaped to fit the cone so that there is a plastic layer between cone and socket. This allows both high vacuum and high temperature to be applied without fear of the joint sticking.

To conclude this section there must be a final word of warning against the indiscriminate use of silicone stopcock grease. This material is of great value where very low or very high temperatures are to be used. For work at normal temperatures it is unnecessary and it tends to spread over the surface of glass-ware and is difficult to remove. This is a nuisance, especially in the case of volumetric glass-ware where the total wetting of the glass surface is desirable.

THE CLEANING OF APPARATUS

Glass apparatus is usually easy to clean. The glass is inert and any residues left upon it may be removed with detergents or strong oxidizing agents without the apparatus being damaged.

Apparatus should be cleaned as soon as possible after use before the residues have hardened. Rinsing with tap water may remove particulate matter and strong oxidizing or reducing agents before the cleaning material is applied.

Modern detergents are capable of removing most chemical soiling. In the author's laboratory, where protein residues have to be removed from apparatus, Diversey 'Pyroneg' detergent has been found to be extremely effective and economical for these difficult substances. In other laboratories Shell 'Teepol' and I.C.I.

'Lissapol' have been found to be equally effective. The apparatus is carefully and thoroughly scrubbed in a solution of detergent of the correct concentration. The detergent is rinsed out with several changes of tap water and, after three final rinses with distilled water, the apparatus is dried in a warm cabinet. For large numbers of test-tubes or small flasks it is convenient to construct a manifold of plastic tubing which rests in the sink and is connected directly to the cold water tap. Thirty or forty small test-tubes can be rinsed free of detergent by tap water forced through the jets of the manifold. Hot tap water often contains appreciable quantities of grease, etc. and cold water is to be preferred.

If the apparatus is too large for the drying oven or is required for immediate use it may be rinsed with acetone, drained, and dried with a hot-air jet from a glass tube adapted to a commercial hair drier fitted with a rubber bung.

Where the use of detergent does not remove all the dirt from apparatus 'chromic acid cleaner' is used. This is, in effect, a solution of chromium trioxide in concentrated sulphuric acid. Many recipes for chromic acid have been published and most laboratories have their own preparations. The usual method involves the addition of concentrated sulphuric acid to a saturated solution of sodium dichromate. In other laboratories a small amount (usually about 25 cm^3) of saturated chromium trioxide solution is added to a Winchester quart of concentrated sulphuric acid. Neither of these methods is really satisfactory. Chromium trioxide is rather insoluble in concentrated sulphuric acid and the actual concentration of chromic acid achieved is low. The cleaning (i.e. the oxidizing) action of these solutions resides almost entirely in the sulphuric acid, the chromium trioxide concentration being too low for it to have any real effect. In the author's laboratory chromic acid cleaner is made as follows:

Chromium trioxide (technical grade) is added to 400 cm^3 of water until no more will dissolve. Two dm^3 of 50 per cent sulphuric acid (v/v) is added and about 20 cm^3 of concentrated nitric acid. A little solid chromium trioxide is shaken into the bottom of an empty Winchester quart bottle and the prepared cleaning mixture added. The mixture keeps for some time if the bottle is well stoppered.

The advantages of this preparation are:

(*a*) There is much less heat generated during the preparation and the dangers of adding large quantities of concentrated acid to water are minimized.

(*b*) The concentration of chromium trioxide is high and the mixture may be used for long periods before the oxidizing action begins to fall away.

When the chromium trioxide has been used up the mixture becomes green in colour due to the formation of chromous oxide. Fifty per cent sulphuric acid (for oxygen estimation) can be obtained readily prepared from Messrs. British Drug Houses at a price similar to that for its total sulphuric acid content.

The use of chromic acid for cleaning should be avoided where possible. Not only is it a dangerously caustic material but it is said to leave chromic deposits on glass-ware which are very difficult to remove and which interfere with many reactions especially in biochemistry. Chromic acid should never be rinsed off with tap water as insoluble precipitates may be formed from the dissolved salts which will cling tenaciously to glass surfaces. After rinsing off the chromic acid with distilled water, cleaning is continued as with detergents above. It is clear that mechanical scrubbing cannot be employed when cleaning with chromic acid and prolonged soaking is the only method that can be used.

For some purposes alcoholic KOH can be of use especially where heavy deposits of grease remain on glass-ware. This cleaner should be used with the greatest care; it is not only very caustic but it attacks glass fairly vigorously. It should only be allowed in contact with glass-ware for a few minutes at a time before being thoroughly rinsed off.

In some laboratories stubborn deposits are removed from flasks by the addition of concentrated nitric acid to methylated spirit. The vigorous reaction which ensues has excellent cleaning effects but it may develop so rapidly that an explosion occurs. The alcohol-acid reaction may also have a long induction period where it appears to be inactive followed by a sudden explosive reaction. This method of cleaning should be discouraged.

To summarize: cleaning with water and detergent is the method of choice whenever it is practicable. Soaking in chromic acid should be limited mainly to volumetric glass-ware where a completely grease-free surface is essential. Even so, many laboratories find that strong detergent solutions suffice for this.

The cleaning of graduated glass-ware follows a slightly different procedure from the general directions given above. Automatic pipette washers are now a common sight in most laboratories. A version of this useful piece of apparatus is shown in *Figure 1.6.*

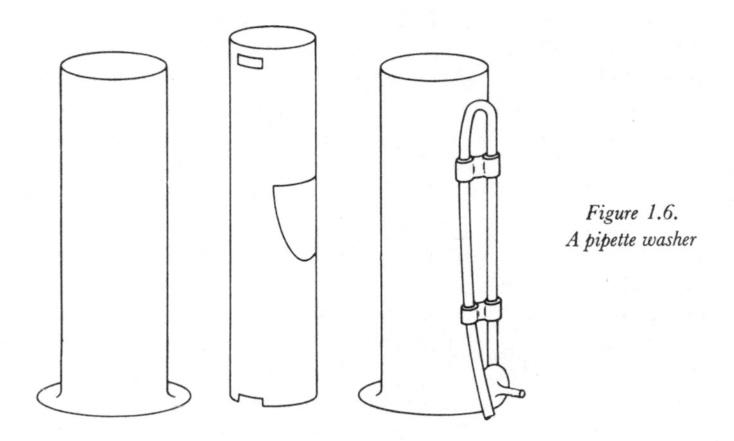

Figure 1.6. A pipette washer

The basket is filled with pipettes with the tips pointing upwards and inserted in a polythene cylinder containing cleaning fluid. After soaking for some hours the basket is removed and drained and inserted into the washer. A flow of water from the bottom of the washer fills it up to the top when it siphons over to empty. The constant filling and emptying washes the pipettes thoroughly. Finally the tap water is stopped and the washer is filled and emptied three times with distilled water. Graduated glass-ware should not be dried by direct heat. One way of drying pipettes is to attach one end to a water pump and draw acetone through. When the pipette is full the tip of the pipette is withdrawn from the acetone and air passes through until the pipette is dry.

Rubber bungs and plastic articles are best washed in detergent solution and thoroughly rinsed.

There can be no hard and fast rules for cleaning apparatus and a knowledge of the nature of the dirt to be removed is obviously useful. If, for instance, a reaction has been carried out in an organic solvent, rinsing out the reaction vessel with small quantities of the pure organic solvent before proceeding to other cleaning methods is a wise move.

The criteria of cleanliness are simple. Dirty glass-ware is usually greasy glass-ware and the presence of grease on glass surfaces is shown by the behaviour of a water film on the surface. A film of water on a surface will be continuous if the surface is clean. The presence of grease will cause the film to break up into large clinging droplets. A clean piece of glass-ware will not show surface droplets when drained after the final distilled water rinsing.

The cleaning of glass-ware has been described at length as the provision of clean glass for the work of a laboratory is usually under the control of the head technician. It is necessary to ensure that those responsible for washing glass-ware follow a rigid routine so that the standard of cleanliness of apparatus remains consistently high.

Finally, mention must be made of automatic washers which have achieved some popularity in recent years. The straightforward jet washer is rather expensive for normal laboratory use in the United Kingdom. It is significant that most of the models in use in the U.K. have been imported from the U.S.A. and Germany where the supply of laboratory domestic staff is even more difficult to arrange than it is here. The ultrasonic washer is a recent innovation. This is a small stainless steel tank with two or more transducers set underneath. The tank is filled with a detergent solution and the articles to be cleaned are immersed. The ultrasonic circuit is switched on and a high-frequency wave is set going in the tank from the transducers. This produces an excellent cleaning effect. After about 5 min the current is switched off and the articles rinsed free of detergent in the normal way. Although the tanks are fairly expensive, the method is invaluable where very scrupulous cleaning is required in such items as tubes used in radioactive scintillation counting, hypodermic syringes, or the grid stages used in electron microscopy.

THE SUPPORT OF LABORATORY ASSEMBLIES

All but the most simple of laboratory assemblies require firm support from stands, bosses, clamps, etc. These are sold in a variety of sizes to support apparatus from micro-analytical to pilot-plant capacity. When using bosses and clamps the right method shown in *Figure 1.7* should be used, i.e. the apparatus should rest on the fixed claw of the clamp and the moving claw should be screwed on to it.

Laboratory scaffolding is very useful, especially where complicated or semi-permanent assemblies are to be constructed. Most of the large laboratory suppliers market a 'system' of scaffolding. All the 'systems' resemble each other closely and none appears to have any particular advantage over the others.

For very large or heavy assemblies slotted angle ('Dexion', 'Handy Angle') is very convenient. It is advantageous to construct

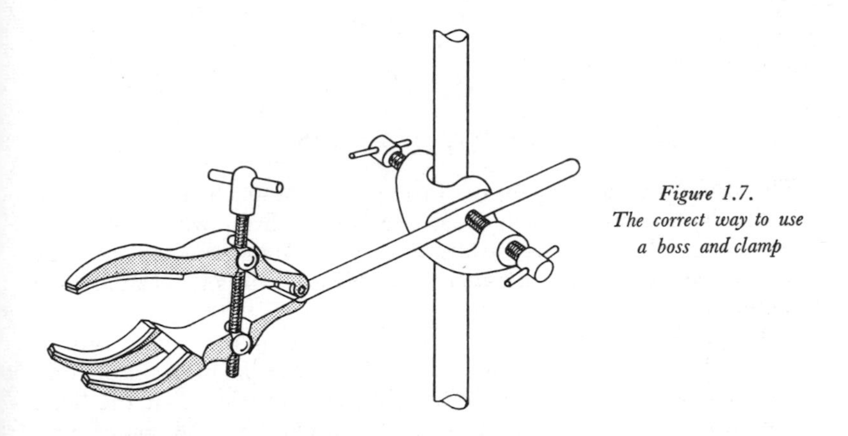

Figure 1.7. The correct way to use a boss and clamp

trollies, tables, etc., which are made the most suitable size for the purpose, and which may not be available commercially.

CONCLUSION

In this chapter an attempt has been made to collect some of the facts about the basic materials which the laboratory worker uses, and some of the manipulations which are common to every branch of chemistry. This information is often omitted from textbooks, but it was felt that some guide would be useful to the technician-in-training.

The subsequent chapters will deal more fully with specific techniques and the more specialized topics.

CHAPTER 2

PURIFICATION

The usual sequence of events in chemistry and chemical research is the synthesis of intermediate compounds from which the final product is obtained, its purification, elementary analysis, a study of its physical nature and then its application involving analytical methods.

The technician must be trained to assist in the whole process but his first responsibility will probably be the purification of compounds after synthesis, where he will gain experience in the basic chemical procedures. Table 2.1 summarizes the procedure,

TABLE 2.1. General Scheme for Analysis of a Sample

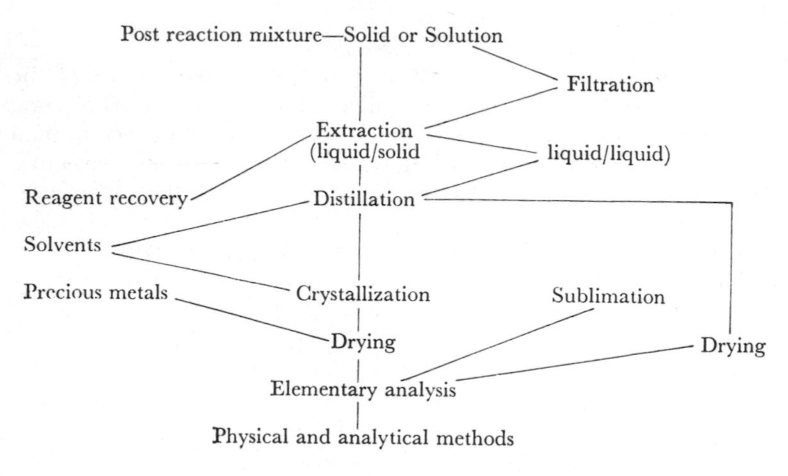

although individual circumstances will need variations, and some of the techniques will obviously repeat themselves, for example, filtration will be required at crystallization and 'recovery' will involve most of the other methods.

In the methods that follow all glass assemblies will be shown with standard ground joints as these have, in general, superseded bungs and corks.

FILTRATION

Separation of Solids from Liquids—Filtration

The method used depends on several variables, such as the physical nature of the mixture, i.e. the volume, viscosity and volatility of the liquid, the amount of solid present and its particle size, and the chemical considerations such as heat stability and light sensitivity.

For most solvents normally employed the 60 degree funnel and a folded filter paper are suitable, the size of these depends on the volume of solution to be filtered and the quantity of precipitate. The volume of the precipitate should be at least 1 per cent. Small

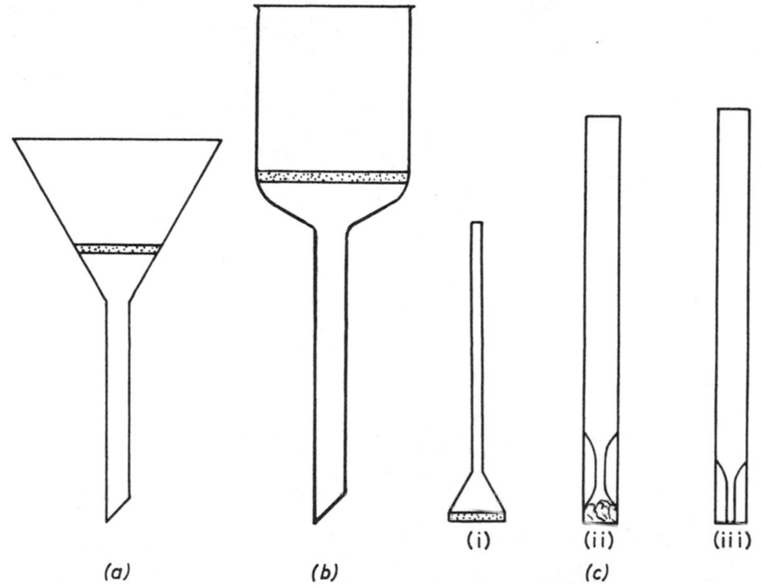

Figure 2.1. Semi-micro filtration apparatus
(*a*) *Hirsch funnel* (*b*) *Sintered glass funnel* (*c*) *Filter sticks*

quantities of precipitates are difficult to recover from the relatively large area of a folded filter paper, therefore a Hirsch funnel (*Figure 2.1a*), sintered glass funnel (*Figure 2.1b*), or one of a variety of filter sticks or tubes (*Figure 2.1c*) may be used. These are all pressure assisted.

Hirsch funnels are of glazed porcelain and have a perforated plate, the diameter of which varies from 10–60 mm. On the perforated disc is placed a circle of filter paper that completely covers the holes but does not turn up the sides of the funnel. The funnel is set into a receiver with a side arm and a mild vacuum applied. The filter paper is then made damp with a few drops of pure solvent which ensures close contact with the plate, and the solution is poured into the funnel using a glass rod to direct the flow. The characteristics of some of the Whatman range of filter papers are given in Table 4.5.

TABLE 2.2. Pore Sizes of Sintered Glass Filters

Porosity grade	*BS specified maximum pore dia. within the range* nm	*Principal uses*
0	150–250	Coarse distribution of gas in liquids
1	90–150	Filtration of very coarse precipitates Support for absorptive layers for filtering gelatinous precipitates Gas distributors in liquids Extraction of coarse grain material
2	40–90	Preparative work with crystalline precipitates Medium gas filters Mercury filters Extraction of medium grain material
3	15–40	Preparative work with fine precipitates Analytical work with medium precipitates Fine gas filters Mercury filters Extraction of fine grain material
4	5–15	Analytical work with very fine precipitates ($BaSO_4$, Cu_2O) Preparative work with finest precipitates Safety valves for mercury
5	Not greater than 2	Bacteriological filtration

The sintered funnel is made of glass or quartz, and the filter disc of glass powder. The latter is produced by careful heat treatment of graded glass powder or grains until they are partially fused giving a porous disc. Various pore sizes are made with a diameter ranging from 0·6 nm to 200 nm. Table 2.2 gives a list of the grades obtainable with their pore sizes and some of their applications. Filtration is performed using a side arm receiver as with the Hirsch funnel.

To improve speed of filtration heat may be used providing the solvent is not too volatile. This is achieved with the larger funnels by placing them in a copper hot water funnel (*Figure 2.2a*), a cone

Figure 2.2. Apparatus for hot filtration
(*a*) *Water jacketed funnel* (*b*) *Copper heating coil* (*c*) *Electric heating mantle*

of coiled copper tubing (*Figure 2.2b*), or a shaped electrically heated mantle (*Figure 2.2c*). The hot water funnel is heated by a Bunsen burner at the side arm. If inflammable solvents are being filtered, a rod immersion heater can be used or an electric heating tape wrapped round the side arm. The copper coil is most suitably used with steam or hot water. The heating mantle has an electric element woven into the glass wall and asbestos jacket, the temperature is controlled by an energy regulator (Chapter 10). When the liquid solidifies or the solute crystallizes in the stem, the whole assembly can be placed in an oven or in an autoclave which is allowed to steam gently.

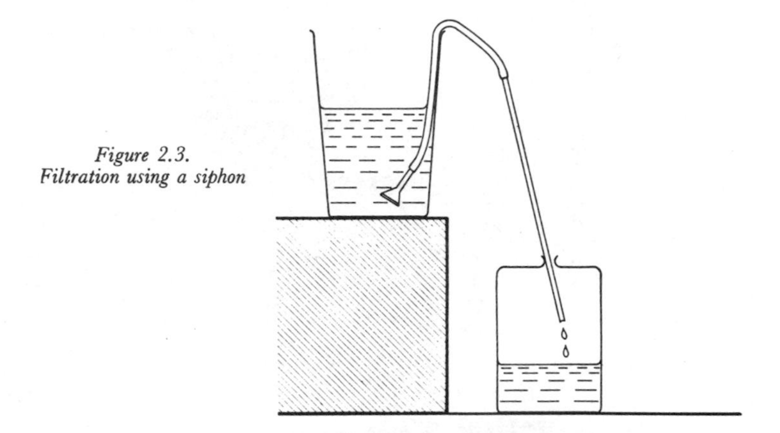

Figure 2.3. Filtration using a siphon

The filter sticks (*Figure 2.1c, i ii*), are plugged with paper pulp, glass wool or asbestos, and can be used for micro-volumes, i.e. 1 cm^3 or less. A rubber bulb or teat is used to give pressure assistance. The filter stick (*Figure 2.1c, iii*), is suitable for clarifying large volumes of liquid by a siphon or suction system (*Figure 2.3*), or the recovery of a faint precipitate.

Vacuum (Pressure Assisted) Filtration

For easier and quicker recovery of material, filtration may be performed, with the aid of pressure, by evacuating the space below the filter funnel. Whenever reduced pressure is used, a trap should be inserted between the pump and the apparatus assembly. The reasons are twofold, firstly, to prevent the loss of material into the pump and secondly, to prevent contamination of the material with the liquid from the pump. A useful addition is a manometer for measuring the pressure within the assembly. The simplest trap is made from a Buchner flask or filter flask which is a thick-walled conical flask with a side arm. The assembly is connected into the top of the flask with thick-walled rubber tubing on a glass tube that is pushed through a hole in a bung. The side arm is connected to the pump. The trap may be modified by the addition of a stopcock for release of the vacuum and a connection for a manometer

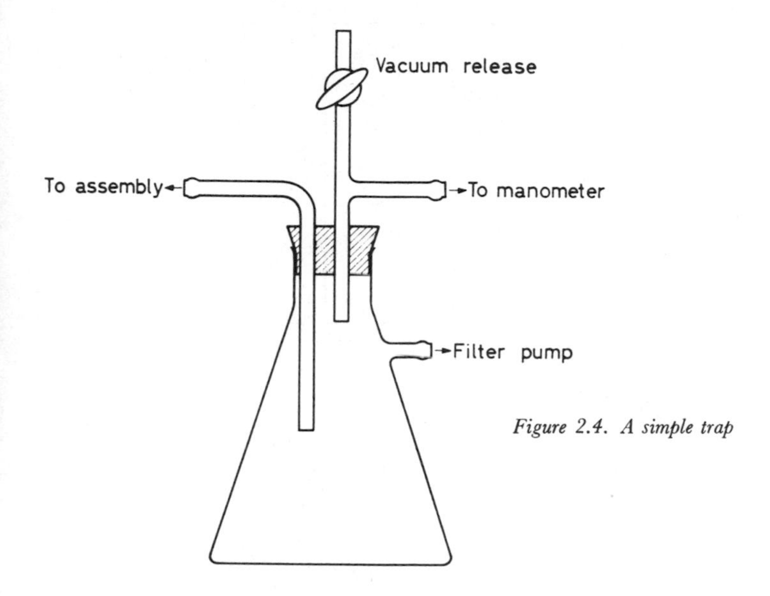

Figure 2.4. A simple trap

(*Figure 2.4*). Most of the vacuum work will not require a manometer of 760 mm length as illustrated in *Figure 2.5a*. More convenient is the combined manometer, vacuum release and non-return valve trap of the designs illustrated in *Figure 2.5b* and *c*. These need be only 15–20 cm tall and the author has found it convenient to attach the trap (*Figure 2.5b*), to the swan-necked water pipe that is serving the water pump. The valves (*V*) prevent liquid from flowing into the assembly if there is a sudden pump failure. The valves, made of glass in *Figure 2.5b* and rubber in *Figure 2.5c* will not prevent the slow seepage of oil or water when the pressure difference across the valves is slight. The pump is attached at (*P*). The manometer in *Figure 2.5b* can be isolated, this is an advantage, as if the vacuum is released suddenly the mercury will return to the top of the tube with such velocity that it may break the glass. Trap *Figure 2.5c* is compact but liquid from the assembly will come into contact with mercury if it enters the trap, and this may be a disadvantage.

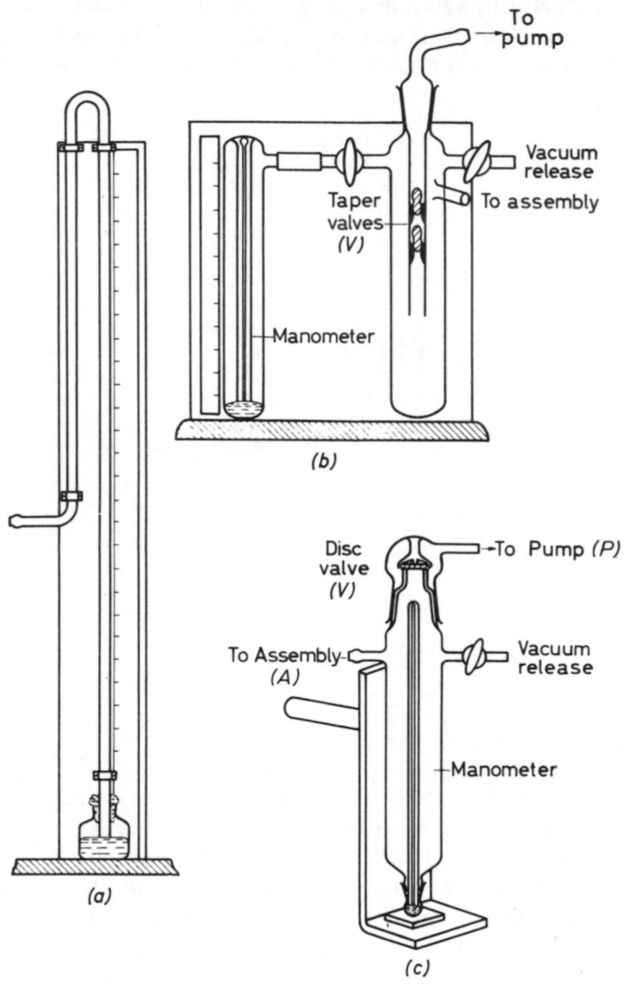

Figure 2.5. Manometer assemblies
(*a*) *Simple manometer* (*b*) *and* (*c*) *Combined traps and manometers*

The water pumps supplied by Edwards High Vacuum Ltd. are widely used. They are made of metal or plastic and have a non-return valve in the vacuum side. These can be attached to the tap by 6 mm connectors. Bourdon type vacuum gauges are available as a pump attachment but these are not as accurate as the mercury type. Glass filter pumps are available but these have the disadvantage of being difficult to unblock and are fragile. All these pumps will produce a vacuum of about 2000 Nm^{-2} (15 mm of mercury) with a water pressure of about 2×10^5 Nm^{-2} (30 lb in^{-2}), but it is recommended that the pressure be boosted to at least about 3×10^5 Nm^{-2} (45 lb in^{-2}), preferably higher, as at the lower pressure, fluctuations due to constantly changing use of the water supply may cause 'suck-back'.

The SI unit of pressure is the *Newton per square metre* (Nm^{-2}) and supersedes all other terms such as mm mercury, torr, lb/in^2, atmospheres, etc. The Newton is the SI unit of force in kg m s^{-2} and the unit of pressure is therefore defined as the force exerted on a unit area. The relationship with the old units is:

$$1 \text{ atmosphere} = 15 \text{ lb in}^{-2} = 760 \text{ mm mercury} = 760 \text{ torr} = 101\,325 \text{ Nm}^{-2}$$

$$1 \text{ mm mercury} = 1 \text{ torr} = 101\,325/760 \text{ Nm}^{-2} = 133 \text{ Nm}^{-2}$$

In the foregoing description the old units are given in brackets after the SI figure. In many cases these are not exact equivalents, the figures having been 'rounded off'. Where the range of a meter or pump is being described this would seem to be a realistic attitude.

If a mechanical pump is used it should be protected against the effects of corrosive vapours by scrubbing the vapours in a wash tower, and then passing them over a cold trap filled with solid CO_2 in acetone. A pump with an air ballast valve is advisable when moisture is likely to contaminate the pump oil. Further information on these pumps will be found in Chapter 9.

Buchner funnels (*Figure 2.6a*) are designed for pressure filtration, they are similar to the Hirsch funnel, are made of stoneware or glazed porcelain, and have a larger capacity. The funnel sits in the top of a filter flask, a rubber cone of suitable size forms a seal. Squares of thin rubber 3–6 mm thick or soft P.V.C. sheet with a hole bored in the middle will also serve. If possible, a funnel should be chosen whose tip projects below the side-arm of the flask and with the bevel facing away from it to lessen the chance of losing filtrate towards the pump. The funnels are heavy and it is necessary to support the assembly. The largest type do not sit directly on the flask and are made in two pieces (*Figure 2.6b*). The upper part is a funnel with a perforated disc, the lower part or base is a bowl so shaped that the filtrate is directed to an outlet in the side. This is connected by flexible pressure tubing to the top of a filter

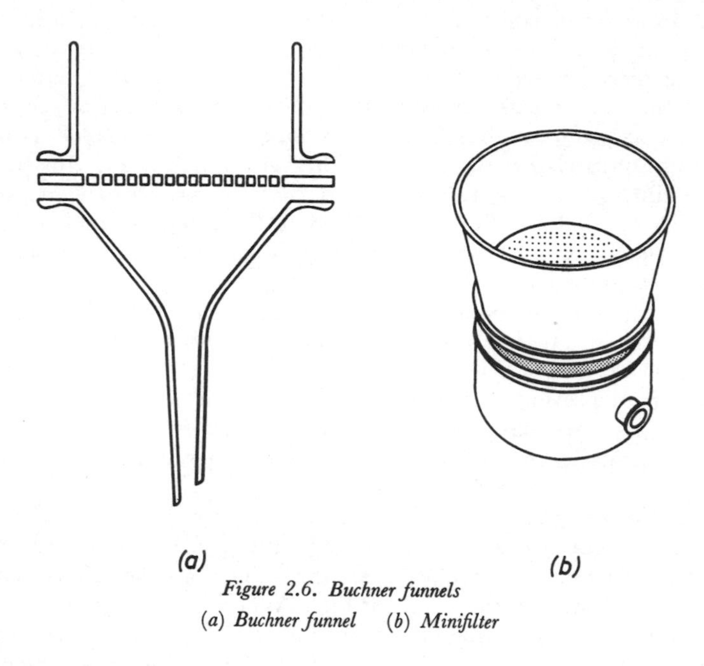

Figure 2.6. Buchner funnels
(*a*) *Buchner funnel* (*b*) *Minifilter*

flask. The type illustrated is the Minifilter manufactured by Messrs Hathernware.

It may be necessary to use two filter papers with large funnels as the plate perforations are often comparatively large and a single paper may burst under the pressure. The vacuum pump is first turned on and the air release on the trap closed, the filter paper circles are then placed in position and made damp with a little of the solvent. This has the effect of drawing the paper on to the perforated plate and covering the holes. The mixture, guided down a glass rod, is then decanted on to the paper and the trap air-release tap adjusted to give a mild vacuum. A large pressure difference at the beginning may cause the filter paper pores to be impregnated by fine material and clog. The vacuum can be increased after a thin filter cake has formed or if the rate of filtration slows down. The cracking of the filter cake when filtration nears completion can be prevented by smoothing the surface with a spatula or the flat top of a glass stopper. Pressure will be released if air is sucked through any cracks.

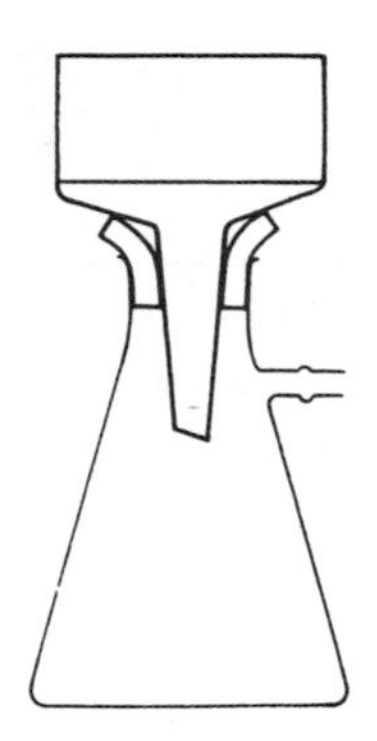

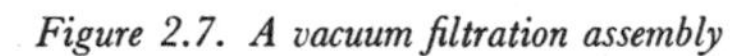
Figure 2.7. A vacuum filtration assembly

Very fine precipitates are difficult to filter, they seep round the edges of the paper or even pass directly through the pores, especially at the beginning. If this is the case, ultra-filtration techniques must be used (see Chapter 9) or, if the precipitate is to be discarded, the mixture may be filtered through a layer of one of the powdered filter aids. These are diatomaceous earths which when dry have the consistency of flour. A pressure filtration assembly (*Figure 2.7.*) is prepared using a medium filter paper circle such as Whatman No. 1. A slurry of filter aid is made with distilled water and this is poured into the funnel and a mild vacuum applied. The filter cake is washed thoroughly with water, the flask is changed and the solution for filtering is passed through. In order not to disturb the cake the mixture is poured on to a small filter paper or watch-glass resting on the filter cake.

Large volumes of liquid are easily clarified by using filter candles. These are tubes of porous ceramic or sintered glass which are immersed in the solution. The candle is connected by flexible pressure hose to a receiver which is in turn attached to a pump (*Figure 2.8*).

Time is often wasted when a filter funnel has to be continuously topped up. It is possible to invert a bottle, containing the solution for filtering, over the filter funnel (*Figure 2.9a*). As the solution level descends, air enters the bottle and more solution pours out. The funnel should not be over-filled as this method stirs the surface of the solution and some may be spilt. Many reactions are performed in a flask and the device in *Figure 2.9b* may be preferred, this

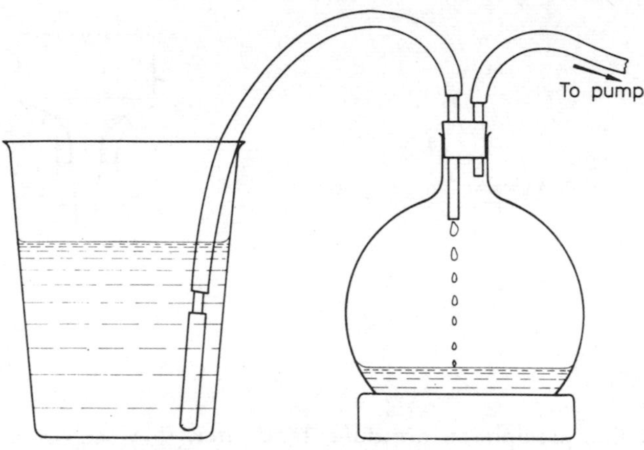

Figure 2.8. Assembly using a filter candle

causes little disturbance of the liquid. The principle of operation is the same as before.

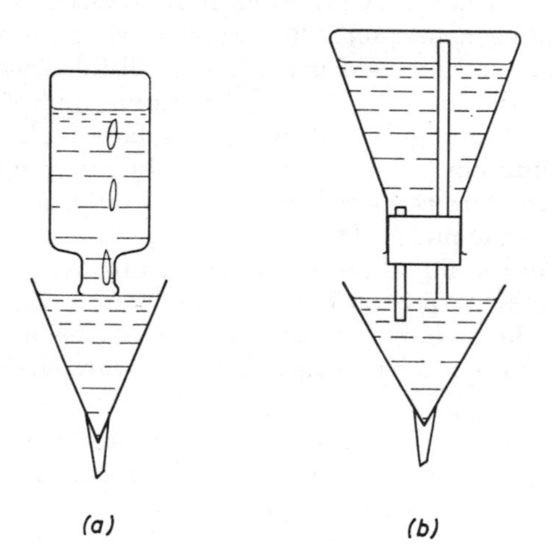

Figure 2.9. Continuous feed systems for a filter funnel

EXTRACTION

Separation and Extraction

In many preparations and experiments it is necessary to extract a dissolved substance from a liquid using another immiscible solvent, and these solvents have to be subsequently separated. This is performed with a *separating funnel*. These may be obtained in several shapes and sizes (*Figure 2.10*). The most common is the pear-shaped funnel (*Figure 2.10a*), the cylindrical type (*Figure 2.10b*), is more easily clamped and can be obtained graduated.

Separation consists of pouring the liquid into the funnel and allowing the phases to separate, then, with the stopper removed, the tap is carefully opened and the heavier liquid drained into a receiver. Occasionally the funnel is twisted to release any of the lower layer that tends to stick up the sides. When nearly complete, the flow is slowed to achieve perfect separation. The final stage is best performed with a funnel that has a short parallel length of tube above the tap in which the interface is made to move slowly.

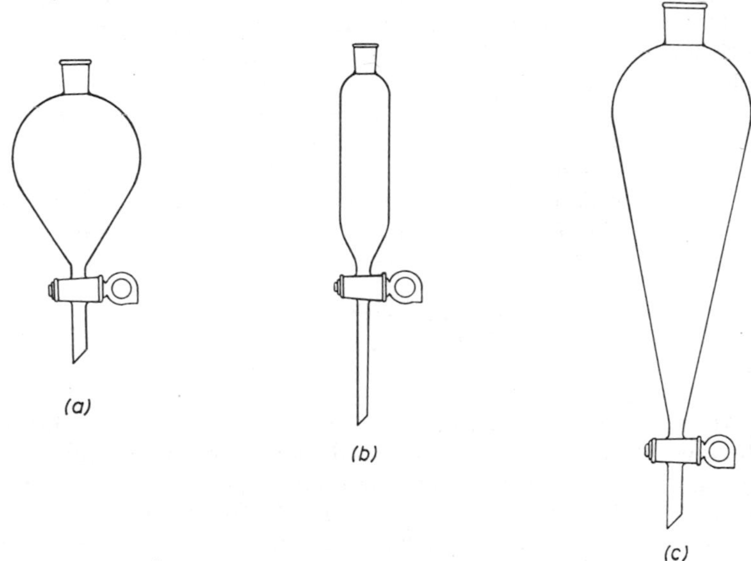

Figure 2.10. Separating funnels
(a) Pear shaped (b) Cylindrical (c) Conical

Funnels can have standard joints and stopcocks with interchangeable plugs. The taps must be lightly greased with Vaseline, rubber grease or a suitable apeizon grease. Silicone grease is not recommended as it may contaminate the inner surface of the funnel, this causes 'non-wetting' of the glass by the solvent and will hinder separation. When using the funnels with fat solvents or for analytical work with lipids, the taps cannot be greased. It is possible to do without a lubricant with some solvents but with others such as ether and the petrols the liquid will creep round the tap. These difficulties may be overcome by using P.T.F.E. taps, or spraying the plugs of glass taps with Fluo-glide (page 16).

Extraction is the transfer of a dissolved solid from one solvent into another immiscible solvent. This is performed by pouring the solvents into a separating funnel, choosing one that will not be more than half-filled. Holding the funnel with one hand securing the stopper and the other the stopcock, shake for a few seconds then invert. Allow the liquid to drain from the tap, then open it to release the gas pressure. Repeat this once or twice, then a prolonged vigorous shake is permissible as equilibrium will have been reached and no further pressure will be generated.

The quantity of solute transferred depends on the *distribution coefficient* or *partition coefficient* of the particular solvent system and solute. The partition coefficient is the ratio of the concentration of solute in one solvent to the concentration in the other, and is a constant at a given temperature. This is easily found by pouring a quantity of both solvents into a separating funnel, adding some solute and shaking as described before. After separation a dry weight is performed on both phases, giving the concentrations in grammes per cm^3. The coefficient is then

$$\frac{\text{g/cm}^3 \text{ in Solvent } A}{\text{g/cm}^3 \text{ in Solvent } B} = K$$

More solute will be extracted from one solvent to the other if the ratio of their volumes is increased, but the partition coefficient does not alter. For example, 6 g of solid A was dissolved in 50 cm^3 of water. With one extraction by 50 cm^3 of chloroform 5 g of A was removed from the water. On extracting an identical aqueous solution with 230 cm^3 of chloroform 5·75 g of A was removed from the water. In the first example the final concentrations are 0·02 g A/cm^3 H_2O and 0·1 g A/cm^3 $CHCl_3$, and in the second 0·005 g A/cm^3 H_2O and 0·025 g A/cm^3 $CHCl_3$. Calculations of the partition coefficient from both sets of figures gives the same result.

$$K = \frac{0{\cdot}02}{0{\cdot}1} = 0{\cdot}2$$

$$K = \frac{0{\cdot}005}{0{\cdot}025} = 0{\cdot}2$$

The efficiency of extraction can be followed using the following formula

$$w_n = w_o \left(\frac{Kv}{Kv + s}\right)^n$$

Where v = the volume of solution being extracted
w_o = weight of dissolved substances
s = the volume of extracting solvent
w_n = weight of substance remaining after n number of extractions
K = the partition coefficient.

It is more efficient to extract with several small portions of solvent than one large one. This is illustrated in the worked examples showing the extraction of acetic acid from water with diethyl ether.

The partition coefficient of acetic acid is 0·58. One extraction of 100 cm³ of 8·5 per cent acetic acid by 200 cm³ of ether removes $8{\cdot}5 - w_n$.

$$w_n = \frac{8{\cdot}5 \times 0{\cdot}58 \times 100}{(0{\cdot}58 \times 100) + 200} = \frac{493}{258} = 1{\cdot}91$$

$$\therefore\ 8{\cdot}5 - 1{\cdot}91 = 6{\cdot}59 \text{ g}$$

Four extractions of the same volume of acetic acid by 50 cm³ of ether removes $8{\cdot}5 - w_n$.

$$w_n = 8{\cdot}5\left[\frac{0{\cdot}58 \times 100}{0{\cdot}58 \times 100 + 50}\right]^4 = 8{\cdot}5\left[\frac{58}{108}\right]^4 = 8{\cdot}5 \times 0{\cdot}0445 = 0{\cdot}378$$

$$\therefore\ 8{\cdot}5 - 0{\cdot}378 = 8{\cdot}122 \text{ g}$$

It will be clearly seen that more is extracted by a volume of solvent if it is divided into portions.

By making use of this law it is possible, by repeated extraction of both upper and lower layers, to separate two or more solutes that have different partition coefficients. The technique requires a number of separating funnels, the more funnels used the better the separation. The process is best illustrated diagrammatically (*Figure 2.11*). The volumes of the solvents are kept constant throughout. In funnel (1) at equilibrium a solute will be distributed between solvents A and B according to its partition coefficient. Solvent A is transferred to funnel (2) and a fresh portion of A is added to funnel (1) and a fresh portion of B added to funnel (2). Equilibrium is achieved in both by shaking. Layer A of funnel (2) is transferred to funnel (3) and a portion of B added. Layer A of funnel (1) is transferred to funnel (2), a fresh portion of A is added to funnel (1). This process is repeated, the number of funnels increasing. It will be seen that the original solvent B always remains in funnel (1) and is repeatedly extracted with fresh solvent A which has the effect of 'moving' the solute from left to right.

If $K = 1$ the percentage distribution after five extractions would be as shown in *Figure 2.11*. This is illustrated graphically in *Figure 2.12a*. With two solutes of different partition coefficients the effect

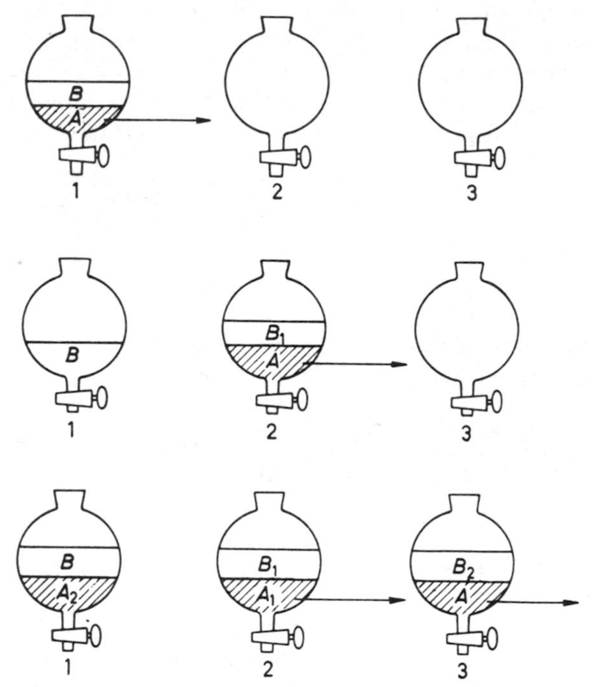

Figure 2.11. The separation of two solutes which have different partition coefficients

shown in *Figure 2.12b* would be expected. Theoretically it is impossible for complete separation to be obtained.

The minimum practical number of separations for this technique to be effective is about 25; 50 or 100 is to be preferred. The use of separating funnels with manual shaking is laborious and therefore the method was rarely employed until in 1949 L. C. Craig designed an ingenious apparatus consisting of glass extracting units linked together. The extractions are carried out simultaneously and then the transfer of solvent layers is made in one movement. In recent years the apparatus has been completely automated (*Figure 2.13*).

An individual unit is illustrated in *Figure 2.13b*. The two solvents are placed in *A* through stopper *B*, an equilibrium is obtained by oscillating the unit about the horizontal. The upper layer is separated off into *D* via *C* by making *A* vertical. On returning the unit to the horizontal the upper layer is transferred to the next unit via tube *E*.

The method is called countercurrent distribution as in *Figure 2.11* solvent *A* moves to the right and solvent *B* to the left. In practice solvent *B* is stationary, the solute being added at one end although

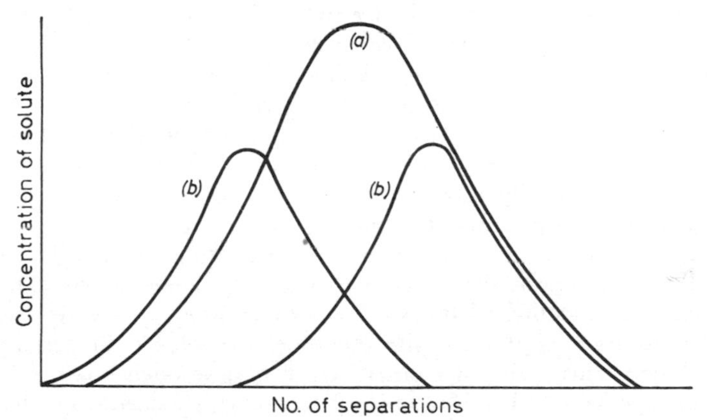

Figure 2.12. Graphical representation of extraction
(*a*) *Distribution of a solute when $K = 1$* (*b*) *Two solutes having different partition coefficients*

this is not necessary. At the end of the train of units solvent *A* is collected in a flask or fraction collector. Distribution trains of 1000 units have been used effecting separations until now deemed impossible.

A very similar apparatus has been devised using glass discs with an eccentric hole. These are separated by P.T.F.E. washers (*Figure 2.14*). A series of chambers is formed and the discs are mounted alternately at 180° to each other so that the holes are not directly opposite. A column of up to 400 chambers is set at an angle of 45° and rotated. The heavier solvent is introduced to each chamber, then the sample is added to the uppermost. A continuous stream of a lighter immiscible solvent is run in at the top. The column is revolved at about 20 r.p.m. and the lighter solvent flow rate is between 10 and 400 cm^3 per hour depending on the partition coefficient of the solute. With low partition coefficients, e.g. $K = 1{\cdot}3$, the lighter solvent may be recycled again thereby doubling the number of extractions.

Occasionally during extraction emulsions occur and separation becomes difficult. Usually, in time, the emulsion clears and this can be hastened in several ways, such as twisting the funnel which tends to burst the emulsion bubbles, and by running out the lower layer very slowly, which has the same effect when the bubbles enter the narrow neck. Sometimes it is permissible to use chemical agents such as alcohol or, when water is the phase being extracted, inorganic salts may be added, such as sodium chloride and ammonium sulphate. These being very soluble in water displace the other compound which is invariably organic. This is called *salting out*. One sure way of breaking up the emulsion is to transfer the solution to centrifuge pots and centrifuge it; 10 min at $2000 \times g$ usually suffices. If it is known that an extraction method has a tendency to emulsify it would be better to use separating funnels that are designed to fit into centrifuge buckets.

Micro-extractions that have a volume of 10 cm^3 or less can be performed in stoppered tubes or preferably tubes with screw caps. After addition of the solvents the tubes are placed in a shaking machine. This is simply constructed from a wooden board with a number of Terry clips fixed to it. This board is rocked or oscillated by a variable speed electric motor and crank mechanism. After extraction, the tubes can be centrifuged if necessary and the layers then separated by pipette. When an assay method requires an extraction, numerous simultaneous extractions generally have to be performed. This work may be lightened if McCartney bottles or Universal bottles are used. These have a volume of 28 cm^3 (1 oz)

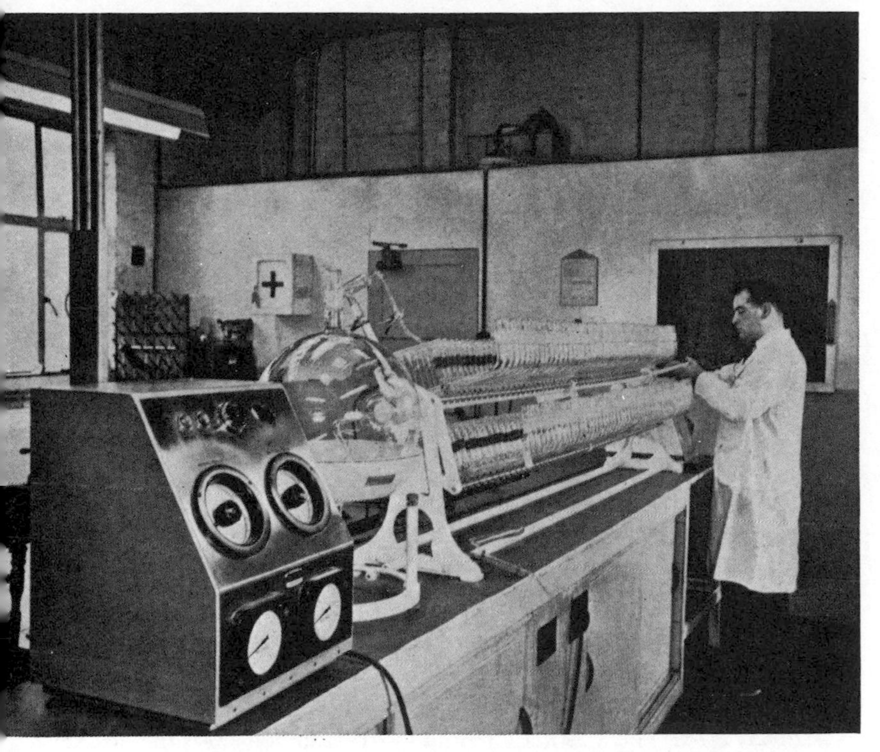

(*a*) [By courtesy of Quickfit and Quartz Ltd]

Figure 2.13.
(*a*) *The Craig counter current liquid/liquid extraction apparatus*
(*b*) *A single unit*

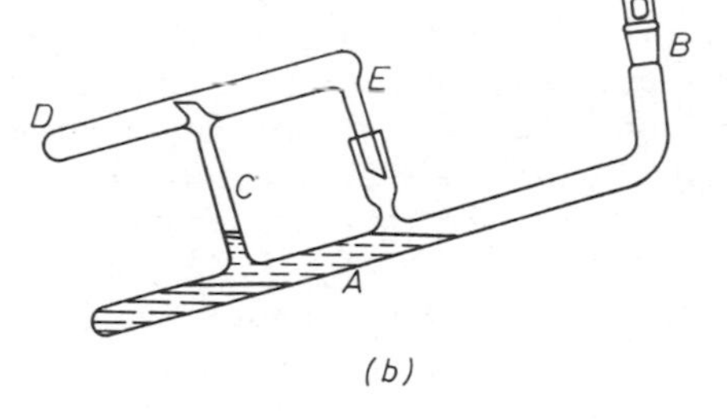

(*b*)

and are used with 10–15 cm^3 of solvent. The bottles are first tested in the centrifuge at $2000 \times g$ and subsequently used below this force. Thin polythene or P.T.F.E. discs, cut with cork borers or a fly press are superimposed over the rubber liner in the screw cap

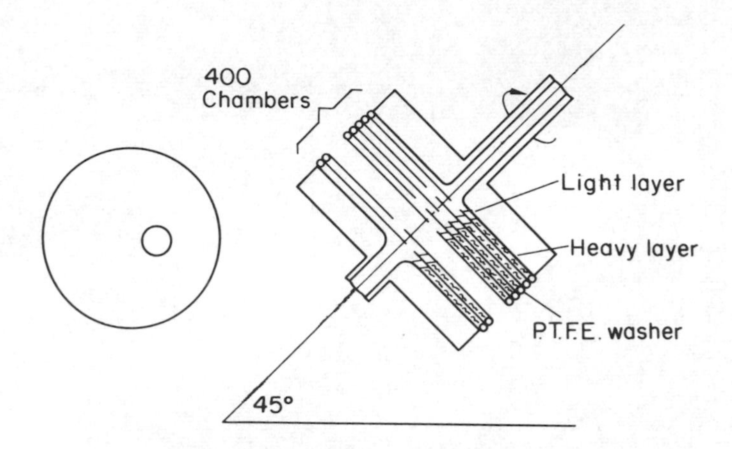

Figure 2.14. Another form of countercurrent apparatus

to prevent contamination of the solvents by the rubber.

For thorough extraction repeated extractions and separations are necessary. This is more conveniently done in a continuous extraction apparatus. These are essential when the substance is only slightly soluble in the extracting solvent. Two basic assemblies are needed, one for when the extracting solvent is lighter than the other (*Figure 2.15a*) and another when the extracting solvent is the heavier (*Figure 2.15b*). In *Figure 2.15a* flask *C* contains the extracting solvent, this is boiled and the vapours condensed in *F*. These pour into funnel *J*, this has to be long to allow for enough hydrostatic pressure to force the solvent through the sintered distributor *S* against the denser liquid. When setting up, the heavy liquid is placed in the assembly, funnel *J* is then put in position with the light solvent being continuously poured through it, otherwise poor distribution will result. The lighter solvent rises to the top and overflows into flask *C*.

In *Figure 2.15b* the denser liquid is in flask *C*, it is boiled and condensed in *F* and is distributed into the lighter liquid through a sinter that is touching the liquid surface. The extracting solvent falls to the bottom and overflows into *C* via tube *G*. Before commencing, some of the denser solvent must first be poured into the extracting chamber otherwise the lighter material may contaminate the solvent in *C*.

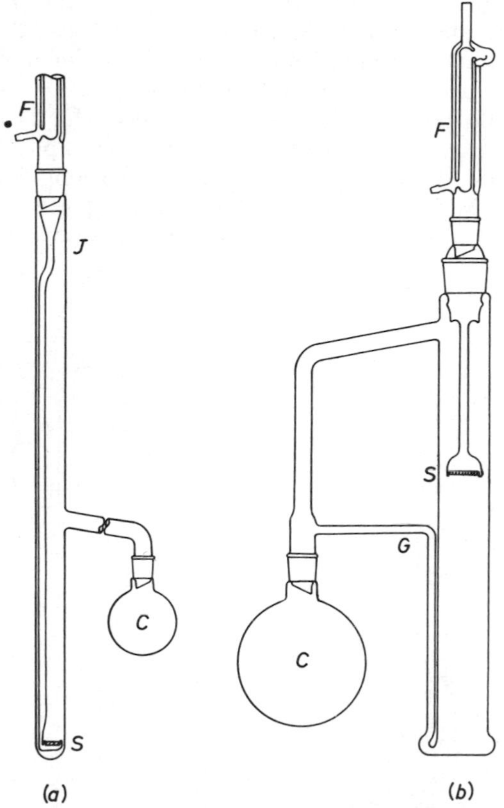

Figure 2.15 Continuous liquid extraction apparatus
(a) for extraction from a heavier solvent by a lighter
(b) for extraction from a lighter solvent by a heavier

There are many designs both for micro and macro liquid/liquid extractions, all using the same principle of operation. In many of these the extracting chamber is either immediately above the boiler or actually suspended in it (*Figure 2.16*). These have the disadvantage that the extracted solution is heated, which is not acceptable with many biochemical preparations.

An all purpose extractor designed by A. Hemmings* may be used for both types of liquid/liquid extractions and for controlled

* *Analyst, Lond.* **76**, No. 899 (1951) 117

percolation or extraction from a solid medium (*Figure 2.17*). It will be seen that by using the extension *H* and funnel *J* and closing both taps the apparatus is similar to that in *Figure 2.15a* and by removing the funnel and opening tap *B* it is similar to *Figure 2.15b*. Both systems are operated as already described. In addition, space *E* can be filled with granulated material which can be either slowly percolated with solvent collecting below tap *A* or continuously percolated by closing tap *A*, partially opening *B* and boiling solvent in *C*. Tap *B* should be adjusted so that there is an overflow down tube *D*, thus ensuring the material is always immersed in the solvent.

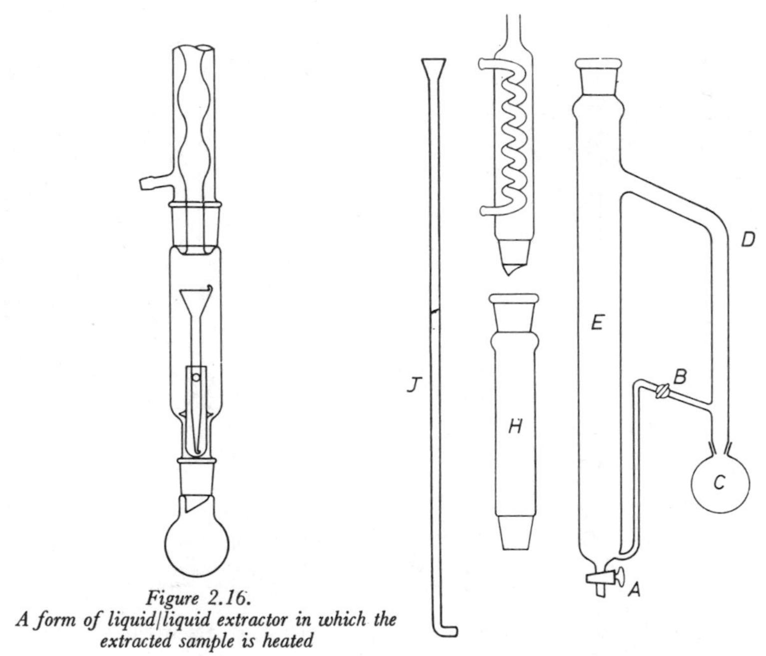

Figure 2.16.
A form of liquid/liquid extractor in which the extracted sample is heated

Figure 2.17.
The Hemmings liquid/liquid extractor

Extraction of Solids

For the continuous extraction of solids the most popular method is the Soxhlet extraction apparatus (*Figure 2.18*). The same principle is employed for both micro- and macro-quantities of solid. The substance is placed in a porous thimble A, usually made of paper but also of porous porcelain or stoneware clay. The solvent is boiled in the flask B and the vapours reach the condenser C via tube D. The solvent falls into the thimble and eventually fills the chamber and siphons out through E back into the flask, this process is repeated continuously. The top of the thimble should be above the level of the siphon to prevent floating particles from passing into B. It may be necessary to prevent splashing of the solid, due to the fall of the solvent, by covering the solid with a porcelain sieve plate, sintered disc or filter paper. This also helps to prevent the formation of channels down the column of solid which will lower the efficiency of extraction.

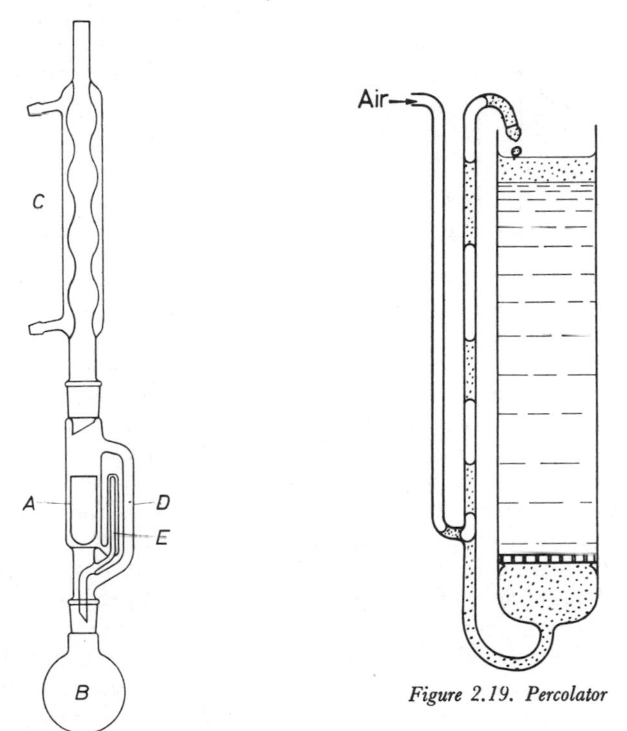

Figure 2.19. Percolator

Figure 2.18. Soxhlet extraction apparatus

The continuous extractions described have the virtue of the extraction always being carried out with pure solvent. A less efficient extraction may be performed by recirculating the solvent with some form of pump. In the continuous percolator in *Figure 2.19* an 'air lift' is used to circulate the solvent.

THERMAL METHODS

A form of micro extraction and separation has been developed by Herbert Weisz mainly for the purpose of carrying out qualitative analytical spot tests (see Chapter 4). This makes possible the separation and concentration of micro-gram quantities. The apparatus

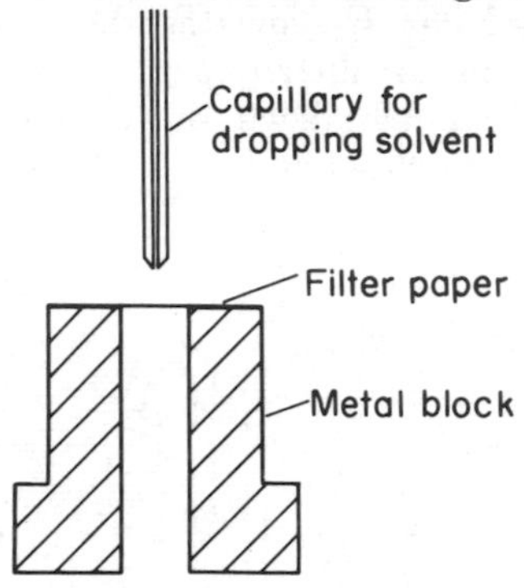

Figure 2.20. The Weisz ring oven

used is called a ring oven (*Figure 2.20*). This is a block of metal having a 22 mm hole bored through its centre. The block is heated electrically and the temperature is arranged to be 5/10°C above the boiling point of the extracting solvent. A disc of filter paper (Whatman No. 40 is most suitable) is placed on top of the oven. To the centre is applied a drop containing the material for extraction. A fine capillary (about 0·5 mm bore) is filled with the solvent and this is held vertically over the centre spot where the solvent runs out onto the paper. The solvent spreads towards the edge of the hole in the ring oven, carrying dissolved solutes from the centre. When the solvent reaches the hot metal it evaporates, leaving the dissolved material in a fine ring on the paper. The extracted material in the centre is removed with a cork borer or punch. Spot tests are carried out on this disc and on the soluble material contained in the ring. The spot may be pre-treated before extraction, for example, metal sulphides can be precipitated. Two sizes of oven rings can be used. After a first separation or precipitation reaction is carried out producing a ring, the paper is then placed over a larger hole and a second extraction made thus forming three distinct areas of separation.

Another purification technique was originated by E. Stahl. It requires the use of the TAS oven and was specifically designed for

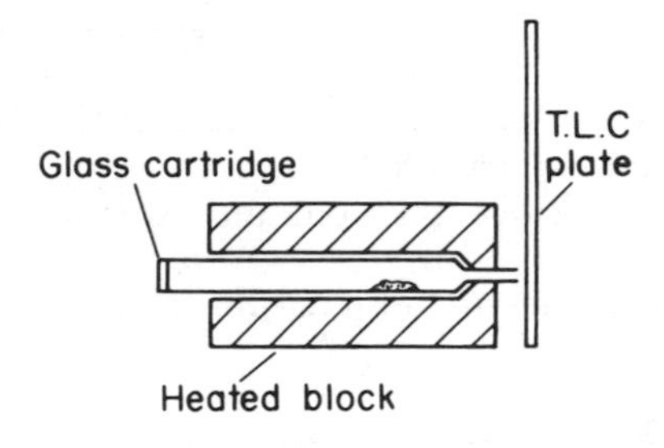

Figure 2.21. The TAS oven

T.L.C. analysis (*Figure 2.21*). The oven is a heated block through which is bored a horizontal hole large enough to contain a special glass cartridge. This cartridge is like a Pasteur pipette having a short capillary. Its simplest use is in the transfer of volatile material from a mixture onto a T.L.C. plate. A few milligrams of the mixture are placed in the cartridge, this in turn is inserted into the oven so that the capillary protrudes to within 1 mm of a T.L.C. plate held vertically. The other end of the cartridge is closed. Heat vaporizes the volatile components of the mixture which impinge onto the T.L.C. plate from the capillary tip. Many variations of the technique are possible, for example, if silica gel is inserted after the sample a form of steam distillation will occur. Other vapour extractions may be done by adding solvent by means of a syringe. Volatilized material from the sample may be pretreated before reaching the T.L.C. plate either by adding reactant to the sample or by using a suitable packing at the capillary end.

Zone Refining or Melting

A technique that was originally developed for the purification of semi-conductor material has been adapted for organic and inorganic compounds. Material that does not decompose at its melting point is packed into a column. This is done by pouring it in the molten state. The column is set vertically. An electric ring or bobbin heater is placed at the top, this is moved by a mechanism that allows it to travel down the tube very slowly. Depending on the velocity of crystallization the speed of traverse will be of the order of one inch per hour. The heater melts a zone and crystallization occurs at the upper zone of the molten area, impurities usually falling. If the reverse is the case the melting must start at the bottom. What occurs is a continuous recrystallization, in fact more than one molten zone can be used provided they are well separated or the heater may simply be returned to the top of the column and the process repeated. A micro zone refiner has been described. This can cope with 0·5 g of material. A tube of 2–5 mm bore by 150 mm long is attached to a mechanical lift that passes it through the focal point of a mirror receiving parallel light from a 100 watt lamp. If the material is light-sensitive the column is surrounded by a metal sleeve.

Similarly, it is possible to perform zone sublimation (*Figure 2.22*).

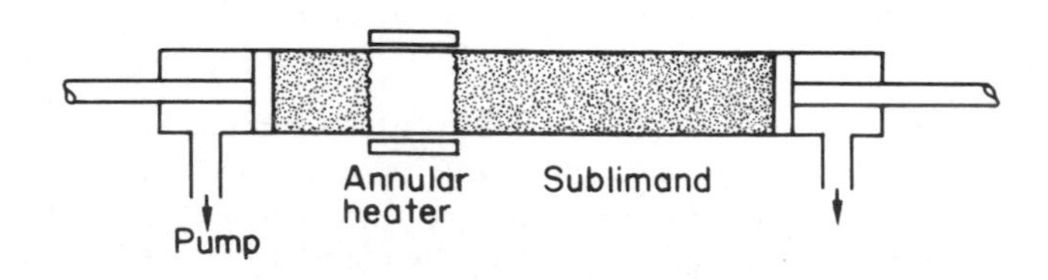

Figure 2.22. Zone refining

The sublimand is packed in a Pyrex tube between two plungers. The tube is evacuated at each end. The plunger is withdrawn by the length of the heater to allow vaporization. Then the heater is moved along the tube. The process is repeated and the impurities will collect at either end.

DISTILLATION

In purification by distillation the liquid is vaporized by heat, the vapours liquefied in a condenser and the distillate collected in a receiver. The vapours over a liquid in a vessel exert a pressure, for an individual solvent this pressure is constant at a given temperature. This vapour pressure rises with temperature and will eventually equal the pressure above the liquid, at this point the liquid boils. The temperature will remain constant with a pure solvent until all the liquid is vaporized, although the quantity of heat applied is increased. The boiling point of a solvent at an atmospheric pressure of 101 325 Nm^{-2} (760 mm Hg) is a reference physical constant called the *normal boiling point*. If the pressure above a liquid is lowered the boiling point is lowered, and distillation under reduced pressure has distinctive advantages where the solvent or a solute decompose at the normal boiling point.

The apparatus used is varied, depending on the volume to be distilled, type of distillation, chemical nature of solution and, not least, the tradition or personal whims of the worker. The liquid is vaporized in a distillation flask, this may be one of four basic shapes, conical, pear-shaped, spherical or round with a flat bottom. All of these shapes are suitable for distillation at atmospheric pressure. The most popular is the spherical flask as this can be used in all types of distillation. The flat bottomed and conical are rarely used as the other types are equally suitable, and more versatile.

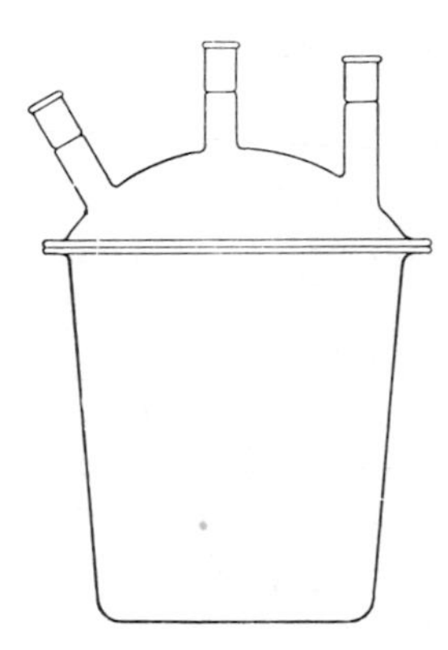

Figure 2.23. Multi-necked reaction flask

The distillation flask may have more than one neck, this is then suitable as both reaction flask and distilling flask, the extra ports being used for stirring, adding of reactants, gassing and thermometer, etc. A very useful flask for this dual role is the wide necked, pear-shaped flask (*Figure 2.23*) with a flat flange joint, the top half having several necks. This type is free standing, the solid or liquid contents can be quantitatively removed and with the correct size of paddle, stirring is very efficient. The plain ground joint is securely clamped by a wire spring rim closure.

Attached to the distilling flask is an adaptor called a distillation head, this has a cone joint to attach the condenser and one or more necks to facilitate taking the vapour temperature before condensation, providing air leaks and the addition of reactants or more solution.

Various types of condensers are in use, these are usually cooled with tap water or refrigerated liquid (see Chapter 10). If the boiling point of the solvent is high the cooling liquid may be omitted.

The commonly used condensers are shown in *Figure 2.24; (a)* is a jacketed tube or Liebig type condenser. *Figure 2.24(b)* is the Davies

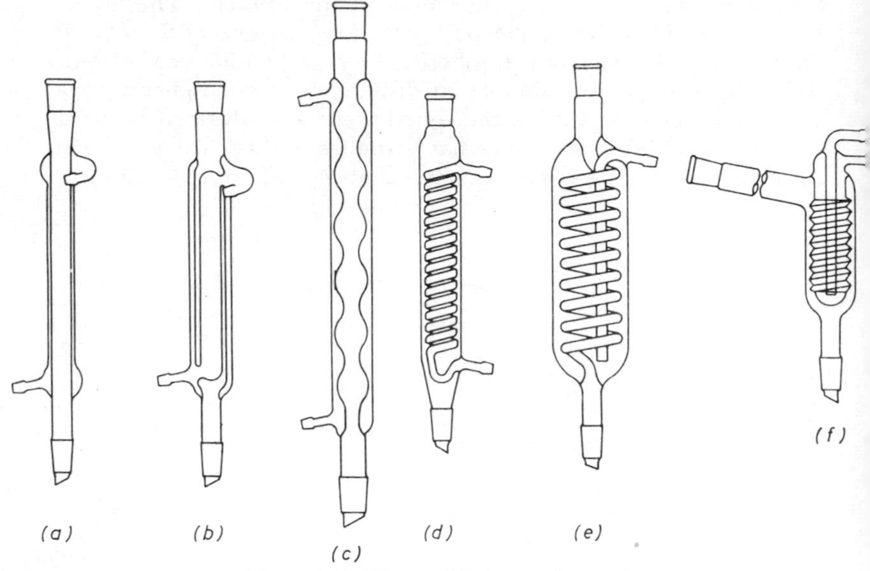

Figure 2.24. Water cooled glass condensers

(*a*) *Liebig*
(*b*) *Davies double surface*
(*c*) *Allihn*
(*d*) *Graham*
(*e*) *Thorpe Inland Revenue pattern*
(*f*) *Friedrich*

double surface condenser and has an obvious advantage that it need be half the length of the Liebig for the same effective condensing surface area. *Figure 2.24*(*c*) is a bulbed or Allihn condenser, (*d*) is a coiled condenser (Graham), the coil taking the cooling fluid. There are many designs of this type, some with two or more integral coils for increased efficiency. (*e*) is the Thorpe Inland Revenue pattern (BS 1848/1952) in which the coil takes the distillate. (*f*) is the Friedrich condenser which has a side entry for the vapours.

Receivers—The condenser leads either directly into the receiver or via an adaptor. The adaptor is designed to facilitate the entry of the distillate into the receiver and also produce a closed system except for a vent so that harmful vapours can be directed away with rubber tubing, or by a connection to a vacuum pump.

Receivers may be any of the vessels used as distillation flasks or, where permissible, beakers or other open dishes.

Methods of Heating—Providing the liquid is not inflammable the flask may be heated by a Bunsen burner. If the vessel has a flat bottom an asbestos gauze should intervene to give even heating. This is not necessary with a spherical flask but the flame should not be a fierce one and should be positioned off centre. For higher boiling points an air bath is used, this is a small circular oven positioned round the flask. It is constructed of metal or better still a section of asbestos ducting used for fume extraction. A wire gauze is pushed into the bottom and a divided lid of asbestos is used to enclose the flask. The oven sits on a tripod and is heated by a Bunsen flame. For temperatures up to 300°C heating mantles can be used, these allow reasonable observation of the flask contents. They wrap around the flask to give even heating, are light in weight and can be more easily suspended in an assembly where the receiver has to be much lower than the distilling flask. The heating element is woven into glass fibre and asbestos but liquids can penetrate to it, and it is not recommended where there is a fire risk.

For most vacuum distillations and temperatures up to 100°C a steam or water bath can be used, the latter heated by an electric immersion heater and temperature controlled by a thermoregulator (Chapter 10). For higher temperatures an oil or metal alloy bath can be used. Suitable liquids are glycerol (150°C), paraffin oil (200°C), or one of the silicone fluids (250°C). The metal alloys, Wood's metal and Rose's metal may be used up to 250°C. The compositions are as follows:

Metal	*Wood's metal* (m.p. 71°C)	*Rose's metal* (m.p. 94°C)
Bismuth	4 parts	2 parts
Lead	2 parts	1 part
Tin	1 part	1 part
Copper	1 part	—

Refluxing

Occasionally it is necessary to have a mixture boiling for a long time, or during the course of a reaction. To prevent loss of liquid through vaporization a condenser is fitted vertically into the top of the flask. This is termed *refluxing*, the vapours condense and return to the flask (*Figure 2.25*).

Atmospheric Distillation

The conventional distillation assembly is shown in *Figure 2.26a*. The most common reason for distillation is to concentrate a solution by reducing its volume, the evaporated solvent being of no further value except for re-use as stock after recovery. A suitable semi-

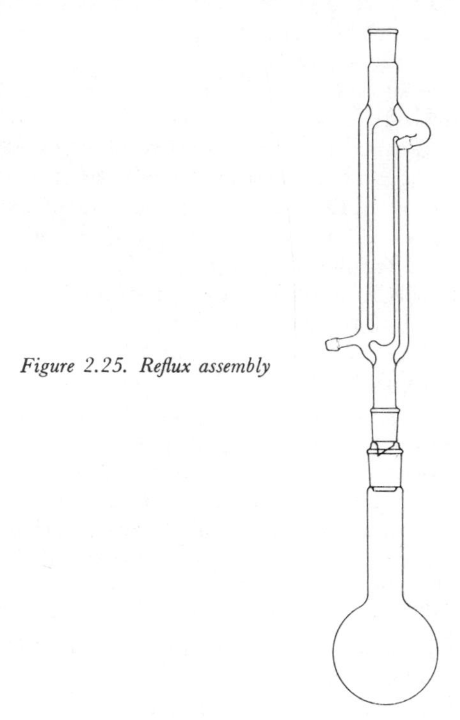

Figure 2.25. Reflux assembly

permanent assembly is illustrated in *Figure 2.26b*, using a Friedrich condenser. A ball joint would give the side arm flexibility. The assembly is conveniently held on one retort stand and takes up less room than other types. The condenser and distillate can be screened from radiant heat by a piece of asbestos board.

Where there is a possibility of crystallization in the condenser the Liebig will be found suitable as this is more easily freed from solid.

Steps should be taken to prevent superheating as sudden boiling may carry some of the distilling liquid into the receiver. Prevention is accomplished by using one or two small fragments of porous pot, pumice, asbestos (Gooch), glass beads or a bumping stick. The latter is made from glass tubing or thin wall capillary tubes depending on the volume of liquid. The tubes are sealed two or three diameters from the bottom and sealed at the top. The bottom edge is best left unflamed. When heating a spherical flask

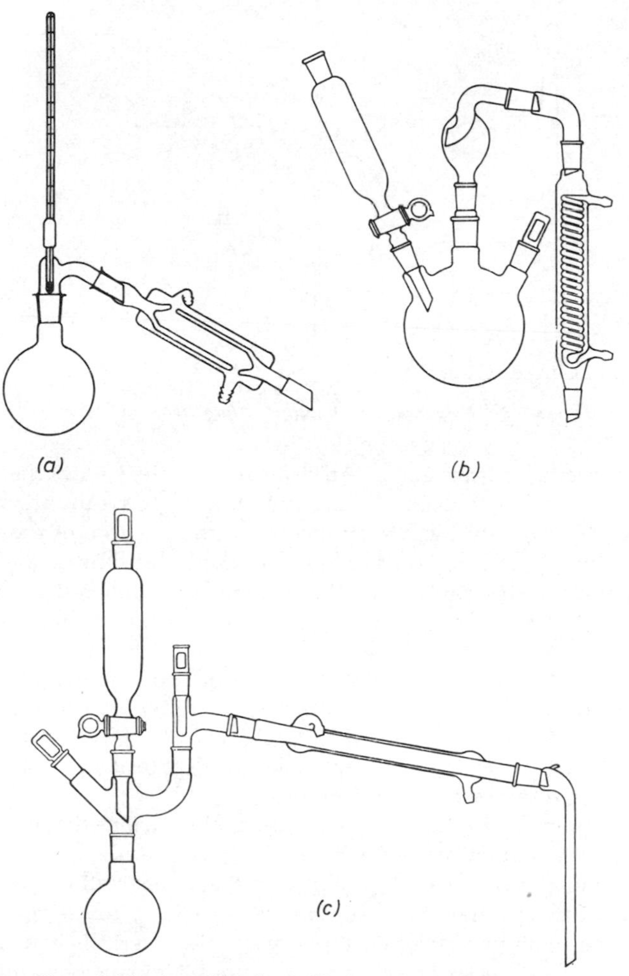

Figure 2.26. Assemblies for distillation at atmospheric pressure

direct with a Bunsen burner as described on page 49 these precautions will not be necessary.

The prevention of frothing is discussed under Vacuum Distillation where the problem is more serious.

Where a large bulk of liquid has to be distilled, that is, 5 dm^3 or more, a continuous feed may be used. This works on the Marriott

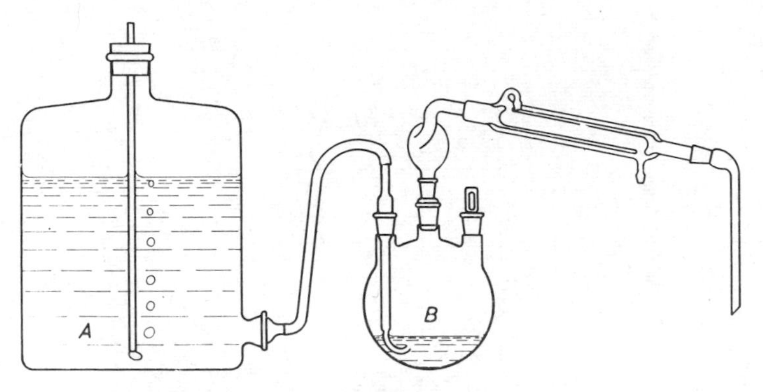

Figure 2.27. Continuous distillation of a large volume of liquid

bottle principle *Figure 2.27*. An aspirator is filled with the liquid and in the neck is placed a bung that has a glass tube through it that extends to within a few millimetres of the bottom of the vessel. The outlet is connected to a multi-necked distillation flask, the liquid entering through a tube that is bent up at the end to prevent vapour bubbles breaking the siphon. The aspirator is positioned so that the bottom of tube *A* is at the desired level of the liquid in the distillation flask *B*. It will be seen that as liquid boils out of flask *B* and the level drops, air will bubble into the aspirator through *A*, and more liquid flow into *B*. The end of tube *A* and the level in *B* will not be exactly the same as the hydrostatic and vapour pressures in the reservoir will affect it.

At this point it will be worth examining the various distilled water assemblies. An apparatus may be used as just described but frequent dismantling and removal of precipitated 'hardness' will be necessary. Where relatively small quantities are required, up to 10 dm^3 a day, one of the ingenious balance types of continuous stills can be used (*Figure 2.28*). These are suspended about a pivot, the outlet for the cooling water from the condenser is connected to the centre of a small manifold. This manifold moves with the assembly and therefore, when the distillation flask is low in water, the assembly is put out-of-balance and the manifold corrects this by directing the water into the flask. In practice, the apparatus oscillates gently and the majority of cooling water goes to waste from the other end of the manifold. With slight modification it is possible to obtain double distilled water by mounting one assembly above

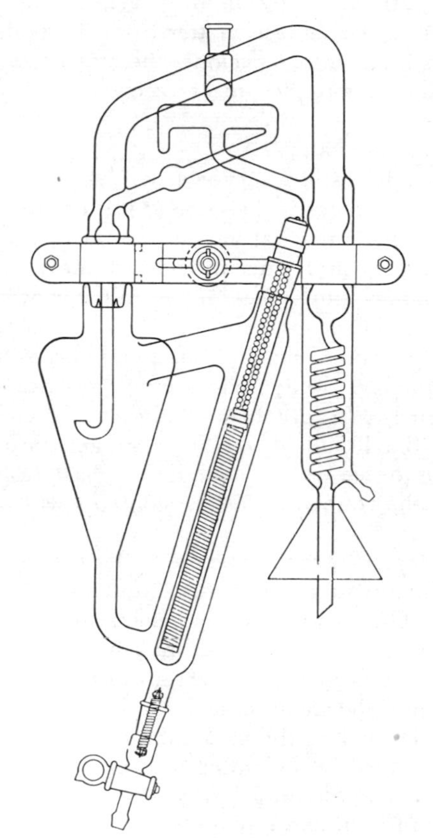

Figure 2.28. Still for production of distilled water

the other. Electrically heated varieties have the protection of a safety device. This is a mercury switch mounted on the pivot arm and when the weight on the flask side becomes too light due to lack of water, the arm tilts and cuts off the electricity supply.

Several glass water stills are available, these have immersion heaters and constant water level devices (Fisons and Scorah).

A robust enamelled steel still with immersion heater is made by Manesty, and is capable of producing 4 dm^3 of distilled water per hour or more. In some varieties the immersion heaters are replaced by a steam coil. Where a laboratory is plumbed with live steam, use may be made of one of the strip-action stills. Such a still manufactured by Fisons is capable of producing 20 dm^3 of distilled water per hour from a 4×10^5 Nm^{-2} (60 lb in^{-2}) steam supply. In this the steam is 'stripped' of particulate matter from the boiler and plumbing system by passing it into a cyclone, the clean steam is then condensed after passing over baffles into a receiver.

Steam Distillation

In a mixture of gases or liquids that do not react with each other the vapour pressure exerted by each gas is the same as if it occupied the space alone. This means that the total vapour pressure is the sum of these individual pressures and thus provides a means of purifying liquids of high boiling points that are immiscible with water and that decompose at or near their own boiling point. When the sum of the vapour pressures reaches atmospheric pressure both liquids vaporize and may be condensed together, providing the temperature is maintained. In practice, steam at 100°C is injected into a mixture of the liquids from a steam generator. Because of this method it is necessary to include a splash head between distilling flasks and the condenser, as the contents will be vigorously agitated.

An assembly illustrated in *Figure 2.29a* may be used. This consists of a steam generator *A*, distilling flask *B*, condenser *C* and a separating funnel *D* as a receiver. The steam may be generated in any suitable flask or a steam can or it may be taken from a steam main. In the latter case some means of cleaning the steam must be used, as in the strip-action still, before the steam enters the flask *B*. The generator is fitted with a safety valve, this is a tube at least 1 m long, one end of which is immersed in the water and the other end bent slightly and pointed in a safe direction away from personnel.

The distillation is carried out by first applying heat to the boiler, clip *E* being open and clip *F* shut. The liquid components are added to flask *B*, the volume of water being only a few cubic centimetres. This flask is also gently heated, to prevent excessive condensation of water. The boiler and flask are connected by rubber tubing at *F*. When steam issues from *E*, clip *F* is opened and clip *E* closed in that order, and distillation commences. At the completion of distillation clip *E* is opened as the burner is removed from under *A*. The boiler is disconnected at *F* and the flask removed. It is unwise to use a 3-way stopcock instead of a T-piece and clips. The stopcock key may jam due to the heat or because the steam has removed the lubricant (a source of contamination in any case) with the result that boiling water is ejected from the safety valve.

A semi-micro steam distillation assembly is illustrated in *Figure 2.29b*. It will be seen that the distilling 'flask' is suspended in the boiler, and is therefore pre-heated by the steam. The bulb in the steam line prevents the flask liquid being sucked back into the boiler section.

'Steam' distillation can also be carried out under reduced pressure and this method will be described later.

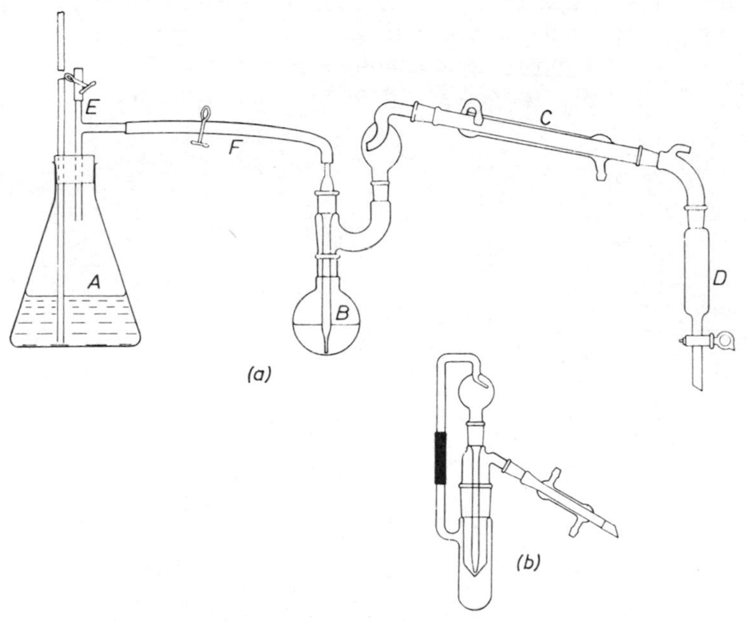

Figure 2.29. Steam distillation assemblies

Fractional Distillation

The purpose of this type of distillation is to separate mixtures of miscible liquids. Consider two miscible liquids (a binary mixture) in a flask. The vapours above the mixture are composed of both components, but the percentage composition will be different from that of the liquid in the flask, it will in fact be richer in the vapour of the more volatile liquid. When distilling such a mixture the vapours of both components will be condensed, but because of the higher proportion of the more volatile component the distillate will have a lower boiling point than the original mixture from which they were distilled. During the distillation the temperature will be continuously rising, the speed at which it will rise depending on the composition. For instance, when recovering acetone (boiling point 56°C) from a mixture containing 90 per cent acetone and 10 per cent water, the temperature will not rise above 58·5°C until the composition of the remaining liquid is 50/50. The acetone content of the distillate at this point will be better than 98 per cent. The temperature then rises more steeply until almost pure water is left in the distilling flask.

A boiling temperature/composition curve may be drawn of a binary mixture *Figure 2.30*. *A* or 100°C is pure water and the

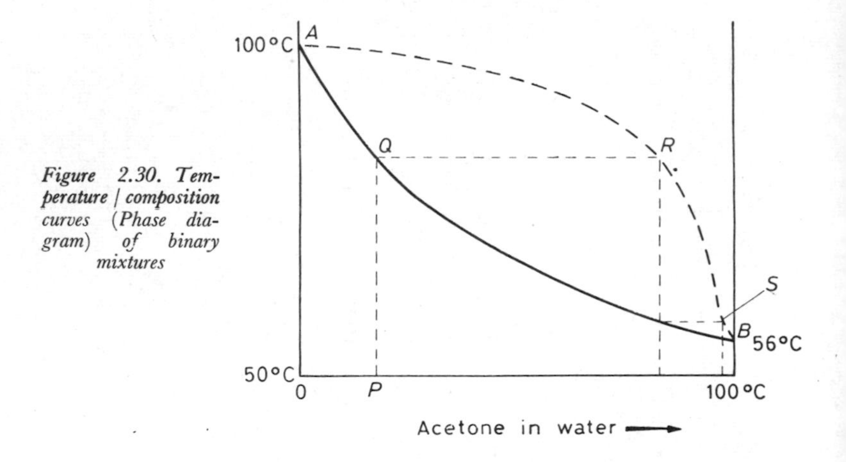

Figure 2.30. Temperature / composition curves (Phase diagram) of binary mixtures

acetone content increases (continuous line) until it becomes 100 per cent at *B* or 56°C. If for the several points of this curve the vapour composition was found, a second curve can be drawn on the same graph 'dotted line' of the vapour condensation temperatures against composition.

A mixture of composition *P* will boil at point *Q* and the vapour produced at that temperature will have composition *R*, and when condensed will have the same composition (*R*) but a boiling point of *S*. This process can be repeated by boiling mixture *R* and obtaining a lower boiling point distillate. If this is done several times an almost pure sample of the more volatile component, in our case acetone, will be obtained. These sucessive distillations are more easily performed in a fractionating column, and the efficiency or enrichment of distillate produced by one of these columns is measured in the number of stages that are indicated on the graph that it is capable of performing. These stages are called 'theoretical plates'. Columns are compared by the length or height that is equivalent to one theoretical plate (H.E.T.P.) for a particular mixture. The H.E.T.P. is found by dividing the column length by the number of theoretical plates.

Some miscible binary mixtures are impossible to separate by fractionation because a particular ratio of components is reached where the vapours are of the same ratio as that of the boiling liquid. The mixture is called an azeotrope and if temperature/composition curves are plotted it will be found that the liquid and vapour curves meet at some point. The mixture behaves normally until it reaches this point, and then a constant boiling point mixture is formed. The ratio of the components in a constant boiling mixture is predictable at a given temperature and atmospheric pressure. Use is made of this characteristic in producing standard hydrochloric acid, and instructions for its production will be found in Chapter 4. Two azeotropic curves are possible.

Fractionating Columns

These columns are interposed between the distillation flask and the condenser. Their design and efficiency has changed considerably over the past 50 years. Vapours from the flask enter the column and are condensed and reflux back; by increasing the boiling rate refluxing is made to take place up the length of the column. This can be finely controlled so that the majority of the vapours are condensed and return to the flask. In all columns the liquid falling back must come into intimate contact with the vapours passing through it. This is to effect an efficient scrubbing of the less volatile component from the rising vapours. All columns are designed to this end.

Several laboratory columns are shown in *Figure 2.31*, (*a*) is the simplest, being a glass tube with several indentations round the circumference repeated at intervals up the column. (*b*) is more complicated, the ball acts as a valve which is lifted by the rising vapours which in turn are scrubbed by the released liquid. The Dufton or Widmer spiral (*c*) is a central rod with a glass or stainless steel spiral wound on it, the whole fitting snugly into the column. The liquid refluxes round the spiral and is in effect, a long, thin, plain column. (*d*) is a column easily made by the technician from a glass rod.

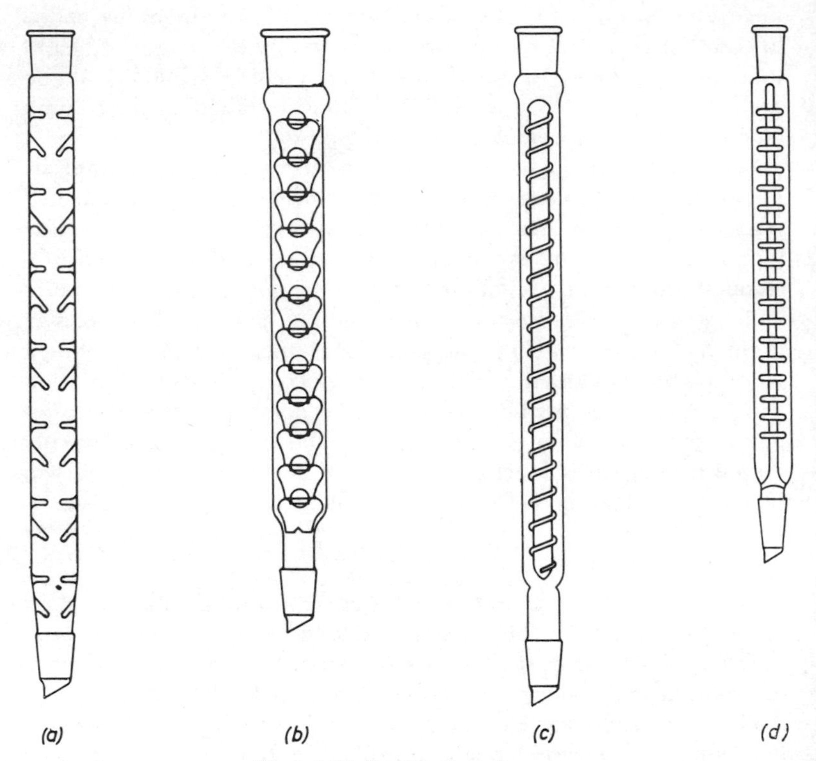

Figure 2.31. Fractionating columns

(*a*) *Vigreux* (*c*) *Widmer* (*Dufton*) *spiral*
(*b*) *Pear bulb* (*d*) *Rod and disc* (*Youngs*)

These columns are only suitable for separating mixtures of liquids with very different boiling points, as the H.E.T.P. is probably no better than 15 cm. More efficient columns have been produced through the research of the petroleum industry. These comprise plain columns filled with specially designed packings, the point of these packings as already explained is to present a large liquid surface area to the rising vapours. The simplest are Raschig rings (*Figure 2.32a*), these are short lengths of glass tubing 6 × 6 mm

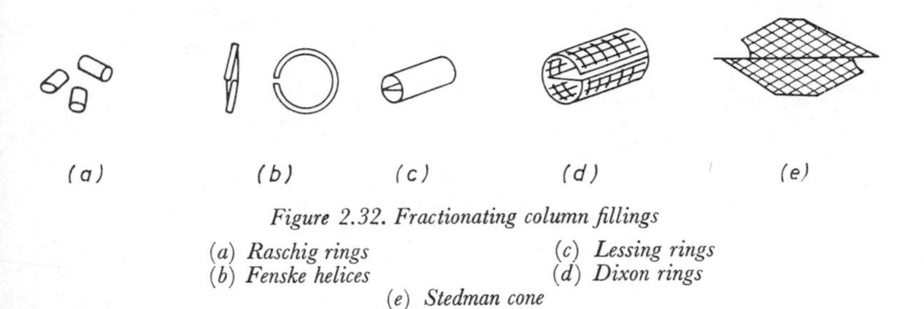

Figure 2.32. Fractionating column fillings

(*a*) *Raschig rings* (*c*) *Lessing rings*
(*b*) *Fenske helices* (*d*) *Dixon rings*
(*e*) *Stedman cone*

or 9 × 9 mm, depending on the diameter of the column to be packed. Fenske helices (*b*) are another popular packing, these are single turns of glass or stainless steel wire up to 0·02 in. diameter. The helices are 2–5 mm in diameter and are made by winding the wire spring-wise over a rod and then cutting across the turns. *Figure 2.32c* and *d* are cylinders with a central division; (*c*) are called Lessing rings and are made of glass or porcelain, (*d*) are Dixon rings and are made of fine stainless steel wire gauze of 100 × 100 mesh.

Filling the columns with packing requires care, first a septum or wire helix is placed at the bottom and then the packing poured in gradually, the column being gently vibrated to make it settle, the smallest helices may be tamped down with a light plunger. If there is any regularity in the way the packing is arranged the column must be emptied and refilled as there is a danger of 'channelling'.

A different type of packing is the Stedman cone *Figure 2.32e*. These fit 25 mm diameter columns, they are made of stainless steel wire mesh and are cells produced by two cones, one inverted on the other and welded together. There is a hole top and bottom set 180 degrees apart. They are packed one on top of the other and are held in by friction. The H.E.T.P. of these columns may be 2 cm or better. The packed columns must be lagged, those up to 25 mm in diameter are usually fitted with a vacuum jacket, and the larger ones may be electrically heated. The temperature is carefully controlled throughout the column length.

It is essential that the column be mounted vertically to prevent the possibility of 'channelling'. The still head above the columns may be the normal type with provision for a thermometer, or a Claisen head may be used. With this a cold finger type of condenser can be inserted above the column to control the reflux rate. More

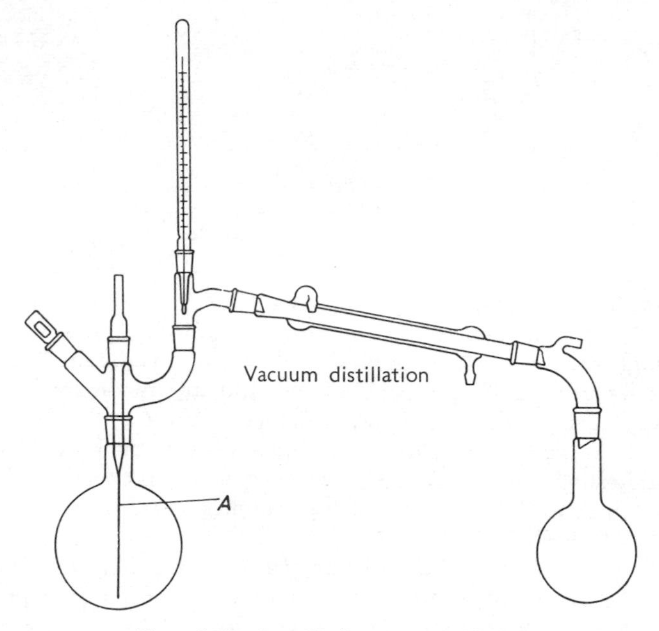

Figure 2.33. Assembly for vacuum distillation

complex still heads are available with valves so that the ratio of the amount returned to the column to the amount condensed into the receiver is controlled. This is called the reflux ratio and a suitable figure has to be found and maintained for each mixture for efficient operation of the still.

After assembly, the condenser water is turned on and then the flask contents boiled. This is usually done by electrical heating but a Bunsen burner may be used if the solvents are not inflammable. It is essential that the rate of boiling can be finely controlled. The vapours are allowed to reflux to the top of the column without collection of distillate. The purpose of this is to 'wet' the column, this may also be done by boiling vigorously and flooding the column, then removing the heat and allowing the liquid to fall back into the flask. For the latter method a still head with a valve is necessary. It has been shown that such 'wetting' may increase the column efficiency by 100 per cent. After the 'wetting' process distillation can commence, more heat is gently applied until one or two drops are distilled every few seconds. The temperature is noted, this should remain steady in ideal conditions until distillation ceases when more heat must be applied for the next fraction. If a fractionating still head is used, the reflux ratio can be adjusted to suit the mixture. This ratio is found by counting the number of drops returned to the column from the reflux condenser during a given time interval, and the number of drops distilled over the same period. If the boiling points of the components are within a few degrees of each other the ratio should be high, e.g. 50 or 100 to 1. Where the difference is large it may be 10 to 1.

For complex mixtures the receiver may have to be changed often to obtain reasonably pure samples. Unless elaborate precautions are taken the condensate temperature must be continually observed, and the receiver changed when the temperature rises 2–10°C according to the mixture and column efficiency.

Perhaps the most sophisticated fractionating system is the spinning band column. This is simply a glass tube having inserted into it a metal strip that is twisted about its axis (*Figure 2.34*). During distillation this is revolved at speeds up to 4000 r.p.m. Its great virtue is its low hold up volume which can be as small as 2 cm^3. The H.E.T.P. is usually better than 1 cm. Commercially available apparatus is very expensive and is designed for use with or without a vacuum. The column is vacuum jacketed and silvered. The spinning band, of which there are several designs, is connected to the motor drive via a magnet, thus eliminating possible air leaks around a shaft seal. Also supplied is an automatic fraction collecting system.

Figure 2.34. The spinning band column

Distillation under Reduced Pressure

Solutions boil at a lower temperature when the pressure above them is reduced. The primarily reason for 'vacuum' distillations is to make possible the purification of substances that decompose below their normal boiling point, or the concentration of solutions of substances that are heat sensitive and may suffer denaturation, for example, certain enzymes. Vacuum distillation is quicker provided the assembly is at hand, and perhaps safer especially when distilling water miscible inflammable solvents, as the fire hazard is greater in a laboratory than that of implosion. Implosion risk is minimized by the careful examination of the glass-ware before assembly. Cracked glass-ware should be repaired or destroyed and any that has scratches or scour marks or appears to be of uneven thickness should not be used.

The assembly is a closed one, and has one main modification to the normal atmospheric distillation assembly. This is the use of a Claisen distillation head which allows the use of a capillary air bleed *A*, *Figure 2.33.* During distillation air is sucked through the liquid via this capillary in a fine stream of bubbles from the bottom of the flask, and minimizes the chance of 'bumping'. This capillary tube can be made in two ways, in the first, a heavy walled capillary tube of 1–2 mm bore and 7–8 mm O.D. is pulled in a brush flame. It is fitted into a rubber bung for insertion in the Claisen head, the position of the bung is adjusted to bring the end of the capillary about 1 mm above the bottom of the flask. The second method is to use a cone joint having a long stem. The stem is pulled out in an oxygen gas flame into an ordinary Pasteur type pipette. This is presented to the assembly and the length required noted, then a final fine capillary is pulled. Both methods give capillaries of strength to withstand the buffeting of the boiling liquid.

To give finer control of the air bleed a 50 mm length of 6–12 mm rubber tubing can be inserted over the other end of the capillary, a length of fuse wire is threaded down the bore, the other end of which is secured to a screw clip gripping the rubber tube. The clip is tightened until the fine stream of bubbles is as required. The wire need not necessarily be used, but if the distilling time is extended the rubber tube may tend to close under its own volition.

The type of distillation flask, condenser and receiver is chosen to suit requirements. The receiver adaptor must of course have a side arm for connection to a trap, manometer and vacuum pump.

The systems described on page 28 are suitable. The flask is usually heated by an oil or water bath or an electric mantle.

The apparatus is assembled on stands, the liquid is added to the flask to half its depth only and then the flask is immersed in the bath fluid. Insert the air bleed and a thermometer, then with a screw clip and the stopcock on the trap open, the pump is turned on. The stopcock is closed and the screw clip adjusted to give a fine stream of bubbles. When the maximum vacuum to be used is reached the bath may be heated, this is brought to a temperature a few degrees above the boiling point as indicated by the thermometer in the assembly. Excessive heating may induce 'bumping'.

The closing down process after completion of distillation is the reverse of the above. First the heat is turned off, then the air bleed is opened to prevent the remaining liquid rising up the capillary, the stopcock is then carefully opened to release the vacuum and finally the pump is turned off. If the assembly is re-evacuated to obtain a further fraction after changing the receiver, it may be that the liquid will distil vigorously or even violently for a while due to the transfer of excess heat from the bath fluid to the liquid in the flask whilst the latter was at atmospheric pressure. This difficulty is overcome by using a system in which distillation does not stop. This is achieved by using either a form of Perkin triangle (*Figure 2.35*) or a multiple receiver adaptor containing several flasks which may be presented in turn for distillate by twisting the adaptor at the appropriate time.

The Perkin triangle has a receiver portion *A* which is attached to the condenser. It is connected to the pump at side arm *P* and the receiving flask *R* is attached. The commencement of distillation is the same as before, taps *B* and *C* being open and the 3-way tap *D* turned allowing vapour to flow between tubes *Q* and *S*. Distillate flows through *A* to *R*, when the second fraction is to be collected tap *C* is closed to retain further distillate in *A*, and *D* is turned such that the vacuum in *R* is released through *Q*. Flask *R* is changed, then *D* is turned to connect *Q* and *S*, thus equalizing the pressure in *A* and *R*. Tap *C* can now be opened. The volume of flask *R* should be kept to a minimum as on connecting it into the evacuated system the pressure is bound to rise, and boiling will cease momentarily. This may cause bumping if the pressure has risen too much.

One of the great problems in distillation especially with biological

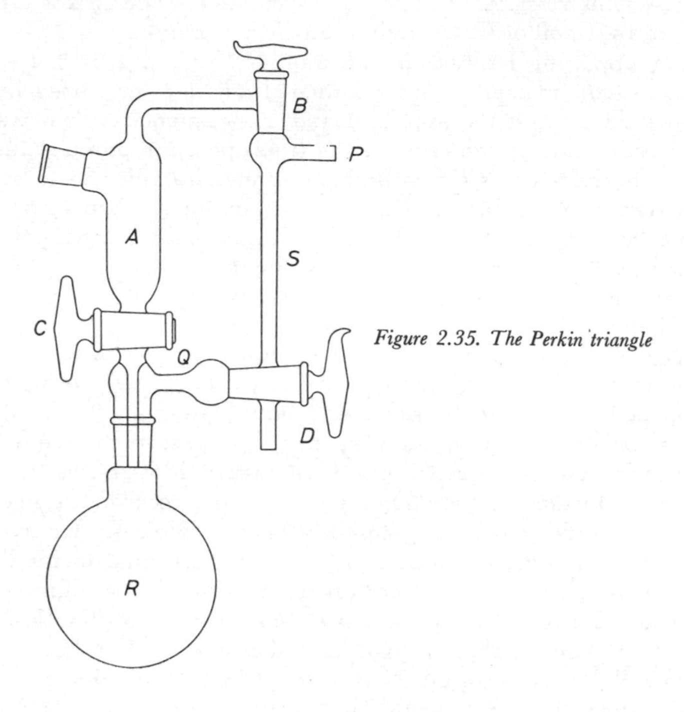

Figure 2.35. The Perkin triangle

material is foaming, and this is aggravated in vacuum distillations. It can be minimized by heating the distillation flask slowly to boiling point, and maintaining gentle distillation, and by heating the flask above the liquid level. The addition of capryl alcohol eliminates frothing completely, but has to be added continuously as it rapidly boils off. A more chemically inert substance is silicone fluid but it has the disadvantage of contaminating the glass-ware making it 'non-wetting'. Where such additions are not permissible a small volume of solution can be distilled in a large flask, for example, 100 cm^3 in a 2 dm^3 flask. More solution can be added dropwise on to the foam, which also tends to break it, and if a thermometer can be dispensed with, a second air leak can be inserted in its place bursting any bubbles that rise above the flask neck. A sure way of preventing foam leaving the flask is to introduce a red-hot wire or coil into the neck, but it must be remembered that its temperature is very high and it may char or change any dissolved material present in the foam. The coil can be of nichrome or platinum which is sealed into an extended cone joint and connected to a transformer of which the type used for hot wire glass cutting is suitable. Finally an anti-foam distillation head is available commercially.

Where large volumes of liquid have to be reduced by distillation under vacuum the cyclone or climbing film type of still can be used (*Figure 2.36*). They may be steam or hot water heated as well as electrically and are therefore suitable for solvent recovery. The solution boils in the heat exchanger arm and vapour and liquid are forced tangentially into the conical section creating a vortex, and all droplets and particles are stripped from the vapours. Under vacuum the liquid can be fed continuously through the tap below the cyclone. Where foaming is a problem a second cyclone fitted between the first cyclone and the condenser can be used. Liquid collected in this is fed back to below the first cyclone. The model manufactured by Quickfit and Quartz Ltd. is capable of distilling 5 dm^3 of water per hour.

Another form of still much favoured by biochemists is the rotary film evaporator (*Figure 2.37*). The original was designed by L. C. Craig but has since been modified and mechanized. The solution is held in a flask that is revolved in a horizontal position continuously exposing a thin film of the solution for evaporation. At the boiling point concentration is extremely rapid, and where large volumes have to be reduced a constant feed is incorporated The solution need not be taken to its boiling point or even above ambient temperature, but it is necessary in this case to use refrigerated fluid in the condenser for greatest efficiency so that the temperature difference is at least 20°C.

Molecular Stills

For the purification of high boiling point, high molecular weight (up to 1000) substances a molecular still can be used. This apparatus requires very low pressures down to about $1{\cdot}25 \times 10^{-2}$ Nm^{-2} (10^{-4} torr), and is based on a different principle as at this pressure the normal relationship between vapour and liquid does not apply. In this case the distance between the evaporative and the condensing surface must be less than the mean free path of the molecule. Within this distance most of the molecules leaving the liquid surface do not return and therefore must condense on the cooler surface. There are two basic types of molecular stills, pot stills and flowing film stills. A simple pot still is illustrated in *Figure 2.38*. An improved

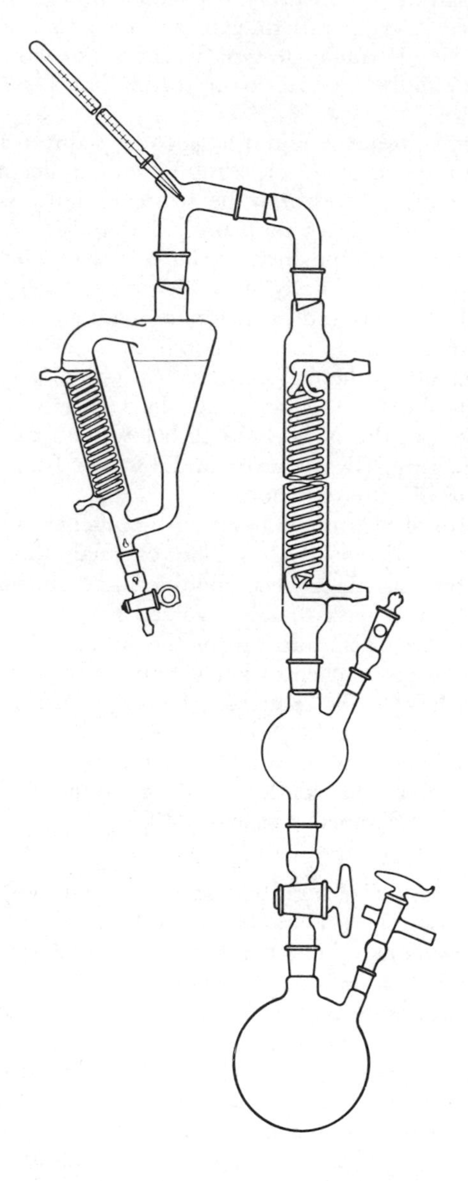

Figure 2.36. Evaporators: Cyclone evaporator

design by Gilson allows for fractionation (*Figure 2.39*). The horizontal boiler is heated by a resistance coil wound on a glass tube. The cold finger type condenser slopes so that the condensate runs back

Figure 2.37. Evaporators: Rotary film evaporator

and drips off a rim into a fractionating adaptor to the receiver. The boiler holds approximately 100 g of material which is inserted by removing the condenser and turning the boiler to the vertical. It is

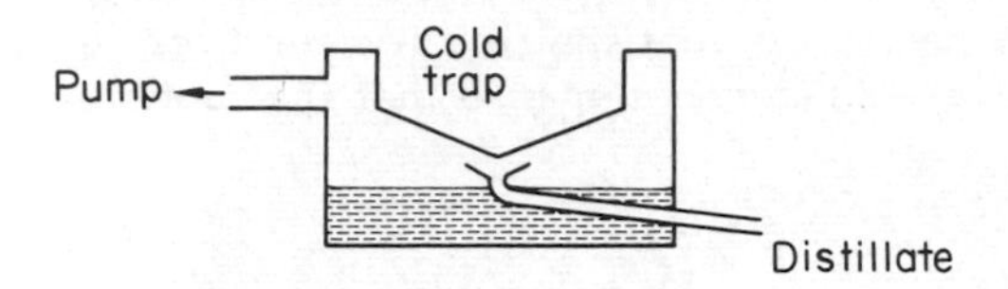

Figure 2.38. A simple pot molecular still

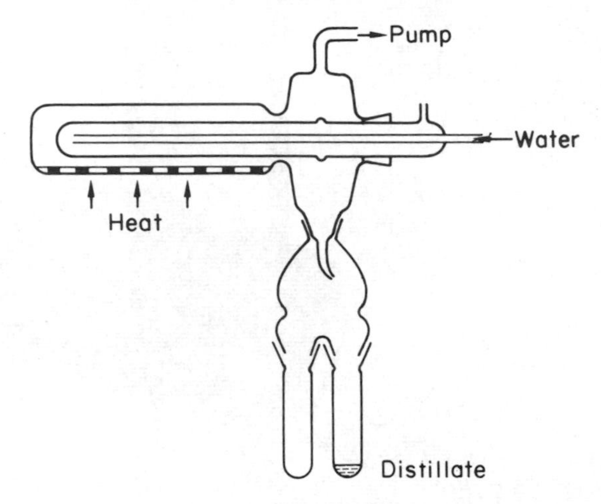

Figure 2.39. An improved molecular still with fraction collector

placed under vacuum in this position and gently heated to remove trapped gases, the vacuum is then released, the condenser replaced and the whole returned to the horizontal for the distillation. A high vacuum is obtained by using a vapour diffusion pump, these are described in Chapter 9.

Various multi-stage types have been designed based on the principle illustrated in *Figure 2.40*. Distilland is placed in trough *A*,

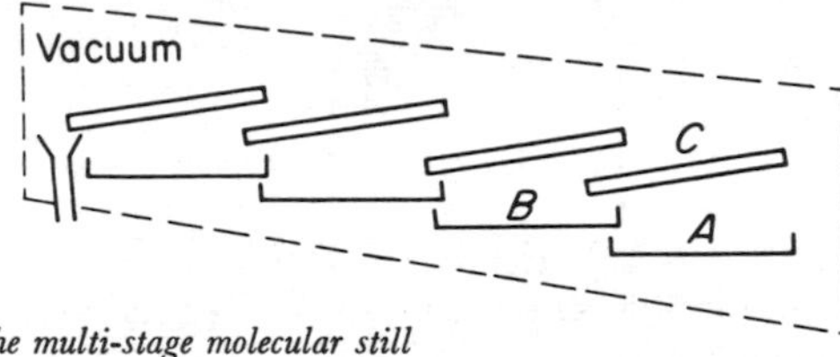

Figure 2.40. The principle of the multi-stage molecular still

this is heated and the vapour condenses on surface C from where it runs into trough B. This process is repeated through all the stages. A temperature gradient may be arranged over a number of troughs. The whole process is of course carried out under high vacuum.

In flowing film types it is generally easier to obtain the condenser close to the evaporating liquid. Three variations are in use. In

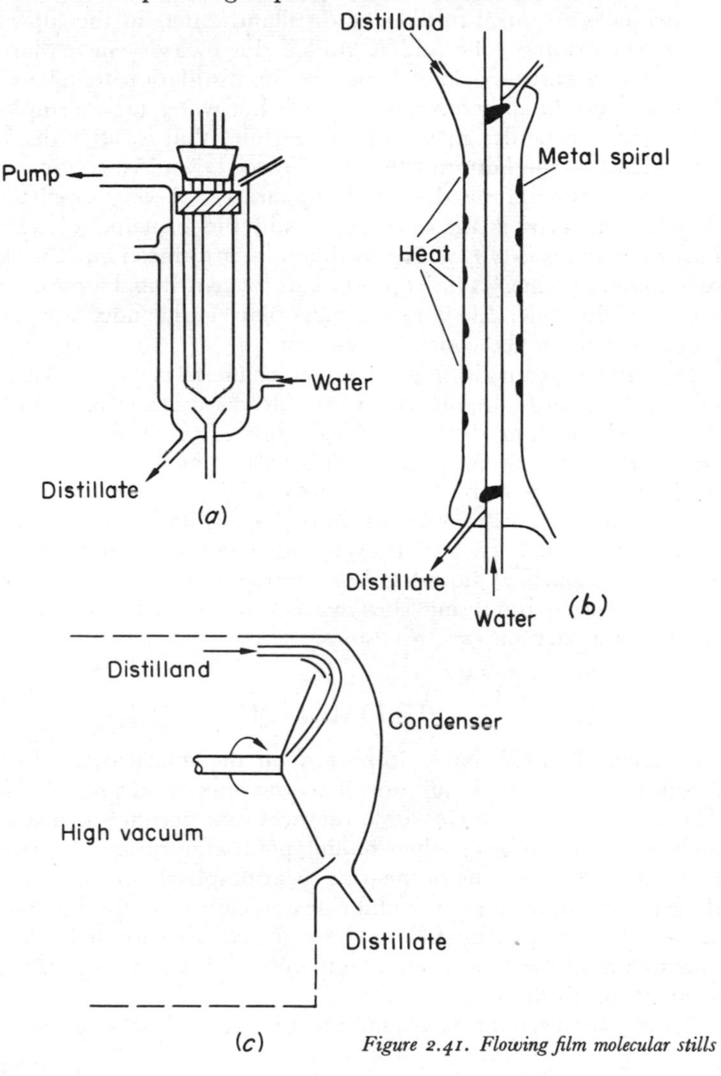

Figure 2.41. Flowing film molecular stills

Figure 2.41a the solution is introduced onto a wire mesh baffle placed around the top of a warm finger. By this means a film flows down the tube in close proximity to the condenser. Remaining distilland is collected into a reservoir from where it is returned to the feed vessel or is pumped directly onto the gauze. The apparatus in *Figure 2.41b* is designed the opposite way to the previous one. The condenser is a central tube. The distilland enters at the top of the outer evaporating tube and is guided downwards via a platinum spiral. This ensures a long exposure for distillation to take place. Heating may be done electrically or a hot water jacket employed. This type can be designed to be reversible, that is, after distilland has passed to the bottom the role of receiver and reservoir can be reversed by up-ending the whole apparatus. A very sophisticated flowing film type is the centrifugal still the principle of which is illustrated in *Figure 2.41c*. The distilland is introduced into the centre of a spinning cone. A fine film spreads by centrifugal force over the surface of the cone. Evaporation takes place and condensate collects on the surface of the concave condenser.

A form of steam distillation under reduced pressure is possible by injecting water vapour down the bleed capillary instead of air. A normal vacuum distillation assembly is preceded by another boiler which has an air bleed. This flask is heated to about 40°C so that the water in it boils under reduced pressure. The stream of air and water vapour passes into the distillation flask which is heated to about 37°C and prevents an increase in water content. Water and purified liquid collect in the receiver. The heating may be done by using small beakers as water baths. This technique is only necessary on rare occasions.

SUBLIMATION

This form of distillation is important in the purification of solids. A substance is said to sublime if its vapours condense as a solid. The method is most useful for substances that decompose near their melting point and before their boiling point at atmospheric pressure. Few substances sublime on heating at atmospheric pressure, among these are iodine, ammonium chloride and camphor. In the majority of cases it occurs by heating under reduced pressure or by heating in a stream of inert gas such as nitrogen, which has a similar effect as in steam distillation.

For sublimation at atmospheric pressure a flask with a Liebig condenser attached can be used, or simply a long-necked flask, the sublimate collecting in the neck or condenser. The latter should be closed with a loose plug of cotton or glass wool. An inert gas can be introduced through an adaptor or by using a 2-necked flask.

Sublimation under reduced pressure is most easily carried out in the apparatus shown in *Figure 2.42*. This is simply a cold finger condenser fitted into a flask via a straight receiver adaptor. For larger scale sublimation an apparatus similar to the molecular still can be used with a stainless steel mesh sleeve fitted on the

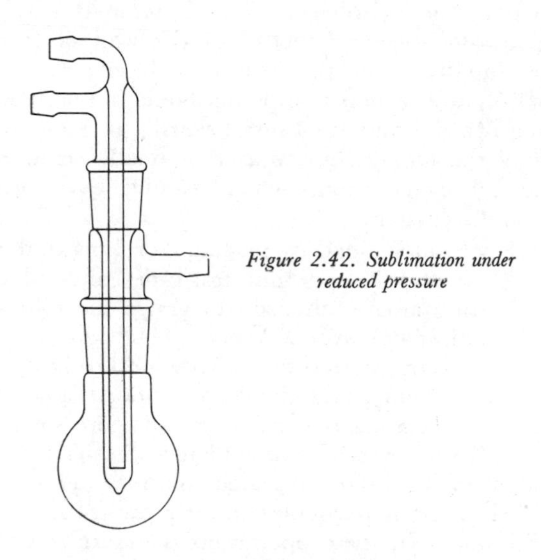

Figure 2.42. Sublimation under reduced pressure

condenser to retain the sublimate. Micro-sublimations are performed in a capillary tube or melting point tube. The substance which is at the bottom is heated in a metal block with the upper part of the tube being cooled with damp filter paper.

CRYSTALLIZATION

The next step in the purification of a compound is the separation from the remaining solid impurities. This may be achieved simply by direct crystallization, but before this is attempted further purification may be necessary by methods to be found in the ensuing chapters, such as adsorption chromatography, column electrophoresis and dialysis.

In the ideal case purification is brought about by producing a hot saturated solution of the impure compound and allowing it to cool, whereupon crystals of the compound begin to form. In organic chemistry this is rarely the case. A suitable solvent must first be found. This should have a gently rising solubility curve with the temperature; if solubility increases too much with temperature, crystallization takes place so rapidly on cooling that impurities are trapped, and if solubility varies little with temperature, crystals do not form. Very volatile solvents should not be chosen as the solid may dry out on the sides of the container through rapid evaporation at the periphery. This is sometimes referred to as 'creeping' as the solvent 'creeps' up the wall of the vessel and evaporates rapidly. For this reason such solvents as ether, low boiling petrols and carbon disulphide should not be used, especially when there is a fire and explosion hazard. It is essential that the solvent is of the highest purity as it is finally removed from the compound by evaporation, which would leave any dissolved impurity on the crystals.

A suitable solvent is found by placing 2 or 3 mg of the compound into each of several 75 × 9 mm test-tubes and adding solvents dropwise, heating after each drop to gauge the solubility and to see if it rises with temperature. When a choice has been made, the impure material is transferred to a narrow-necked flask (to minimize evaporation of solvent), and the solvent added slowly, heating at the same time. Heating is continued and more solvent added as necessary until the material is completely dissolved at or near the boiling point, giving a hot saturated solution. This solution must be quite clear and any particulate matter present must be removed by hot filtration. In these operations due care must be taken if toxic or inflammable solvents are used. In these cases a reflux condenser, is fitted or a fume cupboard used. The vessel is covered and the clear, hot, saturated solution is then allowed to cool.

One of four things will occur, either crystals will slowly form, or the substance come out of solution in a few seconds, an oily, non-crystalline deposit forms, or the solution remains clear and homogeneous. In the first case the solution is allowed to stand until crystallization is complete, it is then filtered. The crystals may be recrystallized one or more times until the melting point is constant. The filtrates (called the *mother liquors*) are usually collected together, and if the preparation is to be repeated the bulked liquors can be evaporated to dryness, a saturated solution prepared and a bonus crop of crystals collected. The formation of crystals may be helped by cooling in a freezing mixture or refrigerator. It may take up to 48 h for complete crystallization.

In the second case of too rapid crystallization, purification is poor. This is overcome by cooling the solution very slowly, and may be done by placing the flask of hot saturated solution into a bath at the same temperature, which in turn can be lagged.

Where oils are formed, the mother liquors are decanted and the oil redissolved in fresh solvent, and crystallization attempted again using the slow cooling method above. If this is still unsuccessful other solvents may be tried, and if these fail the impurity content is still too high and must be reduced by other techniques before further crystallization attempts are made.

If a super-saturated solution is formed crystallization may be induced in several ways, the most common being to scratch the wall of the flask below the liquid surface. This releases microscopic particles of glass which form nuclei for the crystals to 'grow' around. With 'difficult' substances it may be necessary to scratch for a long time before tiny crystals appear. The flask may become permanently marked and should not be used for other processes or assemblies. A tiny crystal of the pure compound may also be used as a nucleus. This is called 'seeding' and the temptation to drop in more than one minute particle must be resisted. Recrystallization is invariably easier.

Where no results are obtained by the methods described, usually due to the absence of the ideal solvent, mixed solvents can be tried. For this method two miscible solvents are required, one in which the substance is too soluble, and one in which it is insoluble or nearly so. A hot solution is made in the first solvent, then the second solvent also hot is added dropwise until a slight persistent turbidity appears. This turbidity is removed by cautious addition of the first solvent and then the solution is allowed to cool. In some instances at the first crystallization it may be necessary to leave this slight turbidity.

Attempts to crystallize large molecules such as proteins may be made difficult by the denaturation effect of high temperature which causes an irreversible change so that they become biochemically inactive. In this instance, solvents are chosen with low boiling points and the temperature difference needed for crystallization to occur is obtained by use of freezing mixtures and low temperature refrigerators and baths.

Some useful mixtures are acetone or alcohol with water; acetone benzene, chloroform or ether with petroleum ether; and alcohol, ether or water with pyridine.

An excellent aid to crystallization where the contaminant is coloured is the use of decolorizing activated charcoal. This is made from bone or wood charcoal, which has the ability to absorb. It is used by adding a heaped half-inch spatula full to every 100 cm^3 of *cold* solution, then bringing to the boil, simmering a few minutes and filtering hot. A No. 42 Whatman filter paper will effectively retain the finely divided charcoal. The compound to be crystallized may also be absorbed so an excessive amount of charcoal should not be used.

MELTING POINT DETERMINATION

The purity of crystals is most easily checked by determining their melting point. The recording of the melting point as a physical constant is necessary in any case, especially where a new compound has been synthesized. The majority of organic compounds have sharp melting points, that is, they melt over a temperature range of 0·5°C or less.

The sample is usually placed in a capillary tube which is then attached to a thermometer and either heated in a bath of high boiling point liquid or in an electrically heated metal block. The capillary tubes are about 7·5 cm long and 1 mm bore, and can be made by pulling 10–15 mm bore glass tubing after softening in a brush blowpipe flame. The long capillary is then cut using a glass knife by resting the knife on the fragile tubing without pressure, drawing it across and then snapping the length off. One end is sealed by placing it in the edge of a very hot flame, care being taken to prevent a thick glass bead forming.

A sample of the damp crystals is dried on a porous tile, between filter papers or by placing on a microscope slide and heating under an infra-red lamp. The open end of the capillary is then pressed on the crystals, pushing them into the bore. The tube is then inverted and gently tapped with the finger or vibrated by drawing a file across the top until the crystals fall and collect at the bottom to a depth of 2–3 mm. In stubborn cases where the crystals will not fall to the bottom the capillary can be dropped down a 60 cm length of 5 mm bore glass tubing that is held vertically on the bench. The melting point tube bounces several times, the shock of which makes the crystals fall.

The bath method of heating is rarely used in practice except for economy reasons in teaching laboratories. The simplest apparatus is a boiling tube fitted with a cork having three holes (*Figure 2.43a*). The centre hole holds the thermometer in position, a wire stirrer passes through another hole and the third hole ensures against an accidental closed system. About 15 cm^3 of bath fluid is placed in the tube and the thermometer is adjusted so that the bulb is immersed half-way in the liquid. The thermometer is then removed, and the prepared melting point tube laid against the stem with the substance adjacent to the bulb. No ties are required as the tube will remain in position due to capillary attraction. The thermometer is placed in the liquid which is then heated by a microflame and stirred gently by raising and lowering the wire loop.

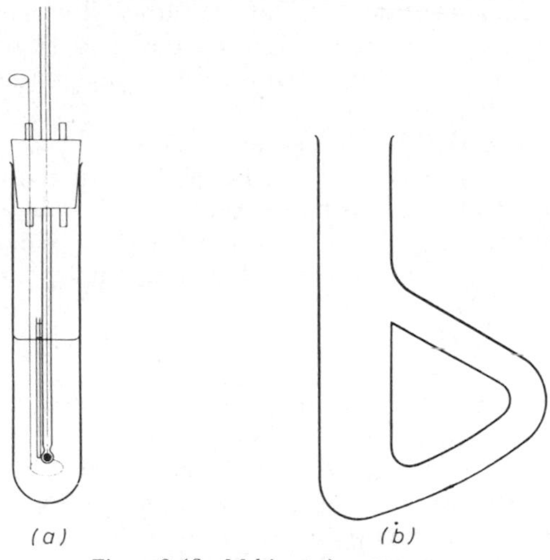

Figure 2.43. Melting point apparatus
(*a*) *Simple form* (*b*) *Thiele tube*

At the melting point the temperature should be rising between 1–2°C/min, at this speed an indication that the melting point is imminent is given by the compound contracting a degree below the actual melt. When the melting point is unknown two determinations should be made, one to find the approximate point and the other to accurately fix it.

The presence of impurities depresses the melting point. If the melting point is known it is easy to check the purity of the sample. It also helps when trying to identify a substance although many compounds have similar or identical melting points. When this is the case the melting point of the unknown is found, then a 50/50 mixture is made with a pure sample of the substance it is believed to be. A melting point is performed on the mixture and if it is the same as the individual compounds, they are identical chemically. If not, the melting point will be lower by several degrees. This technique is called the *method of mixed melting points*.

The ideal liquid to use in the melting point apparatus is silicone fluid that is sold for the purpose, which can be heated to 300°C. Concentrated sulphuric acid is often used but is not recommended for the apparatus just described as pieces of cork will char and colour the acid. Other faults are that the acid fumes above 250°C and the possibility of accidental breakage of the glass container, especially where a score or more melting point sets are being used in a class, will constitute a hazard. Other fluids employed are liquid paraffin, di-butyl phthalate, glycerol and castor oil. They all have limitations usually of temperature range and discoloration from decomposition.

A beaker or a Kjeldahl flask can also be used as a bath. The flask is more convenient as it can be gripped on the stem and the bulb heated directly in the flame, and fuming is less troublesome at high temperatures. Both require stirring loops to mix viscous liquids. Another container is the Thiele melting point tube *Figure 2.43b*. This is a boiling tube with a loop of tubing which is filled with the liquid. The loop is heated at its extremity and mixing occurs by convection.

Electrical melting point apparatuses are designed around two principles, the temperature gradient along a bar heated at one end and an electrically heated drilled block controlled by a variable transformer, resistor or an energy regulator (Chapter 10). The latter type is the more popular in the U.K. The standard models have an aluminium or brass block drilled to accept a thermometer and up to 3 capillary tubes. Two heating elements are clamped in intimate contact with the block, one is connected to the regulating device and the other is switched in for rapid heating to a desired temperature. A built-in lamp and lens system is provided for observation of the melts. A micro-apparatus using the same principle has been designed for use on a microscope stage. The sample is placed on a round glass cover slip and single small crystals can be observed. Where substances sublime the capillary tube must be closed at both ends.

DRYING

When the desired purity has been reached the crystals must be thoroughly dried. Drying in this instance means removal of traces of the solvent from which the substance has been crystallized and not necessarily the removal of water. The method used depends somewhat on the characteristics of the substance itself and the time that can be given to this process.

If the crystals have been gravity filtered the simplest method of drying is to carefully remove the filter paper from the funnel, open it out on another layer of filter paper and allow it to dry by evaporation. The crystals may be covered with more filter paper or with a sheet of glass that is raised a little above them by spacers (e.g. corks).

The speed of evaporation can be increased by gently warming the crystals from above with an infra-red lamp. Heat stable compounds are no problem, the crystals in a shallow dish can be placed in an oven or a vacuum oven (see Chapter 10). Care must be taken to dry below the melting point and not to drive off water of crystallization if present.

Where small samples are to be dried or accurate temperature control is required the vacuum drying pistol can be used or its electrically heated counterpart. A desiccant is used to increase drying efficiency.

The pistol (*Figure 2.44a*), (so called because the original design resembles one) consists of a tube *A* that holds the sample for drying and a flask *B* that holds the desiccant. These fit together and can

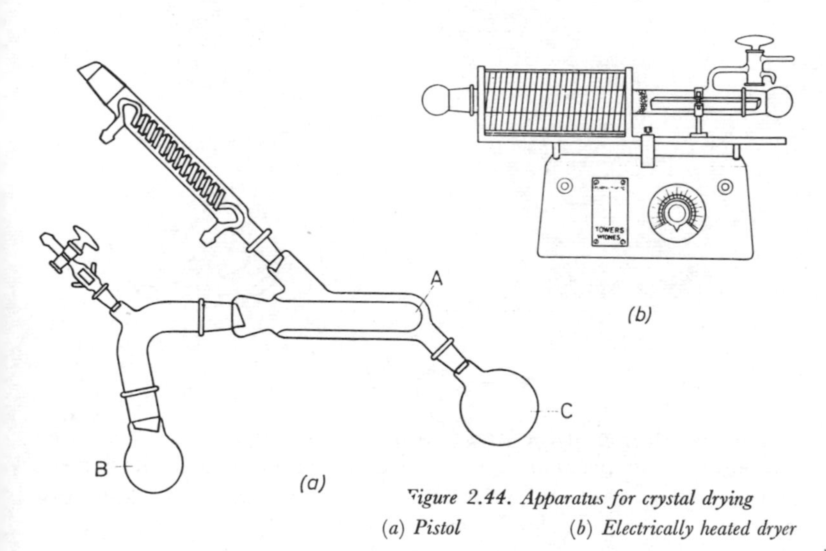

Figure 2.44. Apparatus for crystal drying
(*a*) *Pistol* (*b*) *Electrically heated dryer*

be evacuated. Tube *A* is inserted in a vapour jacket. The temperature of the sample compartment depends on the vapour which in turn depends on the boiling point of the liquid chosen for boiling in flask *C*.

The sample is spread out in a porcelain or platinum boat or on a filter paper which is then inserted in *A*. The desiccant, usually phosphorus pentoxide or fresh paraffin wax shavings, is placed in *B* and the apparatus assembled. The drying chamber is connected to a filter pump and evacuated. The liquid in *C* is then heated until it is gently boiling and refluxing occurs. This is allowed to continue until the sample is dry. The liquid chosen should preferably be non-inflammable.

In the electrical apparatus (*Figure 2.44b*) a Nichrome heating coil is substituted for the vapour jacket. The operation is similar except that the temperature is controlled by a calibrated energy regulator. A similar apparatus but on a micro-scale has been designed by A. C. Thomas*. It features E.C. glass tubing which is glass tubing coated with an electrically conducting layer. Electrical contact is made with silver bands and the coated section is used as a resistance heater, the temperature being controlled by a calibrated variable resistor in series.

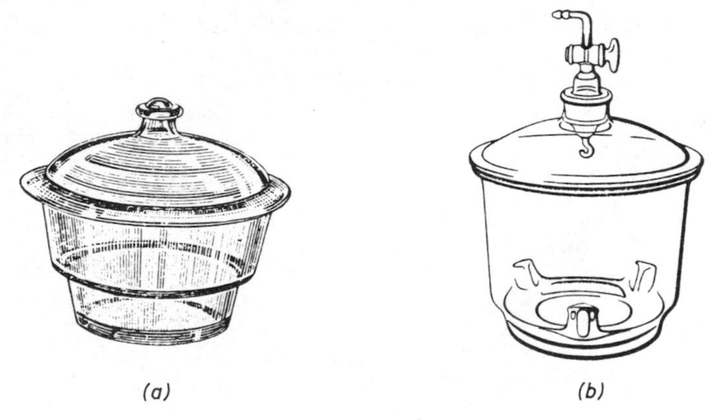

Figure 2.45. Types of desiccator

Vacuum drying in the presence of a desiccant and at an elevated temperature is perhaps the most efficient method if it can be employed but where the substance is not heat stable the desiccator

* *Analyst, Lond.* **85** (1960) 1771

can be used (*Figure 2.45a* and *b*). This usually consists of two halves; the lower contains the desiccant and samples, the upper is a lid that may have a stopcock. An air-tight seal is made between the two halves by ground contact surfaces. The internal working diameter is from 5 to 30 cm. A raised shelf is often supplied made of perforated porcelain, zinc or stainless steel. The desiccant is placed below this or in dishes on it.

The ground surfaces and the stopcock must be evenly coated with a vacuum stopcock grease such as apeizon or Vaseline stiffened with paraffin wax. Alternatively, special rubber seals have recently been introduced. The tube from the stopcock protruding into the desiccator is bent so that inrushing air is evenly dispersed by the curved shape of the lid. This prevents the disturbance of a light fluffy material being dried, also the air flow can be slowed down by placing a filter paper over the stopcock inlet.

Several desiccants are commonly employed in a desiccator. The most efficient is phosphorus pentoxide but this will quickly form a 'skin' so is best used as a thin layer in a shallow dish such as a 10 cm Petri dish. Potassium hydroxide pellets are very efficient and are often used side by side with concentrated sulphuric acid. The latter is placed in the bottom of the desiccator with either glass beads or pumice granules to the same level. This prevents slopping of the acid when the desiccator is moved. The KOH is placed in a dish that will not tip easily such as a petri dish or crystallizing basin, and then stood on the shelf. Calcium chloride granules can be used being as effective as sulphuric acid and safer. For drying off organic solvents silica gel or fresh paraffin wax shavings can be used.

Glass vacuum desiccators are considered by some technicians to be a safety hazard. They believe they should never be used. It is a fact that in most instances they need never be used as an ordinary desiccator is sufficient. When they are used a 'high vacuum' should not be applied, i.e. below 7000 Nm^{-2} (50 mm Hg), a basket type shield must be used and the glass surface be without scratches or blemishes. Some industrial establishments do not allow the use of all-glass vacuum desiccators and insist on the metal type that has a glass observation window in the lid. Sometimes desiccator lids stick or the stopcock will not open to release the vacuum. This should not occur if the ground surfaces are properly lubricated. When ordering desiccators it is worth bearing this possibility in mind and choosing a type that has a chance of being opened without undue danger.

An evacuated desiccator should never be carried. If it cannot be opened, the first step is to apply stopper removing solution to the stopcock and allow time for it to penetrate. A suitable solution is:

Glycerol	5 parts
Chloral hydrate	10 parts
3 M-Hydrochloric acid	3 parts
Distilled water	5 parts

If this is unsuccessful, transfer the desiccator by trolley to an incubator and allow it to warm, or gently heat the stopcock with a 'cool' Bunsen burner flame. Wear an asbestos glove whilst attempting to turn the stopcock; if this fails, the final method is to score the tubing below the stopcock (if there is any) and crack it by applying the molten end of a glass rod. Air will slowly seep through the crack and eventually the whole stopcock can be safely snapped off. Throughout these attempts the worker must be protected from implosion by shields.

Desiccants are often placed in a vacuum oven and therefore should be chosen with care as spillages may corrode the chamber wall. Activated alumina is most suitable as it is readily reactivated *in situ* by heating for several hours at 175°C.

Drying of Solvents

The removal of water from solvents is usually done by adding solid dehydrating agents. These must be chosen with due consideration for the chemical properties of both agent and solvent. The latter must not dissolve the drying agent or react with it. The chemicals used are either oxides, alkalis or dehydrated salts. A list of the common drying agents is given below, the variety is necessary to cope with the different situations that arise. Most of them are equally suitable for use in a desiccator.

Calcium sulphate Magnesium sulphate Copper sulphate Sodium sulphate	Suitable in the majority of cases as they are neutral and inexpensive. Calcium sulphate is the most rapid but has not the absorbing efficiency of the others.
Calcium chloride	Has great absorbing power but forms addition products with many solvents, e.g. acetone, alcohols.

Potassium carbonate
Aluminium oxide (Alumina)
Calcium oxide
Magnesium oxide
Barium oxide
Boric oxide

Phosphorus pentoxide	Where this can be used it is the most effective. A preliminary drying by one of the neutral agents is recommended.

Potassium hydroxide
Sodium hydroxide

Sodium metal	A preliminary drying by another agent is recommended before use.

Another agent that is on a par with phosphorus pentoxide for efficiency is magnesium perchlorate, but is rarely used as there is a real danger of explosion with many solvents and for this reason is not recommended, although it is comparatively cheap.

In purification procedures solutions of substances may have to be dried, therefore the drying agent must not react with, or salt out the substance or have any catalytic effect.

In practice many of the common solvents are finally dried over sodium wire. This is performed by using a sodium press (*Figure 2.46*), this is essentially a cylindrical die with a fine orifice at one end and

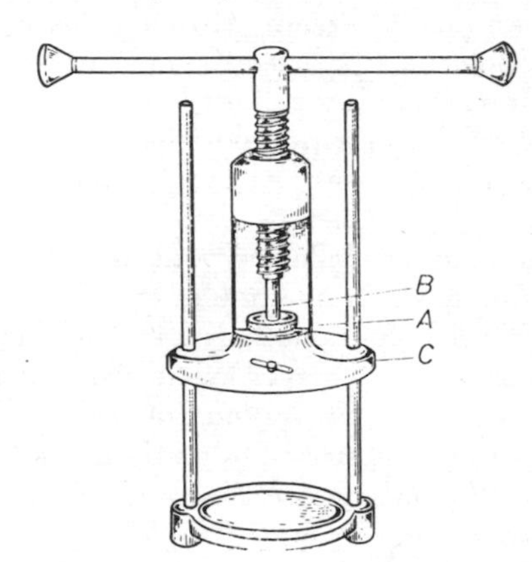

Figure 2.46. The sodium wire press

a plunger *A* and *B*. The die is fixed in a casting *C* and the plunger is actuated by a screw thread. The solvent, which may be in a bottle of up to Winchester quart sized is clamped to an adjustable table under the die. Lumps of sodium are placed in the die using tongs or a knife, about 10 g/dm^3 of solvent to be dried. Sodium stored under paraffin must be wiped with filter paper first. The plunger is screwed down into the die and the sodium is extruded into the solvent as bright unoxidized wire. Hydrogen will be evolved as the sodium reacts with the water, therefore leave the bottle cap loose or close with a bung fitted with a calcium chloride tube. When no more hydrogen is evolved, decant the solvent into a clean dry bottle and add a few small flakes of fresh sodium.

Immediately after using the press, clean the die and plunger by rinsing with ethanol or methanol. If this is not done the plunger will become jammed and the die badly scored. The residual sodium left in bottles after the dry solvent has been used must be treated with alcohol until all the sodium has disappeared and washed out before returning the bottle to stores or the suppliers.

PURIFICATION AND RECOVERY OF SOLVENTS

It must be pointed out that pure solvents (analytical grade) are obtainable from the chemical manufacturers and after drying are suitable for most purposes, and it will normally be found uneconomical for the laboratory to purify the technical grade. If it is desired to do so, excellent methods are to be found in *Practical Organic Chemistry* by Vogel. Where large volumes are regularly being used the recovery of solvents may be a worth while procedure but even this is questionable except with solvents such as chloroform or pyridine. Solvents for spectroscopy have to be especially pure and are sometimes ten times the price of the analytical agent, even so they are best purchased considering the difficulties involved and that only relatively small quantities are required.

Solvents for recovery have to be collected and stored until a reasonable bulk has accumulated. Note should be taken of the regulations where the material is inflammable. If no suitable store is available, it is again better not to recover but dispose of them. It must also be borne in mind that the vapours of many solvents are dangerously toxic, for example, dioxan and carbon disulphide.

The recovery methods that follow can obviously be varied where the contaminating substance is known, for example if the solvent has been used in an acidic extraction process the acid can be neutralized and washed out.

Where two or more solvents miscible with each other are mixed it will be found difficult to separate them sufficiently for re-use in the laboratory unless expensive fractionating apparatus is employed. It is a common experience that chemists and research workers distrust recovered solvents and will not use them, in fact many insist on analytical grade solvents which are in many cases unnecessary.

The following methods are in note form as the details of the procedures used, i.e. distillation, separation and extraction, have already been given earlier in the chapter. *All* recovery procedures entail a distillation for separation from dissolved solids and partial separation of liquids.

Acetone b.p. 56°C

Distil residues collecting the distillate up to 60°C. Dry over anhydrous potassium carbonate or calcium sulphate. Decant and redistil collecting the 55–60°C fraction which is suitable for re-use for crystallization and apparatus drying. Higher purification requires careful fractionation collecting at 56–56·5°C.

Amyl alcohol b.p. 131°C

Distil under reduced pressure. Dry over anhydrous potassium carbonate or calcium sulphate. Distil using an oil bath collecting 125–135°C fraction.

Benzene b.p. 80°C

Distil residues collecting up to 81°C. Wash with M sulphuric acid, 50 cm^3/dm^3 until washings are colourless. Wash once with water then once with 2 M sodium hydroxide and finally wash with water until washings are neutral. Dry over calcium chloride and then sodium. Distil, collecting between 79·5 and 80·5°C.

Sulphur-free benzene may be obtained by distilling analytical grade solvent from phosphorus pentoxide.

Butyl alcohol b.p. 117°C

Distil at atmospheric pressure using a mantle, dry over anhydrous potassium carbonate or calcium sulphate. Distil, collecting the 117–118°C fraction. The higher alcohols are treated in a similar manner.

Carbon disulphide b.p. 46°C

Distil on water bath of the immersion heater type having a

true thermostat set at 60°C, i.e. not a thermoregulator. (Vapours may be ignited by steam pipes and electric light bulbs.) Dry over calcium chloride, redistil as before collecting between 46 and 47°C.

Carbon tetrachloride b.p. 77°C

Distil then wash twice with concentrated hydrochloric acid (ca. 25 cm^3/dm^3), once with water, then twice with 2 M sodium hydroxide. Wash with water until washings are neutral. Dry over calcium chloride, then distil collecting between 75–78°C.

Chloroform b.p. 61°C

Recovery is identical to the carbon tetrachloride method above.

Dioxan b.p. 101·5°C

Forms explosive peroxides and is toxic.

If it is dry it may be refluxed for 2 h over sodium then distilled using a Vigreaux column, collecting the 100–103°C fraction. Wet residues can be dried with two batches of potassium hydroxide, the second being left 24 h. Reflux and distil from sodium as before.

Ether (*Diethyl ether*) b.p. 35°C

Forms explosive peroxides. Distil on water bath. If the distillate is neutral or acid, stand over sodium hydroxide pellets for a day or two. Distil from fresh sodium hydroxide. If the original distillate is alkaline, wash with concentrated hydrochloric acid several times and then with water. Dry over sodium hydroxide and distil. (Never to dryness.)

Often the residues are only wet and all that is necessary is to wash with freshly made saturated ferrous sulphate solution (20 cm^3/dm^3) then with water, finally dry over anhydrous calcium chloride for 24 h, then over sodium wire.

Ethyl acetate b.p. 77°C

Distil, collecting between 70–85°C. Wash with an equal volume of 5 per cent sodium carbonate and then with saturated calcium chloride solution. Dry over anhydrous potassium carbonate or magnesium sulphate. Distil using a fractionating column (Fenske helices) collecting at 77°C.

Ethyl Alcohol b.p. 78·5°C

Distil using a briskly boiling water bath. Redistil over sodium hydroxide pellets (50 g/dm^3). Measure the specific gravity and repeat the distillation if this is above 0·81 which is equivalent to 95 per cent.

Absolute alcohol (not more than 0·8 per cent water) is 'super dried' by dissolving 7 g of sodium per litre of absolute, add 30 g (26·9 cm^3) ethyl phthalate per litre and reflux 2 h. Distil using a long Vigreaux column. Storage under dry conditions is essential.

Methyl alcohol b.p. 65°C

Distil over sodium hydroxide pellets (25 g/dm^3). Dry and reflux over barium oxide. Distil over fresh barium oxide (BaO). Absolute methyl alcohol is dried by the same procedure as for absolute ethanol.

Methyl ethyl ketone b.p. 79·5°C

Distil, collecting the 79–81°C fraction. Reflux over potassium permanganate (2 g/dm^3) for 2 h. Distil using a Vigreaux column. Stand over ignited potassium carbonate for 12 h. Reflux for 2 h over fresh carbonate then distil using a long Vigreaux column collecting at 80°C.

Propyl alcohol b.p. 82°C

Distil, then dry over sodium hydroxide pellets (25 g/dm^3). Decant and distil over fresh sodium hydroxide collecting 81–83°C fraction.

Petroleum ether

Distil collecting over the boiling point range of the original. Dry over calcium chloride and redistil.

It is worth purifying commercial petroleum ether for some purposes. This means the removal of unsaturated hydrocarbons by washing with concentrated sulphuric acid (100 cm^3/dm^3) until acid washings are colourless. Wash with a solution of potassium permanganate in M sulphuric acid until no reduction in colour occurs. Wash with water then with sodium carbonate solution. Dry over sodium sulphate, distil and store over sodium wire.

Pyridine b.p. 115·5°C

Distil on an oil bath collecting between 90–120°C. Add sodium hydroxide sticks, separate aqueous layer. Reflux over a little barium oxide. Distil, collecting the 112–116°C fraction.

Toluene b.p. 110°C

Distil from a water bath using reduced pressure. Wash with M sulphuric acid followed by water, then with 2 M sodium hydroxide and finally with water until washings are neutral. Dry over calcium chloride and distil from an oil bath.

Xylene b.p. 138°C

Proceed as above for toluene.

Greater details and other methods will be found in volumes listed in the bibliography.

RECOVERY OF MERCURY

Mercury is very useful in the laboratory both for its chemical and physical properties. Because of its high density and the fact that it is the only liquid metal at room temperature it is extremely useful for calibration purposes and for thermometry and manometry.

Mercury is obtained and stored in stoneware or polythene bottles, the latter have a removable plug with a fine hole drilled in the centre making it possible to dispense small quantities. Small polythene wash bottles are also suitable.

As mercury is so heavy it is more easily spilt when being poured so it is advisable to carry out this operation over a tray. Spilt mercury must be collected up and stored in a bottle labelled 'mercury for recovery'. This should be done without undue delay as mercury vapour is very poisonous, its effects being accumulative. Where mercury cannot be recovered after a spillage, for example, from cracks and crevices and between floorboards, it must be rendered harmless by sprinkling flowers of sulphur liberally over the area and leaving it for 24 h. This converts it to the sulphide which can then be safely removed with a vacuum cleaner. When mercury is used in open dishes or in spindle seals it should be covered where possible with water or oil to prevent its evaporation.

Globules of mercury can be picked up from the bench by several methods:

1. By using mercury tongs. The tongs are shaped like a deep teaspoon cut in two. The halves when closed round a drop of mercury will pick it up with unsuspected ease.
2. By mercury pipettes. These employ suction, the mercury being

drawn through a capillary into a receiver. A simple apparatus is illustrated in *Figure 2.47a* made from a filter tube, a small Buchner flask is also suitable.

3. By sponge collector. This is a plastic container with a screw-capped lid that has a disc of foam plastic attached. This is pressed on to the spilt mercury and retains it in its pores. On screwing the lid back the mercury is expelled into the container.

Mercury Cleaning Methods

The traditional method is to first remove dust and surface scum by filtering through an ordinary filter paper pierced with one or more small holes. A quicker way is to use a filter funnel having a grade O sintered plate and a Buchner flask attached to a pump.

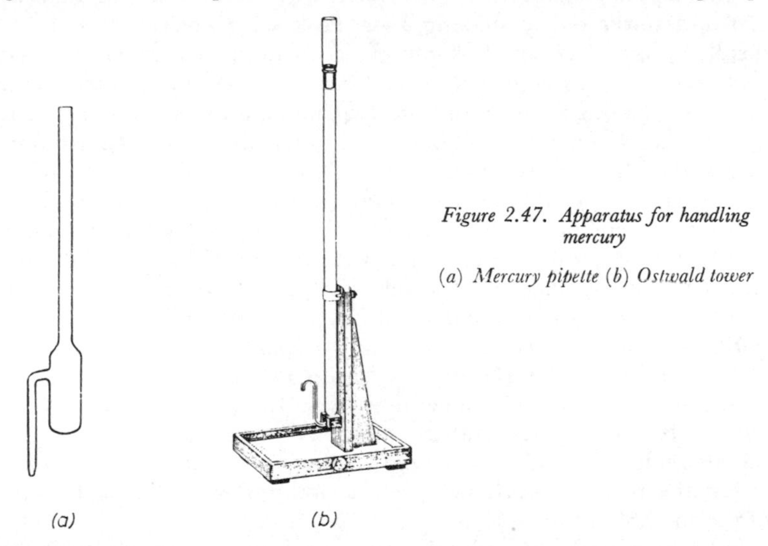

Figure 2.47. Apparatus for handling mercury

(a) Mercury pipette (b) Ostwald tower

Another method is to squeeze the mercury through a chamois leather cloth. The sintered filter will be found the most convenient.

The main contaminants are amalgamated base metals. These are volatile under mercury distillation conditions and must be removed first. This can be done in two basic ways:

1. By oxidizing these metals thus rendering them non-volatile. This is done in a machine which vigorously agitates the mercury causing fine droplets to be formed in air. After several hours of this treatment the mercury is left to stand to allow the oxide to

rise to the surface, this is then filtered off.

2. By passing mercury through dilute nitric acid in an Ostwald tower (*Figure 2.47b*). The glass column, about 1 m tall, has a capillary S-bend at the bottom which is charged with clean mercury. Five to twenty per cent dilute nitric acid is poured on to this to within a few centimetres of the top. A small wide-necked sintered funnel sits in the top into which contaminated mercury is poured. The mercury falls through the nitric acid as finely divided droplets and is washed of its metal contaminations. These are mainly copper, cadmium and zinc. The clean mercury passes into a receiver from the S-bend, dust being excluded by a cotton plug. On no account should concentrated nitric acid be used as a violent reaction will result.

The mercury should then be washed with water either in another Ostwald tower or by shaking in a flask with several changes of distilled water. A simple method of washing in acid is to place the mercury in a Buchner flask and cover it with dilute nitric acid. Fit a bung having a long glass tube protruding into the mercury, then agitate the contents by sucking air through the tube with a pump attached to the side-arm. The nitric acid is then washed out by placing the flask in a sink and connecting the water supply to the tube.

Mercury can be dried by stirring it with a folded filter paper and then filtering through a paper with a small hole in the bottom. This soaks up the majority of water. For perfect dryness the mercury is placed in a distillation flask and heated to 100–150°C preferably under vacuum and certainly in a fume cupboard.

For absolute purity the mercury must be distilled several times. This can be done using a conventional vacuum distillation apparatus with a larger air leak and an aircooled condenser, any volatile metals remaining will be oxidized in the air stream. Before distillation the mercury must be given a preliminary cleaning by one of the methods already given.

Where mercury has to be distilled regularly a specially designed still can be used. The special feature of the design is that the vaporized mercury is condensed into a long, vertical delivery tube which under working conditions acts as a vacuum pump of the Sprengel type.

The apparatus is illustrated in *Figure 2.48*. Mercury for distillation is placed in the reservoir *R* which is suspended by a spring. This operates a simple safety device as when the reservoir is empty its raised position opens a micro-switch that turns off the heater current. Pure mercury is placed in the siphon *S* at the bottom of the delivery tube D. The reservoir is filled and the condenser water turned on and tube T attached to a mechanical vacuum pump. The apparatus is evacuated to a pressure of 100 Nm^{-2}, and the mercury rises into the boiler to within 1 cm of the lip of the delivery tube. Heat is applied gradually to reduce thermal shock. When the mercury boils it falls down the delivery tube after

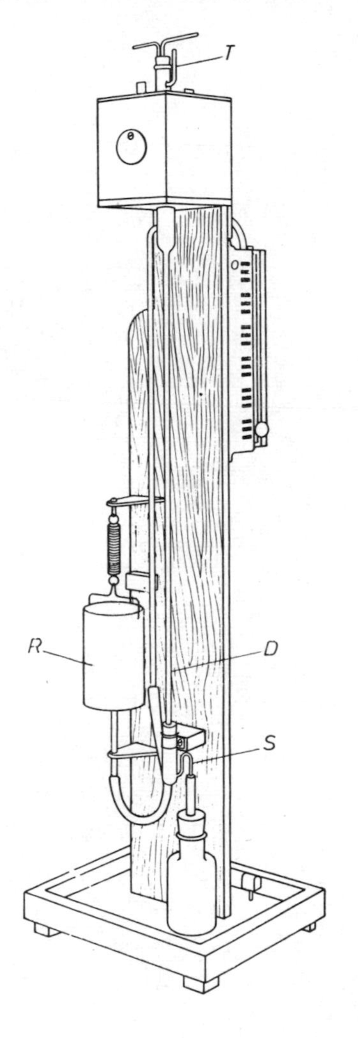

Figure 2.48. Mercury still

condensing on the cold finger, trapping with it air and other volatile materials, thus maintaining the vacuum. The condenser is allowed to run after the heat is turned off until distillation ceases. Provided some mercury is retained in the reservoir the still can be left under vacuum. It is dangerous to distil very impure mercury in this apparatus as volatile contaminants may lower the pressure and cause hot mercury to flow back down the inlet tube and break it.

CHAPTER 3

ELECTROCHEMISTRY

ELECTROLYSIS

Pure water is a very poor conductor of electricity. This can be demonstrated by the circuit shown in *Figure 3.1* in which a battery is connected to a beaker of pure water by two carbon rods. The ammeter shows that a very small current flows through the circuit, indicating the very high resistance of the water. If a little hydrochloric acid is added to the water two interesting results follow.

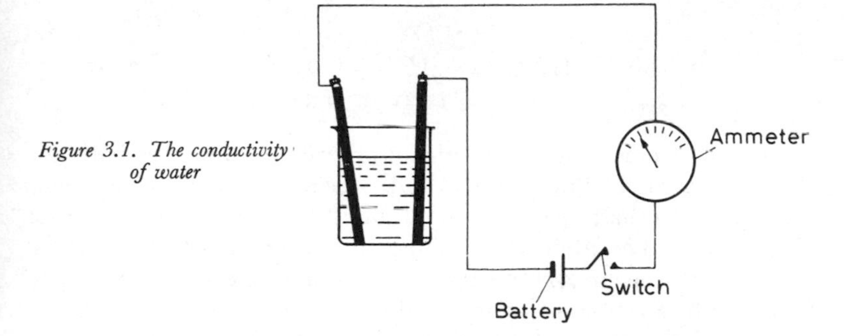

Figure 3.1. The conductivity of water

The current shown on the ammeter increases considerably, showing that a very dilute solution of hydrochloric acid is a good conductor of electricity, and bubbles of gas appear at the two carbon rods. The gas which is evolved at the positively charged carbon rod (or electrode) can be shown to be chlorine, while hydrogen is evolved at the negatively charged electrode. Substances which are good conductors of electricity in aqueous solution, and which cause decomposition of the solution when a current is applied are called *electrolytes*. Electrolytes are usually acids, bases or salts. The process of decomposition by an electric current is called *electrolysis*. The positive electrode is called the *anode*, and the negative electrode the *cathode*.

Towards the end of the nineteenth century the physical properties of electrolytes were investigated. It was found that the freezing points and boiling points of aqueous solutions of electrolytes were different from the figures calculated from the molecular weight and concentration of the solute. (cf. Determination of Molecular Weight, Chapter 5.) To explain the discrepancy Arrhenius postulated his theory of electrolytic dissociation. The theory states that when electrolytes are dissolved in water they dissociate into charged atoms or radicals. Each molecule of electrolyte dissociates into *ions* carrying positive or negative charge, the total number of positive charges being equal to the total number of negative charges so that the solution is electrically neutral, and the number of charges carried by each ion being equal to the valency of the atom or radical from which the ion is formed. Ionization is a reversible process and increases with solution until at ' infinite dilution ' it is complete. The ionization of compounds may be represented by equations, e.g.,

$$HCl \rightleftharpoons H^{+} + Cl^{-}$$
$$H_2SO_4 \rightleftharpoons 2H^{+} + SO_4^{2-}$$
$$CaCl_2 \rightleftharpoons Ca^{2+} + 2Cl^{-}$$

The Arrhenius theory provided a simple explanation both for the abnormal boiling points and freezing points of aqueous solutions that had been recorded, and for the phenomenon of electrolysis. The abnormal boiling and freezing point could be due to each ion in solution having the same effect as a complete molecule of a non-electrolyte. Sodium chloride, which dissociates into two ions, exerts almost double the calculated depression on the freezing point of water indicating almost complete dissociation. It is clear that on this assumption the degree of dissociation of electrolytes in solution can be calculated from accurate measurement of freezing points.

Electrolysis can also be explained upon the basis of the Arrhenius theory. The charged ions migrate to the electrode of opposite charge where they become electrically neutralized. In simple cases such as the electrolysis of hydrochloric acid, the positively charged hydrogen ion would move to the cathode where it would be discharged to become a neutral hydrogen atom. The hydrogen atom would immediately combine with another hydrogen atom to form a molecule and bubbles of the gas would be evolved at the cathode. In a similar manner negatively charged chloride ions would be discharged and chlorine gas would be evolved at the anode.

Discharge involves the addition or removal of electrons, the negatively charged ion of chlorine being equivalent to the chlorine atom with one extra electron attached, the positively charged hydrogen ion being equivalent to the hydrogen atom with its one electron missing. The hydrogen ion simply consists of the nucleus of the hydrogen atom, and is referred to as the *proton*.

In some cases secondary reaction occurs at the electrodes either by reaction of the discharged ion with the electrode or solvent, or because of the potential required for the discharge of the ions of water H^+ and OH^-. In the electrolysis of sodium chloride solution the ions Na^+ and Cl^- are involved. Chlorine gas is evolved at the anode but the sodium ions after discharge at the cathode react immediately with the water

$$2Na + 2H_2O = 2NaOH + H_2 \uparrow$$

and hydrogen is evolved. If a special mercury cathode is used sodium amalgam is formed. This characteristic is employed in the electrolytic manufacture of sodium hydroxide from brine. Sodium sulphate dissociates in solution:

$$Na_2SO_4 \rightleftharpoons 2Na^+ + SO_4^{2-}$$

the sodium behaves in a similar manner to the sodium ion in the electrolysis of NaCl, but the sulphate ion SO_4^{2-} needs a higher potential than the OH^- ion to discharge it. Thus OH is discharged according to the equation:

$$4OH^- = 4e + 2H_2O + O_2 \uparrow$$

The effect of this electrolytic reaction is to decompose the solvent and the solution of sodium sulphate becomes more and more concentrated as electrolysis proceeds.

Faraday's Laws of Electrolysis

The number of molecules in a gram-molecular weight of any substance is a constant denoted by the Avogadro number N. It follows from this that the number of ions in a gram-ion of any ionic substance is also N. Every ion discharged at an electrode alters the number of electrons on the electrode by a definite quantity Z, which is equal to the valency of the ion. It follows that for the discharge of each gram-ion of any substance a change of ZN must take place in the number of electrons on the electrode. To discharge 1 g equivalent weight of any substance the number of electrons involved is the constant N.

The pioneer in the study of electrolysis was Faraday, perhaps the greatest of British scientists, who, incidentally, spent a large part of his working life as 'technician' to Sir Humphrey Davey. Although his work preceded that of Arrhenius he postulated the laws of electrolysis:

1. During electrolysis the amount of any substance liberated is proportional to the amount of electricity passed.
2. The amounts of different substances liberated by the same amount of electricity are proportional to their equivalent weights.

The quantity of electricity which liberates one gram equivalent of a substance is termed the Faraday (symbol F).

The rate of flow of electricity is measured in amperes, and the amount of electricity which passes in one second when a current of one ampere is flowing is one coulomb. The liberation of one gram equivalent of a substance requires 96 500 coulombs. Hence one Faraday equals 96 500 coulombs.

Voltameters

In any quantitative study of electrolysis it is essential to know exactly the quantity of electricity which has passed through the electrolyte during the course of an experiment. For this purpose it should be possible to include an ammeter in the circuit and multiply the current by the time for which the current passes. In fact this is a very unsatisfactory method as chemical changes in the electrolytic system may cause variations in current. The insertion of an accurate *voltameter* or *coulometer* in the circuit allows the estimation of the amount of electricity which has actually passed. The voltameter consists of an electrolytic cell which is connected in series with the experimental cell, and in which one of the substances liberated can be accurately measured.

The instrument used for most accurate work is the silver voltameter. A platinum dish forms the cathode; silver nitrate solution in the dish is the electrolyte, and a silver wire dipping into the solution forms the anode. The platinum dish is carefully washed, dried, and weighed before setting up the apparatus. When a current passes through the system, metallic silver is deposited on the dish. This can be weighed after the experiment and from the equivalent weight of silver (107·9), the number of Faradays which have passed through the experimental cell can be calculated. The silver can be removed from the platinum with nitric acid.

Another common voltameter, which is more simple to use than the silver voltameter, uses potassium iodide as the electrolyte and platinum wires as electrodes. Electrolysis causes decomposition of the KI, and free iodine is formed. The free iodine can be estimated by titration with sodium thiosulphate solution.

For class experiments reasonably accurate results can be obtained by using a copper voltameter. Two copper plates in a glass trough are used as electrodes and a slightly acidified solution of copper sulphate to which a little alcohol has been added to prevent oxidation is the electrolyte. The copper cathode is carefully cleaned, dried, and weighed before and after the experiment, and from the equivalent weight of copper (31·77) the amount of electricity passing during the experiment can be calculated. When great accuracy is not required a reasonable estimate of the amount of electricity passed can be made in a Hofmann voltameter in which slightly acidified water is electrolysed to give hydrogen and oxygen. The volume of gas evolved is measured, corrected for temperature, pressure and vapour pressure, and the calculation based on Avogadro's hypothesis that one gram molecular weight of either gas occupies 22·4 dm^3 at S.T.P. A mole of hydrogen atoms would thus occupy 11·2 dm^3 at S.T.P.

THE HYDROGEN ION EXPONENT, pH

In all aqueous solutions there are free hydrogen ions and free hydroxyl ions present. In pure water:

$$H_2O \rightleftharpoons H^+ + OH^-$$

and from the Law of Mass Action at a fixed temperature:

$$\frac{[H^+] \cdot [OH^-]}{[H_2O]} = \text{a constant } K$$

(The square bracket is a notation meaning 'the concentration of'.) As the concentration of un-ionized water is so large compared with the concentration of ions it may be considered to be a constant, and

$$[H^+] \cdot [OH^-] = K_w$$

where K_w is the ionic product of water.

At 20°C K_w is about 10^{-14} so that $[H^+] = [OH^-] = 10^{-7}$. In all dilute solutions of electrolytes $[H^+] \cdot [OH^-] = 10^{-14}$, thus if $[H^+]$ is increased $[OH^-]$ is decreased, the product being almost constant.

If a 0·1 M solution of hydrochloric acid is prepared it contains 0·1 g replaceable hydrogen per dm³, nearly all in ionic form, and for this solution $[H^+] = 10^{-1}$ and $[OH^-] = 10^{-13}$. Similarly a 0·1M solution of sodium hydroxide will contain 0·1 g moles of the hydroxyl ion per dm³, $[OH^-] = 10^{-1}$ and $[H^+] = 10^{-13}$. This clumsy notation for defining the hydrogen ion concentration of solutions was simplified by Sørensen who devised the hydrogen ion exponent (pH value). Mathematically it may be expressed as the logarithm to base 10 of the reciprocal of the hydrogen ion concentration.

If
$$[H^+] = 10^{-x}$$
$$\frac{1}{[H^+]} = 10^x$$
$$\log_{10} \frac{1}{[H^+]} = x = \text{pH}$$

A less mathematical definition has been stated. The pH of a solution is the logarithm of the number of litres of solution which contain 1 g of hydrogen ions.

It is clear from this that a decimolar solution of hydrochloric acid has $[H^+] = 10^{-1}$ and a pH of 1, a decimolar solution of sodium hydroxide has $[H^+] = 10^{-13}$ and a pH of 13, and pure water has $[H^+] = 10^{-7}$ and a pH of 7.

It must be noted that the greater the acidity of a solution the lower the pH will be and vice-versa. If pure water is 'neutral' a solution with pH < 7 will be acid and one with pH > 7 will be alkaline. A change of 1·0 pH units indicates a tenfold change in $[H^+]$.

The importance of the concept of pH will be seen in many of the topics to be discussed in this book, but it should be emphasized that although Sørensen based his definition upon what were thought to be valid mathematical theories, these theories have since been shown to be only very approximate. The modern idea of pH is based upon a purely notional definition published by B.S.I. in 1950. This definition will be discussed in the section dealing with the electrometric determination of pH. For the theoretical discussion which follows, however, it will be convenient to use the Sørensen definition.

DISSOCIATION CONSTANTS

If several solutions of a weak electrolyte are prepared at different concentrations and the conductivity of each is measured, a graph concentration/conductivity being drawn, it should be possible, by extrapolation, to find the conductivity at infinite dilution of the electrolyte ($\lambda\infty$).

According to Arrhenius it is possible to find the proportion of molecules of an electrolyte dissociated in a solution of known concentration by measuring its conductivity at that concentration. Assuming that the electrolyte is completely dissociated at infinite dilution then

$$\alpha = \frac{\lambda c}{\lambda\infty} \qquad \ldots . (1)$$

Where α = degree of dissociation, and λc conductivity at concentration c. The salt AB dissociates according to the equation

$$AB \rightleftharpoons A^+ + B^-$$

and applying the Law of Mass Action

$$\frac{[A^+]\ [B^-]}{[AB]} = K \qquad \ldots . (2)$$

Where K = dissociation constant of the electrolyte. Often this term will be found in the form $pK = \log_{10} \frac{1}{K}$ by analogy with pH.

If the original concentration of the salt is c g mol/dm^3 there are $(1 - \alpha)c$ g mol of AB per litre and A^+ and B^- each have a concentration of αc g equiv./dm^3.

Substitution in equation 2

$$\frac{(\alpha c)^2}{(1 - \alpha)c} = K$$

$$\frac{\alpha^2 c}{1 - \alpha} = K$$

This is Ostwald's dilution law.

When dealing with a weak electrolyte α is small compared with unity and the law becomes

$$\alpha^2 c = K$$

$$\text{or} \quad \alpha = \sqrt{\frac{K}{c}}$$

Thus for weak electrolytes the degree of dissociation is proportional to the square root of the dissociation constant.

The relative strengths of acids depend upon the hydrogen ion concentration.

Substituting equation 1 in equation 2

$$K = \frac{\lambda c^2}{\lambda\infty(\lambda\infty - \lambda c)}$$

This applies to weak monobasic acids such as acetic or hydrocyanic acid. Dibasic acids (e.g. carbonic acid) have two dissociation constants, tribasic acids have three and so on.

$$H_2A \rightleftharpoons H^+ + HA^-$$
$$HA^- \rightleftharpoons H^+ + A^{2-}$$

In the case of weak dibasic acids the concentration of the anion with double charge is approximately equal to the second dissociation constant. In the case of polybasic acids the stages of ionic dissociation after the first may therefore be ignored.

MODERN CONCEPTS OF IONIC THEORY

It took some years for the Arrhenius theory and its implications to become accepted by chemists, but after the controversies which raged around it had abated it was found to be unsatisfactory on a number of points, although none of these points had been noted by the scientists who opposed its introduction so fiercely.

Although the Ostwald dilution law works fairly well for weak electrolytes in dilute solution there are very large deviations from theoretical calculations when stronger electrolytes are examined. A striking blow against the theory was the provision of evidence that crystalline salts (e.g. sodium chloride) were in ionic form even in the solid state, and that solution in water merely separated the ions. When the absorption spectra of solutions of strong electrolytes failed to show the presence of any undissociated molecules, the validity of a theory which depended upon dissociation varying with concentration was evidently only very approximate.

In 1923, Debye and Hückel postulated a theory which is accepted as the basis for modern concepts of the behaviour of electrolytes. They assumed that strong electrolytes such as inorganic salts, caustic alkalis, and mineral acids are totally ionized in solution. Any deviation from ideal behaviour is not due to incomplete dissociation but to the effect of ions in solution upon each other. As the concentration of electrolyte increases the ions affect each other more and more and the conductivity, which relies upon the movement of ions towards the electrodes of the conductivity apparatus, does not measure the degree of dissociation but the extent to which the ions interfere with each others' movement. Simple interference is not the only factor, for instance highly charged ions produce greater deviation than ions of lower charge.

If a strong electrolyte is completely ionized in solution its apparent concentration determined by physical methods will be less than its actual concentration. The apparent concentration is called the *activity* of the electrolyte (a) and the ratio $\frac{\text{Activity}}{\text{Concentration}}$ is called the *activity coefficient* (α). The activity coefficient is unity at infinite dilution and usually falls as concentration increases.

Activities can be determined by several methods such as the depression of the freezing point of solutes (see Chapter 5) and osmotic properties of solutions.

The implications of the Debye-Hückel theory are of great importance and full discussion of them would be more appropriately covered in an advanced textbook of physical chemistry rather than a textbook of laboratory techniques.

MODERN DEFINITIONS OF ACIDS AND BASES

The Arrhenius definition of an acid is a substance which yields hydrogen ions in solution, and a base as a substance which yields hydroxyl ions in solution. Although this was an adequate theory for much of the early work on acids and bases it becomes insufficient when the properties of electrolytes in non-aqueous solutions are considered.

In 1923, Brønsted and Lowry independently postulated a new and much broader definition of acids and bases which proposed that an acid be defined as a substance which tends to lose or donate protons, and a base as a substance which tends to gain or accept protons. The proton is the positively charged atom of hydrogen which Arrhenius called the hydrogen ion (H^+). This, in fact, is a hydrogen atom which has lost its one electron so that only the nucleus which is positively charged remains. In practice, the proton is never found free in solution but is always in association with one or more molecules of the solvent. In water it is generally assumed that one molecule of the solvent will attach to the proton so that Arrhenius' 'hydrogen ion' becomes not H^+ but H_3O^+. In liquid ammonia the hydrogen ion would be NH_4^+ and in liquid sulphur dioxide HSO_2^+.

According to the Brønsted-Lowry theory a monobasic acid H would ionize:

$$\begin{array}{c} HA + H_2O \rightleftharpoons H_3O^+ + A^- \\ \text{Acid}_1 + \text{Base}_2 \quad \text{Acid}_2 + \text{Base}_1 \end{array}$$

The positively charged ion is an acid because it can donate a proton, the negatively charged ion is a base because it can accept a proton. H_3O^+ is called the conjugate acid to water, A^- is called the conjugate base to HA, and there is a transfer of proton between the two acid-base pairs. Taking acetic acid as an example:

$$\begin{array}{ccc} CH_3COOH + H_2O & \rightleftharpoons & H_3O^+ + CH_3COO^- \\ H_2O + CH_3COO^- & \rightleftharpoons & CH_3COOH + OH^- \\ \text{Acid}_1 + \text{Base}_2 & & \text{Acid}_2 + \text{Base}_1 \end{array}$$

Water can act as an acid or as a base depending on the conjugate system with which it is reacting.

In 1938, G. N. Lewis proposed an even more general theory which defined an acid as a substance which could accept a pair of electrons from a bond and a base as a substance which could donate a pair of electrons to a bond. This advanced theory linked acid-base reactions with oxidation-reduction reactions. Both acids and oxidizing agents are said to accept electrons and are electrophilic. Bases and reducing agents, which tend to donate electrons are said to be electrodotic.

Again the implications of modern theory are wide, and it would not be practical to deal with them in a textbook of this type. The treatment of practical applications which follows is based upon the Brønsted-Lowry theory, although it will be noticed that, for convenience, the hydrogen ion is often written as a single proton H^+.

Salts

A salt, which is formed by reaction between an acid and a base, may not necessarily give a neutral (pH 7·0) solution. The salt may be formed from reaction between one of the following groups:

(*a*) a strong acid and a strong base
(*b*) a weak acid and a strong base
(*c*) a strong acid and a weak base
(*d*) a weak acid and a weak base.

If a salt BA is hydrolysed by water the reaction proceeds according to the equation:

(*i*) $BA + H_2O = HA + BOH$
(*ii*) $HA + H_2O \rightleftharpoons H_3O^+ + A^-$
(*iii*) $BOH \rightleftharpoons B^+ + OH^-$

If HA is a strong acid and BOH a strong base, both will be highly ionized and the solution will be neutral. If, however, HA is a weak acid and BOH is a strong base (class (*b*) above), the salt solution will have a high pH. Class (*c*) salts will give a solution of low pH and the pH of class (*d*) salts will depend upon the relative strengths of the acid and base.

The pH of the solution can be calculated by employing the hydrolysis constant of the salt. (K_h)

$$K_h = \frac{[\text{Free acid}]\ [\text{Free base}]}{[\text{Unhydrolysed salt}]}$$

It can be shown that for:

A salt of a weak acid and a strong base $K_h = \dfrac{K_w}{K_a}$

A salt of a strong acid and a weak base $K_h = \dfrac{K_w}{K_b}$

A salt of a weak acid and a weak base $K_h = \dfrac{K_w}{K_a K_b}$

THE NEUTRALIZATION OF ACIDS BY BASES

It is possible to calculate theoretical titrations for the neutralization of acids by bases. A 0·1 M solution of HCl has a pH of 1·0. If 50 cm³ of 0·1 M NaOH is added to 100 cm³ of the HCl solution the total volume of the mixture will be 150 cm³ which contains 50 cm³ of un-neutralized 0·1 M acid. Assuming complete dissociation of both acid and base, and that the ions all have activity coefficients of unity:

$$[H^+] = 5{\cdot}0 \times 1/150 = 3{\cdot}33 \times 10^{-2} \text{ or pH} = 1{\cdot}48$$

When 75 cm³ NaOH have been added

$$[H^+] = 2{\cdot}5 \times 1/175 = 1{\cdot}42 \times 10^{-2} \text{ or pH} = 1{\cdot}85$$

Continuing in this manner a titration curve can be drawn pH/Alkali added (*Figure 3.2a*). It will be noticed that there is a

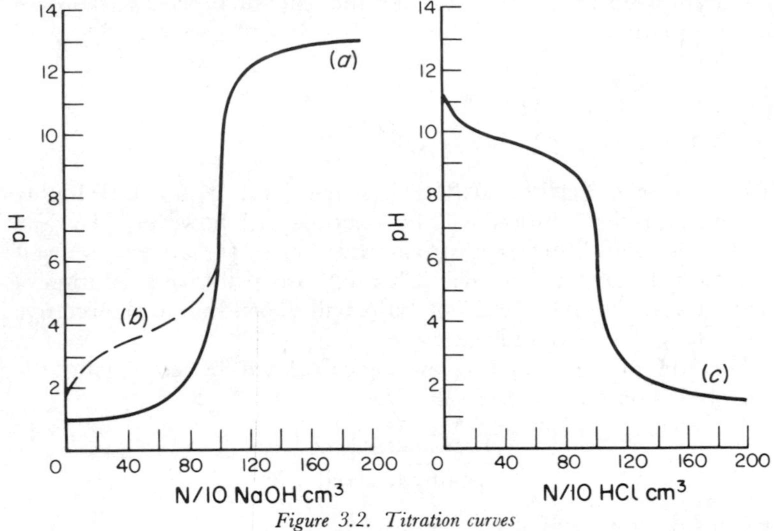

Figure 3.2. Titration curves

(*a*) M/10 NaOH *added to* 100 cm³ M/10 HCl
(*b*) M/10 NaOH *added to* 100 cm³ M/10 *acetic acid*
(*c*) M/10 HCl *added to* 100 cm³ M/10 NH_4OH

very marked and sudden change in pH between 99·9 cm³ NaOH added and 100·1 cm³ NaOH added.

When 99·9 cm³ NaOH have been added

$$[H^+] = 0{\cdot}01 \times \frac{1}{199{\cdot}9} = 5 \times 10^{-5} \quad \text{pH} = 4{\cdot}30$$

When 100·1 cm³ NaOH have been added

$$[OH^-] = 0{\cdot}01 \times \frac{1}{200{\cdot}1} = 5 \times 10^{-5} \text{ pOH} = 4{\cdot}3$$

$$\text{pH} = 14 - 4{\cdot}3 = 9{\cdot}70$$

Thus the change in pH is fairly gradual until the addition of 0·2 cm^3 NaOH in the region of the equivalence point causes a change of 5·4 pH units although the previous 99·9 cm^3 has caused a change of only 3·3 pH units. The calculation of a curve for the neutralization of a weak acid by a strong base is mathematically more complex. There are three factors which must be considered in the calculation:

(*a*) The acid is only partially dissociated to start with;

(*b*) The neutralization of the acid produces a salt which depresses still further the dissociation of the acid remaining;

(*c*) The salt itself is hydrolysed.

Curve *b* in *Figure 3.2* shows the curve for the neutralization of 100 ml M/10 acetic acid by M/10 sodium hydroxide. The first 20 cm^3 of alkali produce a fairly sharp rise in pH as the small number of hydrogen ions present in the slightly dissociated acid combine with the hydroxyl ions of the alkali. The production of sodium acetate suppresses the ionization of the acid and the pH of the mixture rises very slowly for the next 60 cm^3 of alkali added, but as equivalence is approached, the effect of suppression weakens and the pH rises more quickly. The sharpness of the inflection at this point is dependent upon the degree of hydrolysis of sodium acetate. With an acid weaker than acetic acid the degree of hydrolysis would be greater and the inflection would be less sharp. In the case of a strong acid, as has been shown, the inflection is very sharp. At the equivalence point the pH is above 7; when 100 cm^3 of the alkali has been added, the acid has been completely neutralized and the pH of the mixture is that of 0·05 M sodium acetate, i.e. pH 8·7. In all cases of neutralization of a weak acid by a strong base the equivalence point is at a pH greater than 7. The weaker the acid the higher the pH will be at equivalence. After equivalence has been reached the curve continues in a manner similar to that for neutralization of M/10 hydrochloric acid.

The neutralization of a weak base by a strong acid is analogous to the neutralization of a weak acid by a strong base except, of course, that the equivalence point will be below pH 7. *Figure 3.2c* shows the curve for the neutralization of 0·1 M ammonium hydroxide by 0·1 M hydrochloric acid.

BUFFER SOLUTIONS

Aqueous solutions of both sodium chloride and ammonium acetate have a pH of approximately 7. If 1 cm^3 of 0·1 M hydrochloric acid is added to 1 dm^3 of each solution the pH of the sodium chloride solution falls to about 4, but the pH of the ammonium acetate hardly alters. This resistance to change of pH on addition of acid or alkali to a solution is called buffer action. A buffer solution usually contains a mixture of a weak acid or base and one of its salts. The buffer action of such a solution is caused by hydrogen ions being neutralized by the anions of the salt which act as a weak base when an acid is added.

$$H_3O^+ + A^- \rightleftharpoons HA + H_2O$$

whereas if a base is added hydroxyl ions are removed by the neutralization

$$OH^- + HA \rightleftharpoons H_2O + A^-$$

When the buffer solution is a mixture of a weak base and its salt, the equations are:

$$H_3O^+ + B \rightleftharpoons BH^+ + H_2O$$

and

$$OH^- + BH^+ \rightleftharpoons H_2O + B$$

The resistance of the buffer solution to change is called the buffer capacity and is defined mathematically as db/dpH where dpH is the change in pH resulting from the addition of db of base. It can also be defined as the amount of base required to cause a change of one unit of pH (i.e. db when $dpH = 1$). A good deal of information about buffers can be gained from examination of the acid neutralization curves in *Figure 3.2*.

The slope of these curves is dpH/db, the reciprocal of the buffer capacity, so that where the curves are flat the buffer capacity is high and where they rise rapidly the buffering capacity is poor. Strong acids and bases act as good buffers at low and high pH respectively where there is little salt present to repress ionization. Near the neutralization point where the salt concentration is high there is little or no buffering capacity. Weak acids or bases are poor buffers alone as may be seen by the marked inflection at the beginning of neutralization, but when about half the acid has been neutralized, i.e. when equivalent amounts of acid and salt are present, the buffer capacity is greatest. The range of the buffer generally extends to about 1 unit of pH on either side of the pH at this point. Most salts have a poor buffering capacity as can be seen from the sharp inflection at the equivalence point. There are some exceptional cases, e.g. ammonium acetate, where the acid or the base, or both are so weak that there is very little change in pH along the neutralization curve. This is because hydrolysis causes the presence of large proportions of free acid and free base in the salt solution. Such salts may have appreciable buffering capacity.

Buffer solutions are important in many common laboratory processes and will often be referred to in later chapters. To prepare a buffer solution of predetermined pH an acid (or base) should be chosen with a pK_a as close as possible to the pH required. Tables of pK_a values are given in the standard books of reference.

The Henderson-Hasselbalch equation is suitable for designing most buffer solutions.

A monobasic acid HA dissociates:

$$HA \leftrightharpoons H^+ + A^-$$

Then

$$K_a = \frac{[H^+][A^-]}{[HA]}$$

$$K_a \frac{[HA]}{[A^-]} = [H^+]$$

$$\log_{10} K_a + \log_{10} \frac{[HA]}{[A^-]} = \log_{10} [H^+]$$

$$-\log_{10} K_a + \log_{10} \frac{[A^-]}{[HA]} = -\log_{10} [H^+]$$

$$pK_a + \log_{10} \frac{[A^-]}{[HA]} = pH$$

Now consider a mixture of a weak acid and one of its salts in aqueous solution.

The acid itself is hardly dissociated at all and therefore, in the above equation

$$[HA] \simeq [Acid]$$

The salt is almost completely dissociated and all the anions present will be derived from this dissociation

$$[A^-] \simeq [Salt]$$

Therefore we can say:

$$pH = pK_a + \log_{10} \frac{[Salt]}{[Acid]}$$

The lower the pK_a the less accurate this equation becomes, but it fulfils practical needs for pH values between 4 and 10. Outside these limits a more complex treatment is required.

Sometimes buffers are described which consist of two salts representing different stages of neutralization of a polybasic acid. The common phosphate buffers, for instance, consist of a mixture of NaH_2PO_4 and Na_2HPO_4. In this case the salt NaH_2PO_4 is the acid, the dissociation constant being that of the acid $H_2PO_4^-$.

Many formulae for buffer solutions have been published, a few of the common ones being shown in Table 3.1. They may be

TABLE 3.1. Some Useful Buffer Solutions

Buffer			
1. 0·05M *Phalate buffer* 10·21 g Potassium hydrogen phthalate is made up to 1 dm³ in distilled water. This buffer is the primary standard for pH	pH at 15°C = 4·000 pH at 20°C = 4·001 pH at 25°C = 4·008 pH at 30°C = 4·015		
2. *Acetate buffer* The following amounts of M acetic acid are added to 100 cm³. M Sodium acetate and the mixture made up to 1 dm³ in distilled water.	cm³ M *acetic acid*	*pH*	
	350	4·1	
	300	4·2	
	100	4·6	
	80	4·8	
	50	5·0	
3. *Phosphate buffer* Mixtures of M/5 Na_2HPO_4 and M/5 NaH_2PO_4 are made up to 1 dm³ in distilled water.	cm³ M/5 Na_2HPO_4	cm³ M/5 NaH_2PO_4	*pH*
	12	460	5·3
	50	350	6·0
	125	125	7·0
	163·5	10	8·0

prepared speedily and precisely, but for the most accurate work the preparation should be checked with a pH meter. This instrument is described later in this chapter.

The buffer solutions described have a limited range, but 'universal' buffer solutions have been described. An example of these is a mixture of weak acids (citric, diethyl barbituric, boric acids and potassium dihydrogen phosphate). A buffer of any pH from 2–12 can be prepared by the addition of the appropriate amount of alkali. These useful mixtures are obtainable ready-mixed from the laboratory suppliers.

THE MEASUREMENT OF pH

The accurate measurement of pH is, perhaps, the commonest of laboratory procedures. The methods employed may be divided into two main groups:

1. Colorimetric measurement (measurement by indicators)
2. Electrometric measurement (measurement by electronic meter).

Colorimetric measurement

Colorimetric measurement of pH involves the use of neutralization indicators. An indicator is a substance which varies in colour according to the pH of its surroundings. All indicators are organic substances which are either weak acids or weak bases. In the acid form an indicator has one colour and in the basic form it has a different colour. The pH at which the colour change takes place depends upon the strength of the compound as an acid or base.

Consider an indicator of the formula HIn. This would dissociate according to the equation

$$HIn + H_2O \rightleftharpoons H_3O^+ + In^-$$

$$K_{In} = \frac{[H^+]\,[In^-]}{[HIn]}$$

K_{In} is the ionization constant of the indicator. The concentration of water is so large in comparison with the concentration of indicator that it may be considered to be constant and ignored.

$$[H^+] = K_{In} \frac{[HIn]}{[In^-]}$$

The colour of HIn will be different from the colour of In^-, and so the colour of the solution depends upon the ratio

$$\frac{[HIn]}{[In^-]}$$

Thus pH can be related directly to the colour of the solution.

When the two forms of the indicator are present in equal amounts $[HIn] = [In^-]$ and $[H^+] = K_{In}$. Then pH $= -\log_{10} K_{In} = p$In. The factor pIn, the pH when both forms of the indicator are present in equal amounts is called the indicator constant. From the equations it is clear that at any pH both forms of the indicator will be present. In practice however, the eye is unable to detect one colour in the presence of ten times as much of the other colour. The limits of the colour change of an indicator form the colour change interval. This varies as it is purely subjective and depends upon the acuteness of colour vision of the observer, and the published figures are therefore only approximate values. They are based upon the range

$$\frac{[HIn]}{[In^-]} = 10 \text{ to } \frac{[HIn]}{[In^-]} = \frac{1}{10}$$

$$\text{or pH} = p\text{In} \pm 1$$

In practice the range is usually 1·5–2·0 pH units. A list of indicators in common use in the laboratory is given in Table 3.2 together with pH range, colour change, and details of preparation. They are purchased either already in solution or as solids from the laboratory suppliers.

Besides the narrow range indicators listed, some of the laboratory suppliers supply a so-called 'universal' indicator. This is used for approximate readings of pH through the range pH 3·0 to pH 11·0, the colour changes following the colours of the spectrum and giving readings correct to about one unit of pH by comparison with a colour chart supplied with the indicator. Small booklets of universal 'pH papers' are also available. A paper is dipped into the test solution and the resulting colour compared with the colours on a chart. These universal indicator methods are invaluable in deciding which narrow range indicator is appropriate for an accurate determination.

A number of methods in which indicators are used to determine pH have been published. All of them involve the preparation of some sort of standard so that the colour produced by an indicator in a test solution can be compared with a solution of known pH. One of the earlier methods, devised by Bjerrum was the indicator wedge *Figure 3.3.* A rectangular box of glass is divided into two compartments by a diagonal glass sheet. One compartment is filled with an acidified indicator, the other with the alkaline indicator. By looking across the combination from the front a continuous change in colour is seen corresponding to varying ratios of the two colours. As the variation with pH is linear for all practical purposes, a test solution with indicator added can be matched to a similar colour combination on the wedge and the pH read off a linear scale. It is possible to measure the actual ratio of the two coloured species using a colorimeter or spectrophotometer (see Chapter 6) and very accurate pH determinations can be made in

TABLE 3.2. A Range of Acid–Akali Indicators

Indicator	pH *Range*	*Colour in acid solution*	*Colour in alkaline solution*	pK_{In}	*Working solution*
Cresol red (acid)	0·2–1·8	Red	Yellow	—	0·1 g in 13·3 cm^3 0·02 M NaOH made up to 250 cm^3 in water
Thymol blue (acid)	1·2–2·8	Red	Yellow	1·7	0·1 g in 10·75 cm^3 0·02 M NaOH made up to 250 cm^3 in water
Bromo-phenol blue	2·8–4·6	Yellow	Blue	4·0	0·1 g in 7·5 cm^3 0·02 M NaOH made up to 250 cm^3 in water
Methyl orange	3·1–4·4	Red	Yellow	3·7	0·1% in water
Congo red	3·0–5·0	Violet	Red	—	0·1% in water
Bromo-cresol green	3·8–5·4	Yellow	Blue	4·7	0·1 g in 7·25 cm^3 0·02 M NaOH made up to 250 cm^3 in water
Methyl red	4·2–6·3	Red	Yellow	5·1	0·1 g in 18·6 cm^3 0·02 M NaOH made up to 250 cm^3 in water
Bromo-cresol purple	5·2–6·8	Yellow	Purple	6·3	0·1 g in 9·25 cm^3 0·02 M NaOH made up to 250 cm^3 in water
Bromo-thymol blue	6·0–7·6	Yellow	Blue	7·0	0·1 g in 8 cm^3 0·02 M NaOH made up to 250 cm^3 in water
Phenol red	6·8–8·4	Yellow	Red	7·9	0·1 g in 14·3 cm^3 0·02 M NaOH made up to 250 cm^3 in water
Cresol red (base)	7·2–8·8	Yellow	Red	8·3	As for acid indicator
Thymol blue (base)	8·0–9·6	Yellow	Blue	8·9	0·1 g in 10·75 cm^3 0·02 M NaOH made up to 250 cm^3 in water
Phenolphthalein	8·3–10·0	Colourless	Red	9·6	0·1% in 50% aqueous ethanol
Thymolphthalein	8·3–10·5	Colourless	Blue	9·2	0·1% in 80% aqueous ethanol
Alizarine yellow R.	10·1–12·0	Yellow	Red-orange	—	0·1% in water

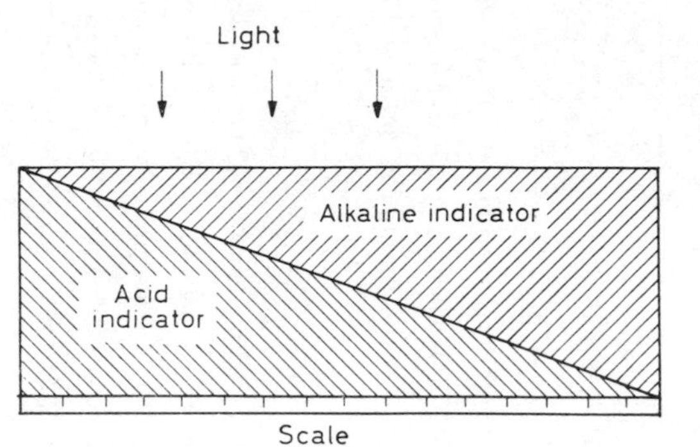

Figure 3.3. The Bjerrum wedge

this way. The most common methods, however, depend upon artificial standards. The Lovibond Comparator (see Chapter 6) can be fitted with discs having glasses corresponding to the colour changes in a large number of common indicators, the range of each indicator being covered in steps of 0·2 units. The method is simple, 10 cm^3 of test solution being placed in the right-hand test-tube, and 10 cm^3 of test solution to which 0·5 cm^3 of indicator has been added in the left-hand tube. An experienced operator can estimate to 0·1 pH units by assessing a colour intermediate between that of two consecutive glass filters.

ELECTROMETRIC METHODS

(1) *Standard Electrode Potential*

If a rod of metal is partially immersed in a solution of its own ions it becomes electrically charged. This arrangement is called a half-cell. The potential exhibited by the rod at 25°C can be calculated from the equation

$$E = E_0 + \frac{0{\cdot}053}{n} \log_{10}(a)$$

Where a = the activity of the ion in solution and n = the valency of the ion and E is the potential. In dilute solution the activity of the ion and its concentration (c) can be assumed to be equal, and

$$E = E_0 + \frac{0{\cdot}053}{n} \log_{10}(c)$$

If the solution used is normal, the concentration becomes unity, $\log_{10}(c)$ becomes zero, and

$$E = E_0$$

E_0 is the Standard Electrode Potential (S.E.P.) of the element and is defined as the electrical potential of a substance in contact with a normal solution of its ions.

It is obvious that it is impossible to measure the value of the S.E.P. of an element directly. If two half cells are connected (*Figure 3.4*) two S.E.P.s can be compared. In the diagram a rod

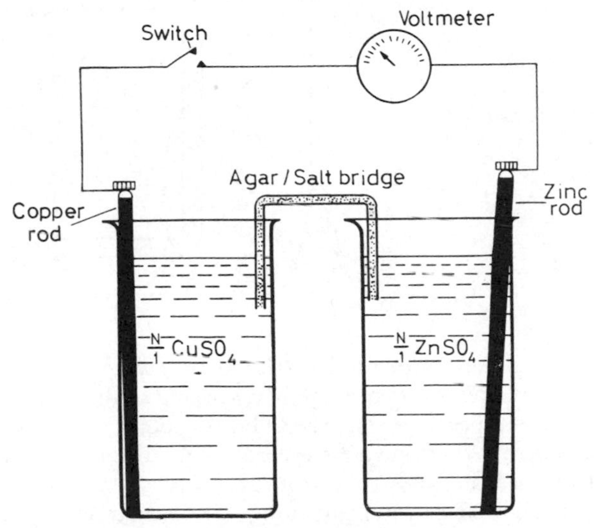

Figure 3.4. Comparison of the standard electrode potential of two half cells

of zinc is partially immersed in a normal solution of zinc sulphate and a rod of copper is similarly immersed in a normal solution of copper sulphate, the two solutions being connected by a potassium chloride/agar salt bridge. The difference in potential between these two standard electrodes can be shown with a voltmeter to be 1·1 V the copper being positively charged with respect to the zinc. This experiment shows that the S.E.P. for copper is 1·1 V greater than the S.E.P. for zinc, but it gives no indication of the actual S.E.P. for either element. The difficulty has been overcome by assigning to the standard hydrogen electrode an arbitrary value of 0·000 V, all other electrodes being ultimately compared with the standard hydrogen electrode.

(2) Salt bridges

In *Figure 3.4* the two beakers are connected with a potassium chloride/agar salt bridge. The salt bridge is used in electrochemistry where two solutions are to be connected electrically without being allowed to mix. Salt bridges are usually made from glass tubing of 5–6 mm O.D. bent to a convenient shape (*Figure 3.5*).

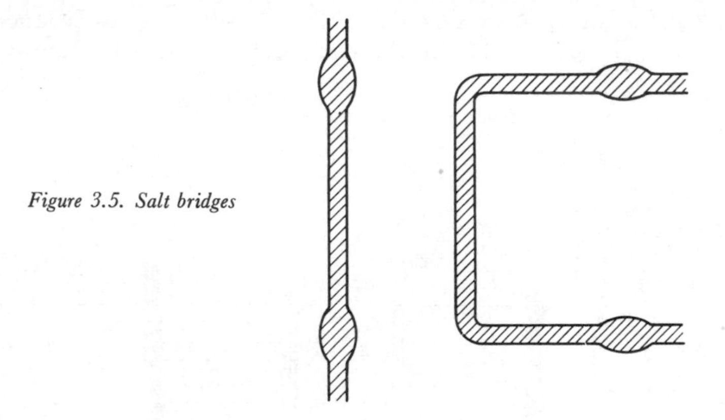

Figure 3.5. Salt bridges

A mixture of 90 cm^3 saturated KCl, 10 cm^3 water and 3 g good quality powdered agar is heated gently, stirring so that the agar does not char. The temperature may rise to boiling point as long as boiling is not prolonged. When the agar is completely dissolved the mixture is sucked up into the glass tubes through a piece of rubber tubing, and the gel allowed to set by cooling. The agar tends to shrink slightly on cooling, and may draw away from the ends of the tube. This can be rectified by cutting away a small piece of the tube so that the bridge reaches the end of the tube. The tendency of the gel to slide out of the glass may be avoided by blowing an 'olive' at one or both ends of the tube as shown in the figure. Bridges may be stored almost indefinitely with both ends dipping into a saturated solution of KCl.

Salt bridges may be used repeatedly. If an end becomes contaminated it may be cut off using a glass file.

Some workers prepare salt bridges in P.V.C. tubing. This allows for flexibility and the contaminated end can be simply cut off with scissors.

The electrolyte in a salt bridge is adequate to connect two solutions and the bridge has negligible electrical resistance.

(3) The Hydrogen electrode

Although the copper and zinc electrodes discussed above are in rod form, it is plainly impossible to have a rod of hydrogen immersed in a solution of hydrogen ions. In fact the hydrogen electrode

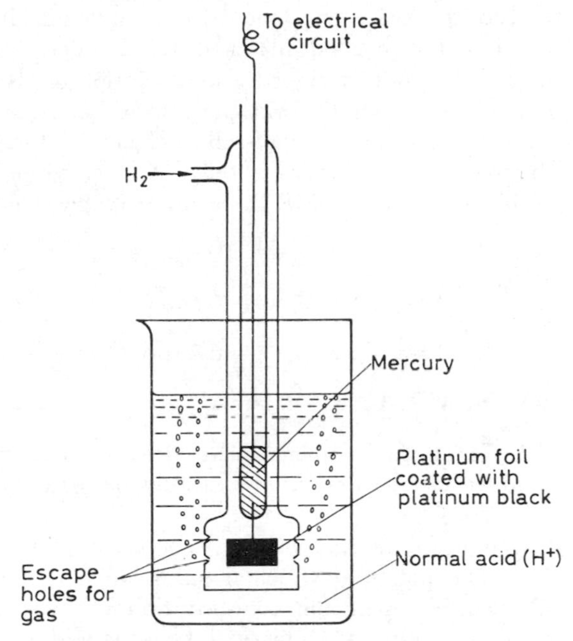

Figure 3.6. The hydrogen electrode

consists of a piece of platinum foil covered in platinum black over which pure hydrogen gas at one atmosphere pressure is passed. A simple hydrogen electrode is shown in *Figure 3.6.*

To prepare a hydrogen electrode a small piece of platinum foil about $\frac{1}{2} \times 1$ cm is cleaned in hot chromic acid followed by a copious washing with distilled water. It is then coated with a layer of colloidal platinum by electrolysis. The platinum forms the cathode, another piece of platinum foil is the anode, and the electrolyte consists of an aqueous solution of 3·0 g chloroplatinic acid and 25 mg lead acetate per 100 cm^3. The current is supplied by a 12 V accumulator regulated by a sliding resistance. Hydrogen is evolved at a moderate rate and in 2–3 min a jet black coating is deposited which must be as thin as possible or the electrode will not give accurate results. The platinum black surface is thoroughly washed with distilled water and replaced in the electrolytic cell where, with M/2 sulphuric acid as electrolyte, electrolysis for about 0·5 h removes any chlorine from the chloroplatinic acid. After washing, it is sealed to the platinum connecting wire which is in turn sealed into the glass envelope which surrounds it in use. The electrode is stored in good quality distilled water.

The hydrogen electrode is dipped into a solution of HCl of unit activity (in practice 1·2 M) to form the *standard hydrogen electrode*. If this is used instead of the zinc half-cell in *Figure 3.4*, the potential difference between the electrodes is 0·340 V, i.e. copper has an S.E.P. of + 0·340 V and the S.E.P. of zinc can be calculated:

$$E_{o_{Cu}} - E_{o_{Zn}} = +1{\cdot}101\ \text{V}$$

$$E_{o_{Cu}} - E_{o_{H}} = +0{\cdot}340\ \text{V}$$

$$\therefore E_{o_{Zn}} - E_{o_{H}} = -0{\cdot}761\ \text{V}$$

i.e. S.E.P. for zinc at 25°C = − 0·761 V

The S.E.P.s of all the elements have been calculated, mostly indirectly, and an electrochemical series may be arranged (Table 3.3).

The hydrogen electrode may, of course, be used for the measurement of pH. The potential which it gains when dipped into a solution will depend upon the concentration of hydrogen ions, i.e. the pH of the solution. This will be discussed in the next section.

The Electrometric Measurement of pH

All electrometric measurement of pH requires two electrodes. One of these, the reference electrode, is at a constant potential if the temperature is constant. The potential of the other varies with the pH of its surroundings. For the most accurate measurement of pH the variable electrode is the hydrogen electrode already described. The reference electrode most commonly used is the calomel cell (*Figure 3.7a*). In the commercial calomel reference electrode illustrated, the inner tube contains the calomel half-cell consisting of mercury in contact with saturated mercurous chloride (calomel) mixed into a paste with potassium chloride. This arrangement produces a potential which varies only with temperature. The reference element is surrounded by a wider tube which

TABLE 3.3. Standard Electrode Potentials

Element	*S.E.P.* (Volts)	*Element*	*S.E.P.* (Volts)
Li	−3·01	Cd	−0·402
Rb	−2·98	Co	−0·27
Cs		Ni	−0·23
K	−2·92	Mo	−0·2
Ba		Sn	−0.140
Sr	−2·89	Pb	−0·126
Ca	−2·84	H	−0·0000
Na	−2.713	Sb	+0·1
Mg	−2·38	As	+0·25
Ti	−1·73	Bi	+0·32
Be		Cu	+0·34
Al	−1·66	O	+0·39
Zr	−1·5	I	+0·54
V		Hg	+0·798
W	−1·1	Ag	+0·799
Mn	−1·05	Br	+1·08
Se	−0·78	Pt	+1·23
Zn	−0·763	Cl	+1·36
Cr	−0·71	Au	+1·7
S	−0·55	F	+1·92
Fe	−0·44		

contains saturated potassium chloride, and which terminates in some form of liquid junction, e.g. a sintered glass or ceramic plug. The outer tube forms the salt bridge which connects the reference element with the solution whose pH is to be measured. A simple calomel electrode which may be prepared in the laboratory is shown in *Figure 3.7b*.

To measure the pH of a solution the two electrodes are dipped into the solution and the potential difference between them is measured either directly with a voltmeter or by comparison with a potentiometer. A simplified circuit for potentiometric measurement is shown in *Figure 3.8*. In all potentiometric measurement the unknown potential is opposed along a wire by a variable known potential of opposite sign. When these two potentials are equal no current flows along the wire, i.e. the value of the unknown potential (E) is equal to that of the known potential, and pH can be calculated from the expression

$$\text{pH} = \frac{E - 0{\cdot}243}{0{\cdot}059} \text{ at } 25°\text{C}$$

The procedure is as follows. With the sliding contact at *D* the double-throw switch is turned to position 1 and the variable resistance *A* is adjusted so that there is a null reading on the galvanometer. This means that the potential drop along the wire *CD* is

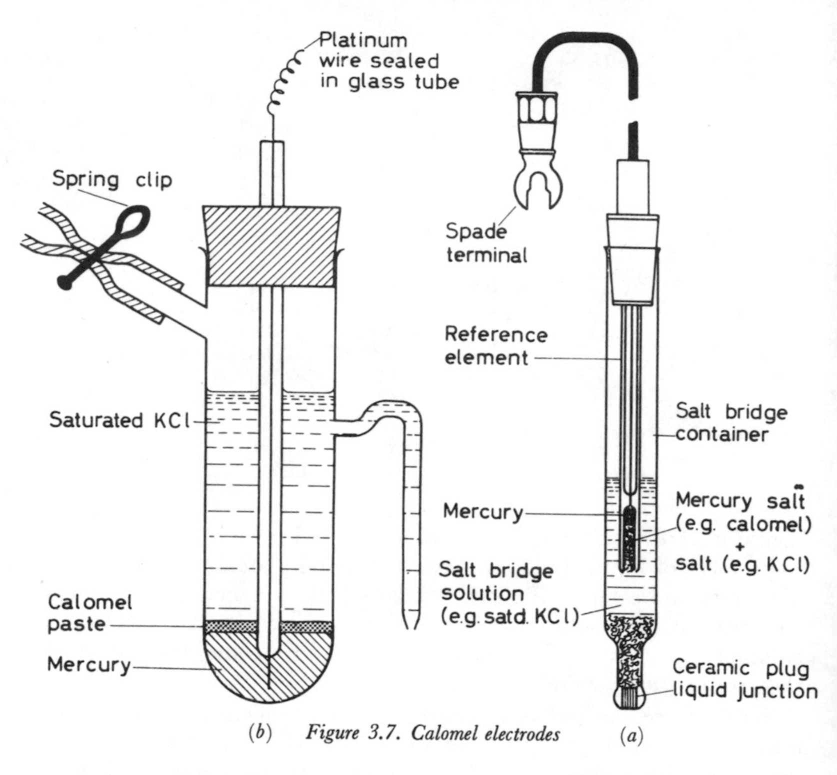

(*b*) *Figure 3.7. Calomel electrodes* (*a*)

1·0183 V, the e.m.f. of the standard Weston cell. The double-throw switch is turned to position 2 and the potentiometer pointer adjusted until there is again a null reading on the galvanometer. If the wire *CD* is calibrated so that 1·0183 V represents its entire length, the reading at *R* represents the potential difference between the hydrogen and calomel electrodes, *E* in the equation above.

A large number of commercial devices which incorporate the necessary components of a pH measuring potentiometer have been marketed. These are compact instruments in which the calibration of the resistance *CD* is in pH units at 25°. The resistance is usually in two parts in the form of two connected potentiometers. One

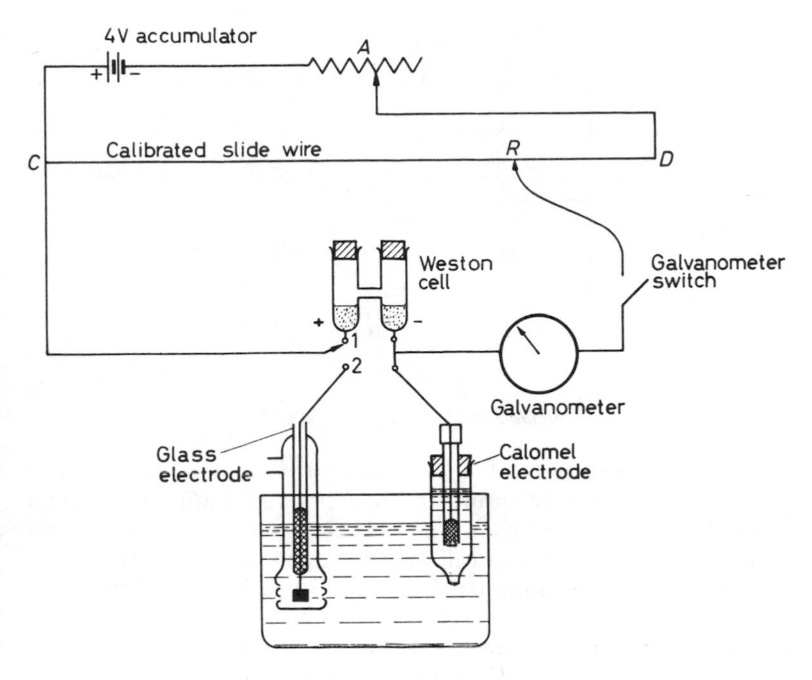

Figure 3.8. Simplified circuit for the potentiometric measurement of pH

of these is of the step type giving readings from pH 0 to pH 14 in steps of 2 units, the other is of the continuous type reading over 2 units of pH with divisions of 0·1 units. They are also provided with a small 'trimming' resistance which is calibrated to adjust the instrument for variations of temperature. This type of instrument can be used to measure pH with an accuracy of 0·05 pH units.

The Standard Weston Cell

In the method for estimation of pH described above, the standard against which the e.m.f. between the two electrodes is measured is the Standard Weston Cell (*Figure 3.9*). The 'positive' limb contains an electrode of mercury covered with solid mercurous sulphate, and connected to the exterior terminal with a piece of platinum wire sealed into the glass container. The 'negative' limb is a 12·5 per cent amalgam of cadmium in mercury with a similar platinum

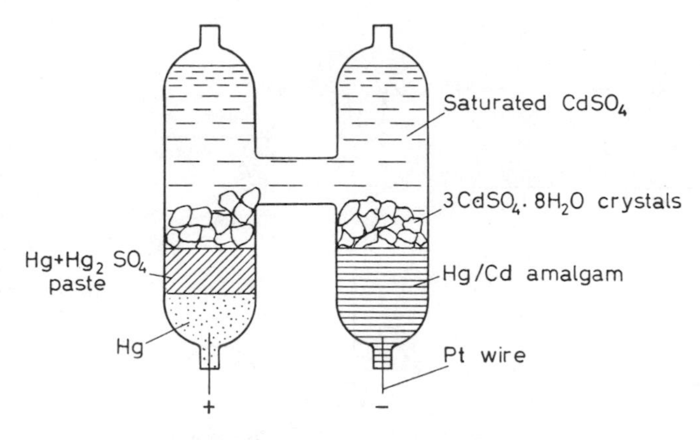

Figure 3.9. The Standard Weston Cell

connection. The electrolyte is a saturated solution of cadmium sulphate which is kept saturated by crystals of cadmium sulphate $3CdSO_4 \cdot 8H_2O$ placed over the cadmium amalgam. The e.m.f. from this extremely stable cell may be calculated for variations in temperature by the formula $E = 1{\cdot}01830 - 4{\cdot}06 \times 10^{-5}(t - 20) - 9{\cdot}5 \times 10^{-7}(t - 20)^2 + 1 \times 10^{-9}(t - 20)^3$. For most practical purposes the temperature coefficient is so small that at room temperatures the value $E = 1{\cdot}01830$ V may be used.

Another type of cadmium cell, the 'unsaturated' Weston cell is often used in laboratory work as a secondary standard. The difference lies in the electrolyte being a solution of cadmium sulphate saturated at 4°C so that at room temperature it is unsaturated. This cell gives an e.m.f. of 1·0186 V and under normal laboratory conditions is virtually independent of the temperature.

The Glass Electrode

For routine measurements of pH in the laboratory the use of the hydrogen electrode as a sensitive electrode is not practical. The electrode is difficult to prepare, difficult to maintain, becoming 'poisoned' very easily, and requires large quantities of pure hydrogen, which is itself rather dangerous. The electrode most commonly used is the glass electrode (*Figure 3.10*). This consists of a thin glass membrane sealed to a stem of high resistance glass containing the stable reference element, which is in contact with the glass membrane. The reference element usually consists of a

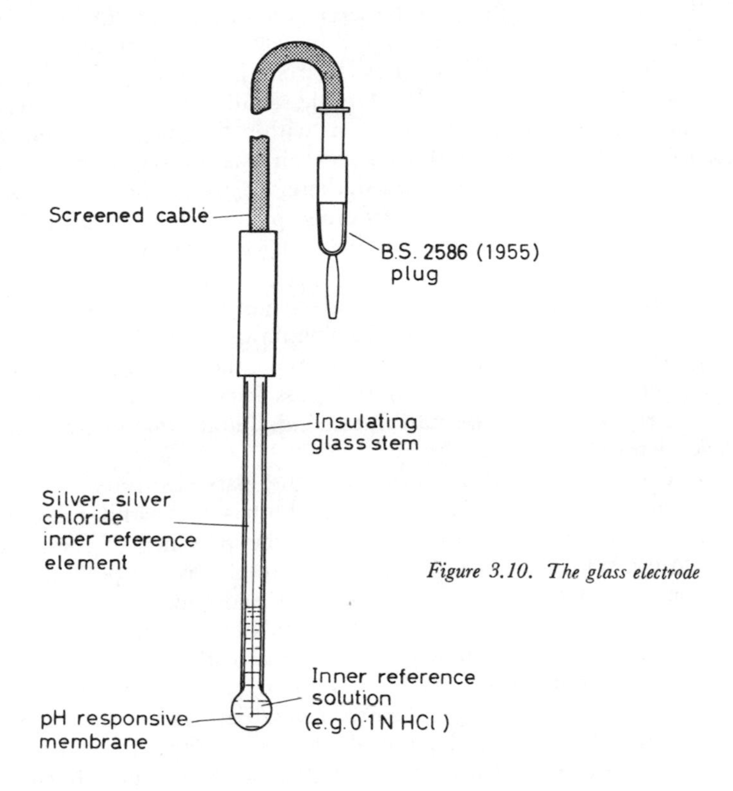

Figure 3.10. The glass electrode

length of silver wire coated with silver chloride, and dips into a reference solution, usually N/10 hydrochloric acid.

It has been found that a glass membrane in contact with a solution of hydrogen ions develops a potential between its walls which depends upon the pH of the solution. If the glass bulb of the electrode is dipped into an aqueous solution a potential will arise between the standard liquid inside the bulb and the test solution outside the bulb. This potential can be compared with the potential of a calomel reference electrode and the pH of the test solution estimated. The arrangement may be represented thus:

Silver | Silver chloride M/10 HCl | Glass membrane | Test solution ||

Saturated KCl salt bridge | Mercurous chloride | Mercury

The potential difference between the silver and the mercury in this system is a measure of the pH of the test solution.

The mechanism of the glass electrode is by no means fully understood. It appears that the thin glass bulb can act as a conductor, partly because of the sodium ions within the glass itself, and partly because the water, and its ions, can pass through the thin glass membrane. It may be assumed, therefore, that if the glass membrane separates two solutions of differing pH, some ionic transfer takes place through the membrane involving a separation of charge, and a potential arises.

One strange feature of the glass electrode is that even if the solutions on either side of the membrane are of identical pH, a potential develops, the asymmetry potential. This is probably due to such factors as strain in the glass itself, and means that glass electrodes have to be calibrated individually against solutions of known pH.

There are several limitations to the glass electrode. Glass itself is an unusual material containing both silicate and sodium ions. It is not surprising, therefore, that the presence of these ions in the test solution can give misleading results as the potential would not depend solely on the presence of hydrogen ions. In fact this error is negligible below about pH 9, but between pH 12–13 the *alkali error*, as it is known, can be appreciable.

The manufacturers have attempted to overcome this difficulty by using glass which gives smaller alkali errors than the soda glass traditionally employed for making glass electrodes, and in recent years electrodes using glasses of potassium, lithium, barium and calcium have been produced which can give reasonably accurate results at the extremes of pH.

Another source of error is found in the fact that the glass electrode will respond accurately only to solutions of high water content, and the presence of large quantities of acid, organic solvent, ionized colloidal particles, and high temperatures can introduce appreciable errors.

The glass electrode can reasonably be expected to detect variations in pH down to about 0·02 of a pH unit with accuracy. The hydrogen electrode can measure to 0·005 of a pH unit, and is still used for the most accurate work.

With a resistance of about $10 \times 10^8\ \Omega$, the glass electrode is unsuitable for the potentiometric measurement of pH and special electrometers have been developed as pH meters, usually of the valve voltmeter type.

pH *Measurement by Voltmeter*

The modern pH meter is in fact a valve voltmeter in which the potential difference between the glass and reference electrodes is amplified several hundreds of times, and the measured voltage read in terms of pH. The direct reading instrument is basically a triode valve, in which the potential difference between the electrodes is applied to the grid and cathode, and the effect on the anode current is noted.

A very popular commercial design of this type is shown in *Figure 3.11*. The Electronic Instruments Ltd. Model 23A pH meter is a compact instrument suitable for laboratory or industrial use. The measurement of pH is simplicity itself. The glass and reference electrodes are mounted on clips on a small stand, together with a platinum resistance thermometer which automatically compensates for changes in temperature. The instrument also has a 'check' position which monitors the condition of the electrodes. These are normally kept immersed in distilled water.

To measure the pH of a solution the electrodes are raised from the beaker of distilled water in which they rest when not in use and 'blotted' very gently with a Kleenex tissue. They are then immersed in a buffer solution of known pH. The control knob is turned from 'Check' to '0–14 pH', and the 'Set Buffer' control adjusted until the meter reading is that of the known buffer pH. The control knob is returned to the 'Check' position, the electrodes again raised, rinsed with distilled water, 'blotted', and immersed in the solution of unknown pH. The knob is once again moved from 'Check' to '0–14 pH' when the meter reading will give the pH of the unknown solution.

An alternative method which may be used with this particular instrument is the 'Δ pH' method. The advantage of using the Δ pH scale is that it covers a range of $-3{\cdot}5-0-+3{\cdot}5$ units of pH over the same length as the '0–14 pH' scale, i.e. each unit of pH is twice as long on the pH scale. The centre zero on this scale may be set to any known pH with a solution of known value, and the difference between the known and unknown solution measured to well within the accuracy which may be expected with the standard glass/calomel electrode system. The Δ pH range has no automatic temperature compensation and temperature correction must be applied. A correction chart is supplied with the instrument.

Most laboratory suppliers market accurately measured buffer solutions for the standardization of instruments. Alternatively, buffer tablets may be used. Each tablet dissolved in water and

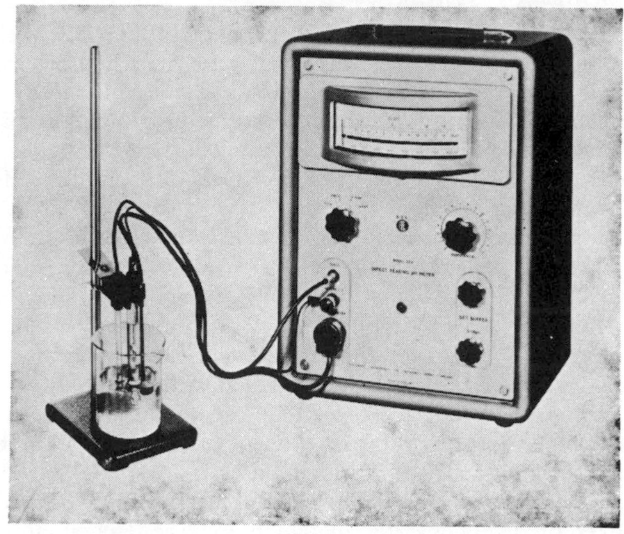

[By courtesy of Electronic Instruments Ltd]

Figure 3.11. A commercial direct reading pH *meter*

made up to 50 cm^3 in a volumetric flask gives a solution of known pH. Of course, suitable buffer solutions may be made up in the laboratory (Table 3.1).

The pH meter does not measure hydrogen ion *concentration* but hydrogen ion *activity*. It is clear, therefore, that the original Sørensen definition of pH is no longer considered valid, and a new definition of pH has been found necessary.

The British Standards Institution has laid down a *primary standard* for pH measurement. A 0·05 M solution of pure potassium hydrogen phthalate is stated to have a pH of exactly 4·000 at 15° C. At any other temperature between 0° and 60°C the pH is given by the formula

$$\text{pH} = 4{\cdot}000 + \tfrac{1}{2}\left(\frac{t-15}{100}\right)^2$$

where t = temperature in °C.

Using this primary standard (S) the pH of an unknown solution (x) is found by the difference in the e.m.f. of a cell using a hydrogen electrode and a reference electrode with the two solutions as electrolytes and the same temperature and hydrogen pressure:

$$\text{pH}\,(x) - \text{pH}\,(S) = \frac{E_x - E_s}{2{\cdot}3026\,RT/F}$$

Where E_x is the e.m.f. of the cell using solution X, E_s the e.m.f. using solution S, R is the gas constant, T the absolute temperature and F the Faraday (= 96 493 coulombs). This purely notional definition of pH is universally used. The National Bureau of Standards in the U.S.A. has laid down a similar definition, the difference in American and British Standards being that the former has four primary standards covering a range from about pH 3·5 to about pH 9·5 and two secondary standards, which may give results of slightly lower accuracy, for the extremes of the scale.

Potentiometric Titration

By using the pH meter described above it is possible to carry out neutralization titrations without using a coloured indicator. The titration is carried out in a cell containing suitable electrodes and the pH measured after each addition of titrant. A graph can be drawn and the equivalence point found. As can be seen from the graphs in *Figure 3.2* on page 102, the change in pH for the addition of a given amount of titrant is greatest at the equivalence point.

In fact, it is not necessary to measure the actual pH of the titration mixture to find the equivalence point, the potential of the system, measured in millivolts, may be measured instead and in fact, usually is.

The preparation of a titration graph may present difficulties as the best curve through a series of experimental observations can lead to errors of personal judgement. The *derivative method* of locating the end-point produces better results. Instead of plotting e.m.f. (E) as ordinate versus Volume (V) as abscissa the graph is plotted with change of E/change of V as ordinate versus V as abscissa ($\Delta E/\Delta V$ versus V). The end-point is shown as a sharp spike on the graph (*Figure 3.12a*). Even better results are obtained by plotting $\Delta^2 E/\Delta V^2$ versus V producing the graph shown in *Figure 3.12b*. Special apparatus is available for the direct measurement $\Delta E/\Delta V$.

Potentiometric titration is an ideal procedure for automation and many designs of automated titration equipment have been marketed. The burette usually has a solenoid operated valve outlet which is linked to a pH meter. The meter is set to the pH of the end-point and the solenoid valve is opened. When the predetermined end-point pH is reached, the valve is switched off and

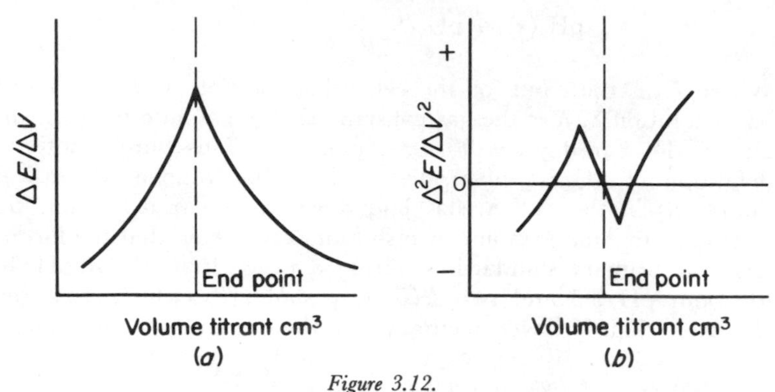

Figure 3.12.

Derivative methods for the location of the end-point in potentiometric titration
(a) By plotting $\Delta E/\Delta V$ against V *(b) By plotting $\Delta^2 E/\Delta V^2$ against V*

the burette read. Minor refinements include such things as visual or auditory indicators when the titration is finished. Many pH meters which can be used for automatic titration have an outlet socket for connection to a chart recorder so that a titration curve is drawn as the titration progresses.

Perhaps the most useful of the commercial instruments for automatic titration is the Radiometer Titrimeter and Titrigraph (*Figure 3.15*). This Danish machine is marketed in this country by Messrs V. A. Howe. It is possible to follow the course of a titration very accurately with this apparatus, or alternatively to use it as a 'pH-stat', i.e. to keep the pH of a reaction mixture stabilized within very close limits.

Ion Selective Electrodes

The idea of a glass electrode for measuring hydrogen ion activity was first postulated by Haber in 1909. Although it should be possible to measure the activities of other ions in solution by the use of suitable electrodes, this extension of the use of what has always been referred to as the 'pH meter' has only recently become possible. In the last few years a large number of *ion selective electrodes* have been developed. These will measure the activity of a particular ion in solution depending upon the characteristics of the sensing membrane of the individual electrode.

Glass Membrane Electrodes

The earliest forms were based upon the glass electrode and were suggested by the alkali error which became apparent when measuring high pHs in the presence of sodium ions. By using glasses of special formulation, electrodes which were particularly responsive to monovalent cations could be produced. The Corning sodium electrode, for instance, is responsive to sodium ions and is 10^3 times more responsive to silver ions. It can be used directly for the potentiometric measurement of silver ions in solution at a concentration (pAg^+) of 0–7. Sodium ions interfere quite considerably and potassium ions very slightly. The electrode, which looks very much like a normal glass pH electrode, is responsive to hydrogen ions and can only be used within a pH range of 4–8. In the absence of silver ions it can be employed to measure pNa^+ in a range of 0–6 and within a pH range of 7–10. In order to minimise the effect of hydrogen ions it is considered necessary to work at a pH at least 2 units above the pION, i.e. the concentration of the measured ion must be at least 100 times greater than the H^+ concentration.

A similar electrode developed by Corning is the monovalent cation electrode which is responsive to Ag^+, K^+, NH_4^+, Na^+ and Li^+ in that order. It is possible to measure potassium ions in the presence of sodium ions only if the comparative concentration of the latter is very low or if a correction is made from an independent estimation of sodium. In fact more accurate results are obtained by flame photometry. An example of the use of this electrode is in the measurement of urea nitrogen by acting upon urine with the enzyme urease and measuring the NH_4^+ produced.

Crystal Membrane Electrodes

Anions can be measured by the use of the appropriate crystal membrane electrode (*Figure 3.13*). The glass membrane which is

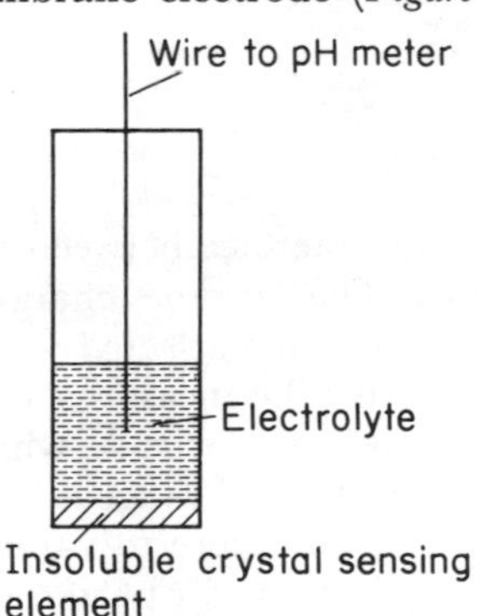

Figure 3.13. The crystal membrane electrode

so commonly used is replaced by a thin pressed pellet of crystals of a sparingly soluble salt of the anion. Chloride, bromide, iodide and sulphide electrodes have been made using the appropriate silver salts as membranes. In the case of the halides mentioned many other cations interfere but fluoride and sulphide may be measured with crystal membrane electrodes with remarkably little interference from other ions present in the test solution. The crystal membrane is in contact with a silver nitrate solution of constant Ag^+ activity in which is immersed a solid silver electrode.

Liquid Ion Exchange Electrodes

None of the electrodes discussed above is suitable for cations with a valency greater than one. It has not been found practicable to formulate special glasses which would respond to the common divalent cations. The problem has been solved for many metal ions by the preparation of liquid ion exchangers (see Chapter 7, p. 271) which are in contact with the test solution through a porous glass or ceramic membrane (*Figure 3.14*), contact being made through the interstices of the membrane. The ion-exchange liquid is usually made up in a solution high in chloride ions. The 'sensitive

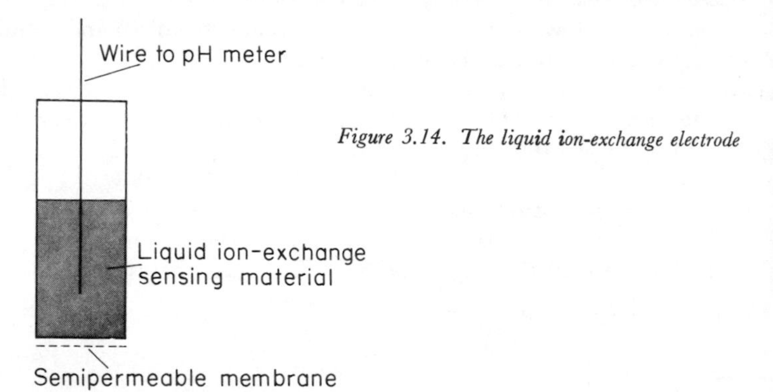

Figure 3.14. The liquid ion-exchange electrode

membrane' in these cases is the actual liquid interface between the test solution and the ion-exchange solution. This becomes charged, the charge depending upon the concentration of the selected ion in the test solution. A large number of cations may be measured with these electrodes. Corning manufacture a calcium electrode which works within a pCa^{2+} range 1–6. The permissible pH range is 7–11 and the other group IIA metal ions and Ni^{2+} interfere to varying extents. Orion Research Inc. offer a similar electrode for pCa^{2+} 1–5 which operates over a slightly wider pH range and which suffers interference from Mg^{2+}. Another interesting Orion product is a general divalent cation electrode which responds to divalent cations of lead, nickel, zinc, iron, calcium, magnesium, barium and strontium in that order. The Corning potassium electrode is about 100 times more responsive to that ion than it is to sodium. Other liquid ion exchange electrodes are available for anions, notably perchlorate and chloride.

Ion selective electrodes are an important extension of instrumental analysis as long as the problems associated with their use are appreciated. Specificity is still a problem and interference from other ionic species must be taken into account. Some types are very susceptible to electrode poisoning especially by proteins and strong reducing solutions which limit their applicability in some fields. However, their introduction is still a comparatively recent event and at the time of writing they form an important growth point in instrument technology. New and improved products are regularly introduced and for the most up-to-date situation in this field the manufacturer's literature should be consulted. In this country Messrs Eel, Beckman, E.I.L., and Pye-Unicam all produce useful electrodes for use with their pH instruments.

Conductometric (Coulometric) Titration

The titration of weak acids with weak bases is complicated by the difficulty in judging the end-point as the change in pH may be very slight. Conductometric titration overcomes this difficulty.

[By courtesy of V. A. Howe Ltd]

Figure 3.15. The Radiometer Titrimeter and Titrigraph

The conductivity of the titration mixture is measured after the additions of successive amounts of reagent from the burette and the figures plotted on a graph. The graph usually consists of two straight lines which intersect at the equivalence point (*Figure 3.16a*).

The conductivity of a solution is controlled in the main by two factors:

1. The number of ions in solution.
2. The speed at which the ions move (ionic mobility).

The hydrogen ion is by far the most mobile of the cations and the hydroxyl ion, similarly, is the most mobile of the anions, and in a neutralization process, where other ions are substituted for H^+ or OH^-, the conductivity will be lessened, and where these ions are added to solutions the conductivity will increase.

Where very weak acids and bases, which are only slightly ionized, are involved, the addition of acids and alkalis may increase the conductivity rather than decrease it, but even in this case there will be a sharp increase in conductivity at the equivalence point (*Figure 3.16b*). It should be noted that the actual conductivity value at the equivalence point has no practical significance.

In order to obtain sharp end points the volume of solution in the conductivity cell should not increase very much or the lines of the graph become so curved that the end-point is difficult to see clearly. This is usually achieved by making the titrant, i.e. the solution in the burette, much more concentrated than the solution in the cell.

If this is not possible, each conductivity reading should be adjusted by multiplying by the factor:

$$\frac{V - v}{V}$$

where V = total volume in conductivity cell
v = volume of last addition of titrant.

Conductometric titration is very useful where such factors as hydrolysis, solubility, and dissociation of the reaction product may cause inaccuracies in potentiometric titration. Large amounts of electrolytes which take no part in the reaction do affect the conductivity of the solution, however, and in this case potentiometric titration is to be preferred.

THE PREPARATION OF CONDUCTIVITY WATER

For experimental work in the field of electrochemistry it is necessary to use water with as low a conductivity as possible. Low con-

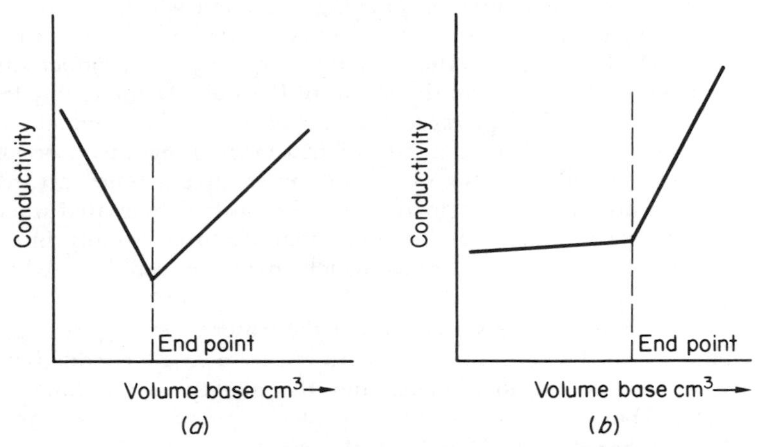

Figure 3.16. Titration curves in conductimetric titration
(a) Strong acid/strong base (b) Weak acid/weak base

ductivity may be equated to purity, especially freedom from dissolved electrolytes.

Many methods for the preparation of *conductivity water* have been described and in the past a great deal of research has been reported on this topic.

Ordinary distilled water, as used in the average chemistry laboratory, has a conductivity of about 10×10^{-6} Siemens (S) and is unsuitable for conductivity work at low dilution. The very purest water obtainable (e.g. that produced by numerous distillations in quartz apparatus) may have a conductivity as low as $0{\cdot}05 \times 10^{-6}$ S. Unfortunately, this is so difficult to prepare and store that it is rarely used except in research work of the very highest refinement. For most purposes equilibrium water is suitable. This has a conductivity of about $0{\cdot}9 \times 10^{-6}$ S. Traditionally it is prepared by redistilling ordinary distilled water over acidified potassium permanganate in a hard glass flask using a hard glass condenser. A rather better product (about $0{\cdot}1 \times 10^{-6}$ S) which can be classed as conductivity water, can be obtained if ordinary distilled water is redistilled, first as above and then over a little barium hydroxide. The distillate is only partially condensed so that the steam which issues from the apparatus excludes air and carries off gaseous and volatile impurities.

Nowadays most conductivity water is produced by percolating distilled water through ion-exchange resins (see Chapter 7). This is really an extension of water-softening and water which is virtually free of foreign ions may be obtained. Several de-ionizers are on the market which produce water of suitable quality for conductivity experiments. The Permutit Mark 6 Portable Deminerolit, for instance is claimed to produce 130 dm^3 of water of a conductivity less than $1 \cdot 0 \times 10^{-6}$ S at a rate of 60 dm^3/h using raw London tap water. Distilled water would, of course, give a much greater volume of product. The apparatus can be easily regenerated when the resin is exhausted. Some other manufacturers supply interchangeable cartridges of resin which may be replaced when exhausted.

An advantage of this system is that the water emerging from the apparatus can be continuously monitored by a small conductivity flow cell connected to a meter and the quality of the water is therefore known.

Jacobs* has described a neat and simple apparatus for the preparation of water with a conductivity of $0 \cdot 1$–$0 \cdot 2 \times 10^{-6}$ S at 25°C. A mixed bed column of Amberlite resins IR 120 and IR 400 in the ratio 1:2 is used, and the apparatus described is fitted with a laboratory-made conductivity cell. This apparatus may be constructed by any technician with a grasp of the normal laboratory arts.

The storage of conductivity water presents a problem in some laboratories. The main hazard is carbon dioxide from the air which readily dissolves in water, raising its conductivity. The water should be stored in small vessels of hard-glass or inert plastic and should be protected from atmospheric CO_2 by guard tubes containing soda-lime or carbosorb granules. Prolonged storage in glass should be avoided as very pure water tends to leach out salts from the glass surface. A cool dark place is recommended for the storage of the bottles or direct sunlight may promote the growth of the microscopic green plants which are very difficult to remove entirely from any water and which are certainly not desirable in conductivity experiments.

* *Chemy Ind.* July (1955) 944–946.

CHAPTER 4

INORGANIC ANALYSIS

SOME GENERAL IDEAS ON ANALYSIS

A large part of the activity in any laboratory comes under the general heading of *analysis*. The dictionary describes analysis as the separation of a system into its component parts, and in chemistry, as in mathematics or grammar or book-keeping, the analysis of substances consists mainly of separatory processes.

When the chemist is given a sample of unknown material to analyse, he first separates it into *fractions*, by physical methods, e.g. the water-soluble from the water-insoluble fraction, or the low-boiling distillate from the high-boiling distillate (see Chapter 2, Table 2.1). The separated fractions are then subjected to *qualitative analysis* by chemical tests to determine what kind of material is present in the original sample, i.e. the components are *identified*. *Quantitative analysis* follows, to measure the amount of each of the components in the sample. All these procedures involve the separation of the components, and in all analytical work the order is the same, fractionation followed by qualitative and then quantitative analysis.

The sample presented to the working chemist for analysis is often extremely impure and may consist of a crude amorphous mass. The soil chemist, for instance, may be expected to analyse a sample of mud; the oil chemist a sample of sludge from the sump of a car engine. Samples of this type may require a great deal of fractionation by physical methods before chemical tests can be applied to the fractions. In the elementary study of inorganic analysis the samples which are provided are usually in a pure state and physical separation is rarely necessary. In qualitative analysis it is possible to proceed directly to the chemical testing of the sample. It might be mentioned that in clinical biochemistry the sample is not specially purified even for elementary students, and physical fractionation is usually necessary.

The amount of sample which is available for analysis is of some importance, and special techniques have been developed so that very small quantities of substances can be identified and measured.

To give some idea of the scale on which analysis is carried out, there is a generally accepted nomenclature which describes this:

1. Macroanalysis requires 0·5–1 g of sample.
2. Semi-microanalysis requires 0·05–0·1 g of sample.
3. Microanalysis requires less than 10 mg of sample.

These limits are by no means rigid but it will be noted that each system is scaled down by a factor of 10 from the preceding one. The teaching of elementary qualitative analysis is almost always on the semi-micro scale, partly because of the greater speed of the technique and partly because of the saving in materials. Students of elementary chemistry rarely have the experience to master the manipulative skills of microanalysis with success.

THE FUNCTIONS OF ANALYTICAL TESTS

In all qualitative analysis there are two kinds of test which may be used:

1. *Group tests* in which more than one substance gives a positive result, but which will distinguish the positive group from all other substances. For instance, if a slight excess of dilute hydrochloric acid is added to a solution of inorganic salts, a white precipitate will show the presence of Group I of the inorganic analysis scheme, i.e. any or all of the three metal radicals, silver, lead, mercury (monovalent). A lack of precipitate will show the absence of these radicals. The precipitate can be separated by centrifugation and its components identified, while the metal ions remaining in the filtrate may be precipitated by further group tests.

2. *Specific tests* in which only one substance will give a positive result. If a solution of barium chloride is added to a solution of inorganic salts slightly acidified with HCl, a white precipitate indicates the presence of the sulphate radical. No other substance produces this result. Specific tests are comparatively rare. They must not be confused with the individual tests which distinguish members of the same group from one another. These are only of value when the group has been separated from all other components.

It is often difficult to decide whether the result of a test is positive or negative. The reagents used may themselves give what appears to be a faintly positive result. It is often useful to perform *control tests* both with an authentic sample of the substance suspected and in its certain absence. The former is called the *positive control*, the latter the *negative control* or *blank*. Controls must be set up in every case in which a test gives equivocal results. Their value cannot be over emphasized.

Every test has a lower limit of identification, a quantity or concentration below which the substance will not give a positive reaction. This minimum is called the *sensitivity* of the test. The sensitivities of the more common tests cover a wide range. A flame test for sodium would be positive with a few microgrammes of sodium salt, and indeed, very minute quantities of sodium salts present as an impurity are often reported by students as positive components of a mixture. The standard wet tests of the inorganic analysis tables require a minimum concentration of about 1 mM. In microanalysis it is essential to know the sensitivity of each test so that the necessary conditions for successful identification of the components are fulfilled. In macro and semi-microanalysis sensitivity does not often give rise to difficulties as the amounts and concentrations used are usually far in excess of the permissible lower limit.

QUALITATIVE ANALYSIS

There are a number of ways in which the qualitative analysis of inorganic substances may be performed. Among the more common, which will be considered are:

(*a*) Systematic wet tests ('analysis tables').
(*b*) Spot tests.
(*c*) Crystal tests.
(*d*) Polarography (see Chapter 9).

Systematic Wet Tests

The elementary course of study of qualitative inorganic analysis always includes the study of the analysis tables. These are a series of systematic tests which separate the more common metal ions into one of seven groups, mainly by precipitation reactions, and the examination of the separated precipitates to identify the individual cations present. The anions are similarly separated and identified. In most systems the wet tests for anions and cations are preceded by preliminary dry tests (i.e. tests on the solid sample rather than an aqueous solution) which give some indication of the ions which may be expected to be found in the more reliable wet tests which follow.

The more complex the mixture the more difficult the separation and identification of its components, and whereas the student is usually introduced to analysis by the identification of a simple inorganic salt, as he gains experience he may be expected to identify a mixture of three or more compounds with radicals in the same group or which require special treatment because they 'interfere' with one another.

Over the years scores of different sets of 'analysis tables' giving the systematic tests for the identification of the more common inorganic compounds have been published. These usually differ from each other in minor details only and the general procedure for analysis is common to most of them. Nevertheless, teachers of chemistry appear to have marked preferences for one scheme or another and most teaching establishments provide their students with duplicated sheets giving precise instructions on the system to be followed. There are also a large number of inexpensive booklets published which give reliable analysis schemes (see Appendix II). For these reasons it has not been considered necessary to reproduce a set of analysis tables in this book.

It should be pointed out that for most practical purposes analysis by tables in this manner is considered of little use in the working laboratory. The method works best when all the components are present in fairly high amounts and when they are in a fairly pure state. The purpose of teaching analysis in this manner is said to be threefold:

1. To teach the main reactions of the more common inorganic substances.
2. To give practice in the main manipulative skills in analysis.
3. To give some training in logical deduction.

The working analyst uses methods which give results of far greater reliability than would be obtained from the analysis tables which form a rather artificial system as they cover only a limited number of the inorganic components which may be found in a genuine sample.

Spot-Tests

Over the last 30 years or so a much more useful analytical technique has been developed, mainly associated with the name of Feigl and his collaborators. Spot analysis is a microanalytical technique which requires a minimum of apparatus and, with a little practice in the manipulative skills involved, excellent results are obtained.

For most practical purposes spot-tests are used to decide the presence or absence of a particular radical, e.g. to check the purity of a material by seeking the possible impurities. They may be used in systematic analysis by separating the components of a mixture into the groups of the analysis tables and identifying the precipitates by spot tests.

The group separation is carried out by techniques which are mainly scaled down from semi-micro methods, and such apparatus as 5 cm^3 beakers and flasks and 0·5 cm^3 centrifuge tubes are obtainable. Skill is needed in using such miniature apparatus, for instance H_2S must be passed into solution through a fine capillary tube with very great care; it is all too easy to blow the solution completely out of the tube. After the precipitates have been produced in centrifuge tubes they are dissolved in a suitable reagent and spot-tests performed on the solutions.

All the reagents used in spot-testing are organic in character, and materials of suitable purity for the technique are marketed by the usual chemical suppliers. The tests themselves may be carried out in several ways:

1. By the production of colours by bringing together one drop of the test solution and one drop of reagent on absorbent paper or a porcelain tile.
2. By placing one drop of test solution on absorbent paper impregnated with reagent.
3. By the production of microcrystals of characteristic shape and colour which may be recognized under the microscope.
4. By the development of colours in solution using micro test-tubes.
5. By the action of the reagent on a small amount of the solid sample.

The apparatus required for performing these tests is very simple. A useful selection of reagents and apparatus is provided in the Spot-Test Outfit marketed by Messrs. British Drug Houses. Although this kit does not provide enough for the comprehensive testing of any sample it contains some 30 reagents together with the apparatus for testing for a wide range of metal ions and radicals including some of the rarer elements not included in the group analysis scheme. Table 4.1 gives a list of the substances which may be detected and the reagents supplied.

TABLE 4.1. Reagents included in the B.D.H. Spot-test Outfit

Reagent	*Reagent for*
Aluminon (*tri*-ammonium aurinetricarboxylate)	Al
Arsenazo	Th
Benzion α-oxime	Cu and V
2,2′-Bipyridyl	Fe^{2+}, Cd
Brucine	Nitrates and Bi
Cacotheline	Sn
Chromotropic acid sodium salt	Cr, nitrate and Ti
3,3′-Diaminobenzidine tetrahydrochloride	SeO_3
Di(2-hydroxyphenylimino) ethane [glyoxal bis-(2-hydroxy-anil)]	Cd, Ca
4-Dimethylaminobenzylidenerhodanine	Ag, Hg and Au
Dimethylglyoxime	Ni, Pd, Fe^{2+}
1,5-Diphenylcarbazide	Cr
1,5-Diphenylcarbazone	Hg, V, Cd, Zn
Dithio-oxamide	Ni, Co, Cu and Ru
Dithizone (diphenylthiocarbazone)	Pb, Cu, Hg and Zn
Hexanitrodiphenylamine	K and Tl
8-Hydroxy-7-iodoquinoline-5-sulphonic acid	Fe^{3+} and Ca
8-Hydroxyquinoline (oxine; quinolol)	WO_4^{2-}, U, V
4-(4-Nitrophenylazo)resorcinol	Mg
1-Nitroso-2-naphthol	Co
3,5,7,2′,4′-Pentahydroxyflavone (morin)	Al, Be, Sn, Ti, Zr
'Phenylfluorone' (9-phenyl-2,3,7-trihydroxy-6-fluorone)	Ge
Picrolonic acid	Ca
Quinalizarin *see* 1,2,5,8-Tetrahydroxyanthraquinone	
Rhodamine B	Sb, Ga, Au, Tl, U
Rhodizonic acid sodium salt	Ba, Ca, Pb and Sr
Salicylaldehyde oxime	Cu
1,2,5,8-Tetrahydroxyanthraquinone	Be, Mg, Al, In and B
4,4′-Tetramethyldiaminodiphenylmethane (tetra-base)	Mn, I
Thiourea	Bi and Ru
Titan yellow	Mg

The apparatus includes:

1. A glazed porcelain tile with twelve depressions for 'spotting'.
2. Absorbent papers for 'spotting'.
3. Microscope slides and cover glasses for crystal tests.
4. Capillary tubes and small teats (micro-dropping pipettes).
5. 0·5 cm³ one mark pipettes.
6. 7·5 × ½ cm test-tubes and test-tube rack.
7. The B.D.H. Spot-Test Outfit Handbook.

The apparatus is shown in *Figure 4.1.* Other apparatus which may be found useful would be :

1. Micro-bunsen burner.
2. Microscope (up to 250×).
3. Nickel micro-spatula (Chattaway pattern 7–10 cm long).

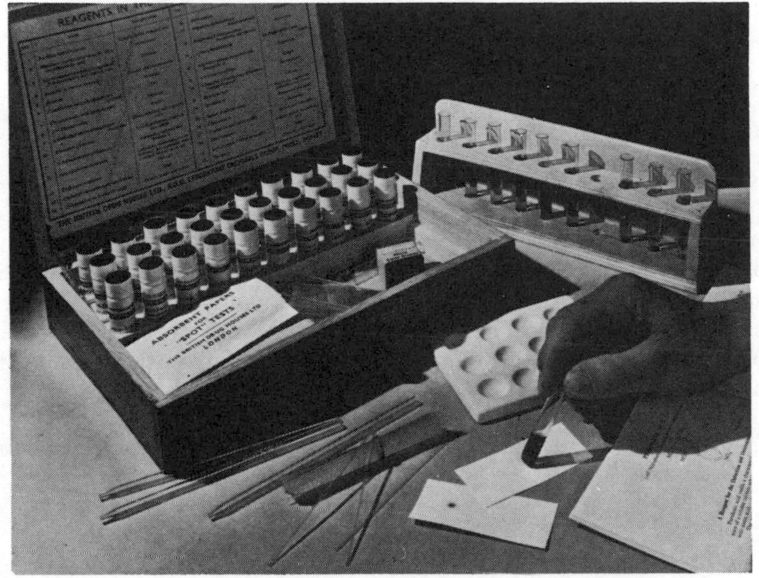

[By courtesy of B.D.H. Ltd]

Figure 4.1. B.D.H. spot-test outfit

4. Fine stainless steel tweezers.

5. Gas reaction assembly for the identification of gaseous reaction products (*Figure 4.2*).

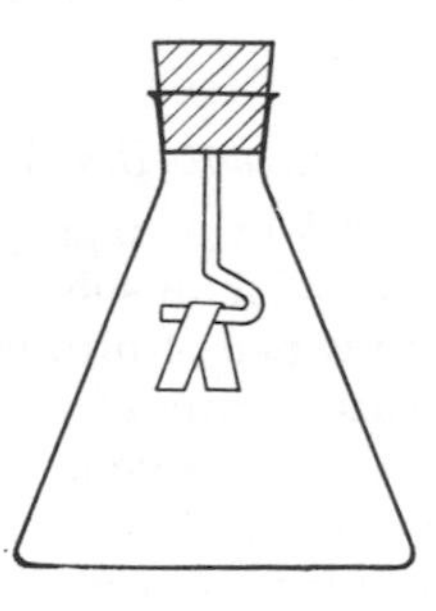

Figure 4.2. Apparatus for the detection of ammonia

Although it would be impossible to provide a complete scheme for analysis by spot-testing the following experiments will give some idea of the scope of the technique and the methods employed.

Experiment 1. The Detection of Nickel, Copper and Cobalt with Rubeanic Acid

$$\begin{array}{l} CS{-}NH_2 \\ | \\ CS{-}NH_2 \end{array}$$

One drop each of 0·1 per cent solutions of cupric sulphate, nickel sulphate and cobalt chloride are placed on separate pieces of absorbent paper. The papers are exposed to ammonia vapour and one drop of 0·1 per cent alcoholic rubeanic acid added to each spot. Nickel produces a blue coloration, cobalt a brown coloration, copper an olive green coloration. The sensitivity of the test is high, 0·1 μg of copper being detectable in the presence of 2 mg of nickel or 0·2 mg cobalt.

Experiment 2. The Detection of Cobalt, Iron and Zinc with Acridine Hydrochloride

$$\left[\text{acridinium (N–H)} \right]^{+} Cl^{-}$$

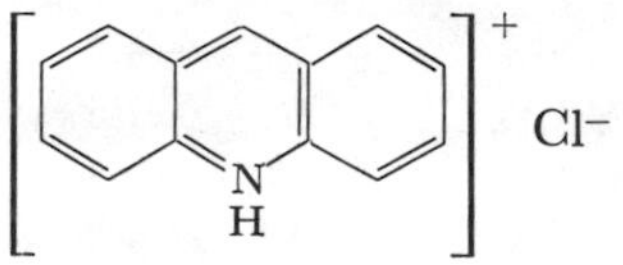

One drop of the solution to be tested is placed on a microscope slide. One drop each of 1 per cent aqueous acridine hydrochloride and 1 per cent ammonium thiocyanate solution are added. Under the microscope the formation of crystals can be seen. Ferric iron produces red crystals, cobalt square green crystals, zinc square yellow crystals.

Experiment 3. The Detection of Nitrates with Brucine ($C_{23}H_{26}O_4N_2$)

One drop of test solution is placed in a depression on the test tile and one drop of 0·02 per cent brucine in sulphuric acid added. A blood red colour indicates the presence of nitrate or nitrite. To distinguish between the two radicals a little of the test solution is treated with sodium azide, acidified, and boiled. This decomposes nitrite while nitrate is unaffected. The brucine test is repeated with the treated solution. Nitrate still produces a blood red colour while no colour is produced if the original material contained nitrite only. This test is extremely sensitive (0·06 μg nitric acid may be detected); great care must be taken to ensure that the reagents are nitrate free, and blank tests must be performed.

Experiment 4. The Determination of Aluminium with Aluminon

The test solution is acidified with HCl to about M concentration and mixed with equal volumes of 3 M ammonium acetate and 0·1 per cent aqueous aluminon solution. A bright red lake indicates the presence of aluminium or chromium. The solution is made slightly alkaline with a solution of ammonium hydroxide to which ammonium carbonate has been added. The chromium lake is destroyed while the aluminium lake persists. 0·02 mg of aluminium can be detected in the original sample.

Experiment 5. The Detection of Tin with Cacotheline

A strip of absorbent paper is dipped into a saturated aqueous solution of cacotheline and blotted. The test solution is acidified with hydrochloric acid and a little zinc powder added to reduce the tin to the stannous form. One drop of solution is spotted on to the prepared paper. A violet coloration indicates tin. The test is sensitive to 2 μg of tin per drop of solution.

Although this is a simple test for tin many substances interfere with the reaction and it cannot be applied in the presence of silver, chromium, copper, ferric iron, mercury, cobalt, nickel, or vanadium. Free nitric acid also inhibits the reaction while titanous chloride reacts to give a similar violet colour.

Many spot reactions suffer from interference by ions other than those being tested for and any possible interfering substances should be sought before applying a test.

Experiment 6. The Detection of Ammonia Released from Ammonium compounds

The apparatus shown in *Figure 4.2* is used. One drop of test solution and one drop of 2 M sodium hydroxide solution are placed in the vessel. A strip of moist red litmus paper is hung over the hook. The apparatus is closed and warmed in a water bath at about 40°C for about 5 min. This test is extremely sensitive and can detect as little as 0·01 μg of ammonia in the test drop, although for concentrations of this order the treated litmus paper should be compared with a strip of moistened untreated litmus paper.

These six experiments are quoted to indicate the type of apparatus and manipulation used in spot and crystal testing. Full treatment of this topic is given in the appropriate volumes listed in the Appendix.

QUANTITATIVE ANALYSIS

After the chemist has decided the kind of materials present in his unknown sample by qualitative analysis, he must determine how much of each component it contains by quantitative methods. There are several basic methods for inorganic quantitative estimation:

1. Volumetric, i.e. the measurement of liquid volumes.
2. Gravimetric, i.e. the measurement of weight.
3. Gas-volumetric, i.e. the measurement of gas volumes.
4. Colorimetric, i.e. the measurement of depth of colour.
5. Spectrographic.

The first two of these methods will be discussed in this chapter, while optical methods are described in Chapter 6.

In volumetric analysis a known volume of a solution containing the 'unknown' is taken and another solution containing a known concentration of some material which reacts with the unknown is added until reaction is complete. This process is a *titration* and the volume of standard solution (*titrant*) is called the *titre*. There are several conditions which must be fulfilled if analysis by this technique is to give accurate results:

1. The reaction must be specific for the particular 'unknown' being determined in the solution, i.e. the titrant must only react with the substance being determined and with no other material present in the 'unknown' solution.
2. The reaction must go to completion, preferably with reasonable speed.
3. The proportions by weight in which the 'unknown' reacts with the material in the standard must be known.
4. There must be some way of knowing when the reaction is complete.

There are several advantages of the volumetric technique over other methods.

1. A great many substances can be determined volumetrically, and physical separation is not necessary. The 'separation' in this form of analysis lies in the specificity of the reaction, i.e. the unknown is separated by being the only component to react.
2. It is usually easy to prepare a standard, i.e. a 'reactive' solution of known strength.

3. Analysis is a speedy process, the measurement of volume is all that is required once the standard is prepared.
4. The apparatus required is comparatively inexpensive and readily available.

The main disadvantage lies in the fact that the measurement of volume is perhaps the least accurate of the measuring processes used in the laboratory and results cannot be expected to be accurate to within less than about 0·5 per cent.

In gravimetric analysis the 'unknown' component is separated from all the other components in a weighed sample by converting it to an insoluble derivative with a known formula. This is physically separated by filtration, washed, dried, and weighed. From the weight of derivative formed the weight of the 'unknown' component in the original sample may be calculated.

In some cases gravimetric analysis can be performed with very great accuracy as the only measuring instrument used is the analytical balance which is by far the most accurate instrument used in the laboratory. This is the main advantage of the technique.

Gravimetric analysis is limited to reactions which go to completion and which are specific for the substance concerned besides forming a derivative which is insoluble and has a precisely defined formula. The difficulty of fulfilling these conditions is one disadvantage of the method. Another disadvantage is that the careful separation, filtration, drying and weighing techniques require skill of a high order and much practice is needed for this skill to be acquired. Further, the method is a slow one, the preparation of derivatives, and the drying process especially, taking hours of laboratory work. The accuracy of the technique when properly performed nevertheless makes gravimetric analysis a common method in the working laboratory.

Gas-volumetric analysis, as the name implies, is concerned with the measurement of gas volumes mainly in the analysis of gas mixtures.

THE ANALYTICAL BALANCE

In both gravimetric and volumetric analysis weighing is a primary skill. It is true that standard solutions may be purchased or may be provided for students in a teaching laboratory, but to prepare the standard solution a weighing must be performed at some point. If the 'unknown' is solid it must be weighed to prepare a solution of known concentration to perform the titration. The analytical balance is used for this purpose. A simple precision balance of the type provided in most teaching laboratories is shown in *Figure 4.3.* A rigid metal *beam* is supported on a central fulcrum or *knife-edge* of optically flat agate or synthetic sapphire. The *pans*, on which objects to be weighed and weights are placed, are hung by stout wires at the ends of the beam, the wires being supported on further knife-edges. A light pointer at the centre of the beam indicates a scale attached to the *pillar* which supports the central knife-edge. In most balances some method of supporting the pan supports and beam off the knife-edges when the balance is not in use is provided to protect the agate from unnecessary wear.

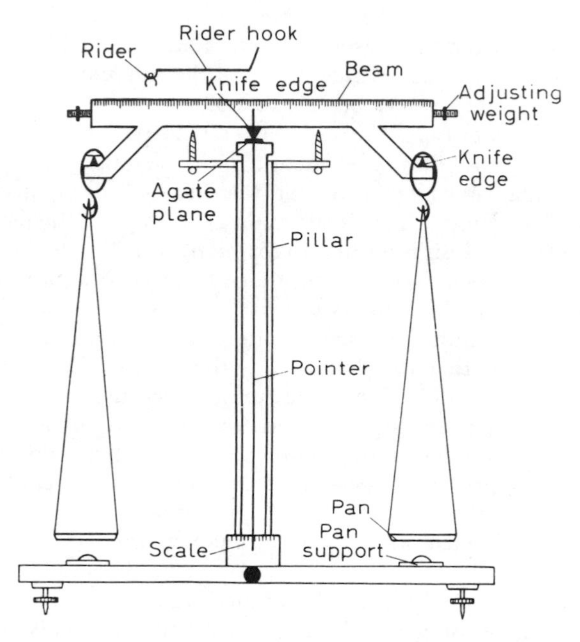

Figure 4.3. A simple analytical balance

The simplest type of balance is used in the same way as the shopkeeper's scales. The object to be weighed is put on the right-hand pan and weights added to the left-hand pan until the beam is horizontal and the pointer vertical. For the analytical balance this is too crude for accuracy and extra refinements must be added in order to weigh substances to an accuracy of less than 10 mg. Weights smaller than this are difficult to handle. The analytical balance is provided with a wire *rider* of 5 or 10 mg. This movable weight may be placed on the beam at different distances from the central knife-edge. The beam is calibrated so that the weight on the pan which is equivalent to the effect of the rider in position may be seen.

The analytical balance most commonly provided in teaching laboratories is intended for a maximum load of 200 g and may be read to the nearest 0·1 mg. The weights provided are of two kinds. Those of 1 g or more are cylindrical in shape with a small knob so that they may be held with forceps. They are usually of brass or stainless steel. The fractional weights are flat sections of aluminium or stainless steel, with one edge raised for forceps.

Care of the Balance

In most laboratories the balances are placed in a separate small room ('balance room') which is protected from the fumes of the main laboratory. Ideally, the balance should rest on a foundation as free from vibration as possible. In a specially built balance room this is usually a heavy stone or slate slab resting on cork and supported on heavy brick piers. Where this is not possible the balance resting on a thick sheet of cork or rubber should be set on a firm table. Alternatively, the laboratory suppliers market anti-vibration balance tables.

It is important that the temperature of the two arms of the beam should be the same. Thus the balance should not be set near a window or radiator, or where it is liable to be affected by draughts.

In many laboratories it is the practice to place a small balance desiccator (a perforated aluminium can containing dry silica gel) inside the balance case so that the humidity is kept low and constant, although it must be said that some reputable balance manufacturers do not approve this, claiming that the placing of a desiccator to one side of the balance case upsets the symmetry of the system.

Clean and tidy working in the balance room is of the greatest importance. A flat camel-hair brush about 2 cm wide should be placed inside each balance case so that any material accidentally spilled may be cleaned up as quickly as possible.

Methods of Weighing

There are three methods of weighing. The most important is *direct weighing* which, in one of its variations, is the method used in most routine chemical weighing operations. The simple method described is adequate for most purposes in elementary quantitative analysis.

1. Check the balance to ensure that the beam and pans are in correct alignment. Check the box of weights to ensure that it is complete.

2. With the rider hook place the rider on the zero position on the beam.

3. With the front window of the balance open and the pans empty, gently release the beam and allow it to swing. If it does not swing one pan may be 'fanned' gently with the hand to set the balance in motion. Lower the window.

4. The pointer may not swing symmetrically around the zero point on the scale. Find the point on the scale around which the pointer swings symmetrically and note it. This is the *rest-point* of the balance. If the rest-point is grossly to one side or the other it may be necessary to bring it nearer the central zero by adjusting the setting nuts at each end of the beam. For most practical purposes it is not essential that the zero is the rest-point as long as the rest-point is known.

5. The weighing vessel is wiped with a chamois leather or other soft lintless cloth and placed on the left-hand pan. Weights are added to the right-hand pan starting with one too large and trying the weights in order down to 0·01 g. *Nothing should be added or removed while the beam is swinging.* The beam should be arrested for every change.

6. After the vessel has been weighted within 0·01 g below the true weight the balance is closed and the rider placed on the 5 mg mark. Release the beam and see if this is 'too heavy' or 'too light'. Continue to adjust the rider until the pointer swings symmetrically around the previously noted rest-point.

7. Take the weights off the scale pan starting with the largest and add up the total weight of the vessel while returning the weights to the box. Note this weight in a practical book (not a scrap of paper).

8. Remove the vessel from the pan.

9. Put the material to be weighed into the vessel and repeat the steps 1–8 above. If a definite weight is needed, guess the approximate amount to put into the vessel and add or subtract as required. The vessel must be taken from the balance when material is being added or removed. For most practical purposes it is not necessary to weigh out a specified amount, as long as the material is near the specified amount and its weight is accurately known.

10. Dust out the balance case if necessary and close it. Check to see that it is at rest and in order before leaving.

As stated, the method above is adequate for most practical purposes in the C.G.L.I. syllabus.

Where very accurate weighing is required, very slight inequalities in the lengths of the two arms of the beam may have a considerable effect. *Gauss's method of double weighing* eliminates this inaccuracy. The substance is weighed as stated above. The vessel is then moved to the right-hand pan and weighed again (care being taken with the rider position). The arithmetic mean of the two weights gives the correct weight of the substance. *Borda's method* is another technique for eliminating beam errors. The substance is placed on the right-hand pan and *tared* using a vessel on the left-hand pan as a *counterpoise*. Lead shot or sand is added until the weight on the two sides is equal. The substance is removed from the right-hand pan and weights added until the same rest-point is obtained. This is a method which is most useful for large or heavy substances.

The Sensitivity of the Balance

Although most student analytical balances are calibrated to 0·1 mg on the beam, they rarely show any difference in rest-point for a change of this amount. This is because the *sensitivity* of the balance is not sufficient to register such a change. Normally this type of balance shows a deflection of one scale division for a change in weight of 0·5 mg, and therefore materials can only be weighed to the nearest 0·5 mg.

The sensitivity of a balance may be defined as the number of scale divisions which the pointer moves for an inequality of weight of 1·0 mg. The N.P.L. has defined sensitivity as the ratio deflection/corresponding change of mass. Sometimes the reciprocal of this ratio is used when the deflection may be needed for very fine weighing. An alternative definition is sometimes used: the sensitivity is the weight needed to shift the pointer by one scale division.

In some catalogues the sensitivity of a balance is quoted merely as a weight value, e.g. 'sensitivity 0·1 mg'. This bald statement is really inadequate to define sensitivity.

It should be noted that in the normal analytical balance the sensitivity decreases as load increases. In some laboratories each balance is calibrated for sensitivity by noting the number of scale divisions of deflection caused by a difference in load on each pan of

1 mg at a number of different loads from 0–100 g. A graph is plotted, divisions per milligramme/load in each pan, and kept inside the balance. The variation in sensitivity is mainly due to the beam being not quite perfectly rigid and the bowing effect of large loads changing the relative positions of the three knife-edges.

Calibration of Weights

Everything in the laboratory which is used to measure should be calibrated when new, and checked at regular intervals thereafter. Weights are no exception. The simplest and most accurate way of calibrating weights is with a set of weights accurately calibrated and certified by the National Physical Laboratory. Borda's method of substitution is used. Three sets of weights are used, the N.P.L. set, the set to be calibrated (*C*) and a third subsidiary set (*S*). The smallest weight of *C* is placed on the right-hand pan and balanced by the corresponding weight of *S*. The rest-point is found. *C* is removed and replaced by the corresponding N.P.L. weight and the new rest-point found. The difference in the two rest-points corresponds to the error in the *C* weight and can be evaluated from the sensitivity of the balance. Every weight in the *C* box can be checked in this manner. The N.P.L. set must be returned occasionally for checking and recertification.

If no N.P.L. set of weights is available a *C* set can be 'internally' calibrated against itself. This assumes that one of the weights in the set is accurate and calibrates all other weights to this using an extra set of weights as a tare. This one weight may be then checked against a single N.P.L. weight to find the true values of the complete set.

It is important to have at least one good calibrated set of weights in the laboratory. Volumetric apparatus is calibrated by weighing as will be seen later in this chapter.

More Refined Analytical Balances

For many years balances of the type described above were in regular use in working laboratories where gravimetric estimations of the highest accuracy were performed. More recently various refinements have been added which speed weighing and which improve the accuracy of the process considerably. Balances possessing one or more of the following devices are usually found in working laboratories.

1. Automatic release is a feature of most modern balances. There are several systems which release the beam slowly and gently no matter how quickly and clumsily the release knob is turned. One (Stanton synchro-release) depends on a geared electric motor which releases the beam when switched on by the turning of the release knob. Oertling Releas-o-matic balances employ a hydraulic

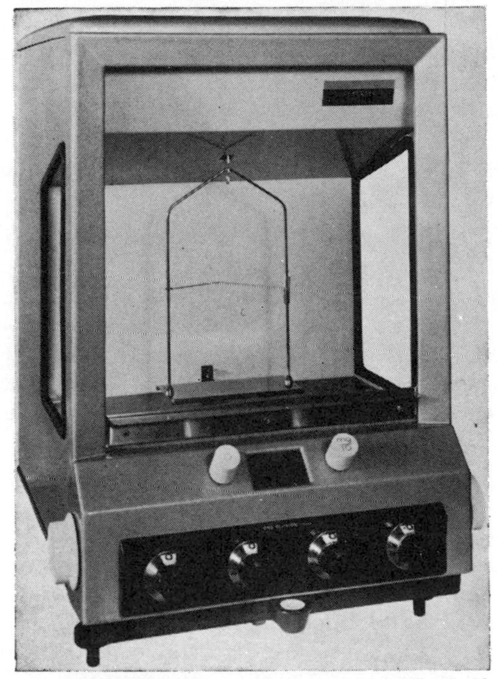

[By courtesy of Oertling Ltd]

Figure 4.4. A modern single-pan analytical balance

piston which gradually releases the beam when the release knob is actuated. Other manufacturers have devised other patented methods. The standardization of the beam release (and arrest) which is achieved by these systems leads to more accurate weighing, and by ensuring low velocity of contact between knife-edges and planes, the agate is protected and repairs are reduced to a minimum.

2. Aperiodic cylinders are often fitted. These are in two halves one attached to the pillar the other to the beam at each side of

the balance. The slow movement of one half inside the other causes hydraulic damping of the swinging beam so that it comes to rest speedily without the tedious swinging of the undamped balance.

3. Projection reading is achieved by fitting a transparent graticule to the lower end of the pointer. A lamp at the back of the balance projects a magnified image of the graticule on to a small screen at the front. The last two or three places of decimal weights may be read off the screen eliminating the tedious adjustment of the rider. The hair line on the screen may be easily adjusted so that the rest-point is at zero, and the sensitivity of the balance is very easily checked.

Projection reading is in fact more reliable than weighing by rider as both rider and beam tend to suffer from wear during use which lowers accuracy over a period of time.

4. Automatic and semi-automatic weight loading is a common feature of good balances. The weights are ring shaped and are placed on individual hooks inside the balance. Dials outside the balance case control the weights which may be lowered on to supports fixed to the right-hand pan stirrup. The weights need never be removed from the balance case and their constant weight is assured as they are never touched. Sometimes the larger weights are placed on the pan manually, in other models all the weights are dial controlled.

5. Single pan balances have recently become very popular. These are mostly of the constant load type. The weight pan is eliminated and equal weights are placed on both halves of the beam. The substance to be weighed is placed on the pan and weights removed automatically from the pan side of the beam until equilibrium is achieved. The constant load on the beam ensures that the sensitivity of the instrument remains constant throughout the weighing range. Many of these instruments have unequal beam arms which makes the zero position more than normally sensitive to fluctuations in temperature. The popularity of these models is mainly owing to the speed with which weighing can be carried out and the sturdy way in which they stand up to clumsy handling.

Semi-micro Balances

These usually incorporate most of the refinements described above. In the U.K. balances with maximum load of 50 g (sensitivity

0·02 mg/div) and 30 g (sensitivity 0·01 mg/div) are manufactured.

Microchemical Balances

Balances with a maximum load of 20 g (sensitivity 0·001 mg/div) are in the microchemical range. These are invariably air damped and have projection reading. The beam and upper part of the pillar is separately enclosed so that changes in atmospheric conditions do not interfere while the weighing is carried out. (These balances are so sensitive that even the heat from the operator's body may cause temperature problems.)

THE TORSION BALANCE

The torsion balance (*Figure 4.5*) is based on a completely different principle from the analytical balance described above. Weighing

Figure 4.5. A torsion balance

[By courtesy of White Electrical Instruments Co. Ltd]

depends on the torsion applied to a spiral spring or wire one end of which is attached to the axis of a shaft and the other end to a pointer. The balance arm, which is attached to the centre of the wire or spring, projects from the case and has a small hook at the end to which a pan may be attached. A separate 'rest-point' is engraved on the dial and a second pointer attached to the balance arm indicates the rest position. A lever moves the pointer along a 32 cm ($12\frac{1}{2}$ in) scale provided with a parallax mirror. There is a separate lever to lock the balance arm when the instrument is not in use.

There may be slight variations in the technique for using torsion balances from different manufacturers. The method described is suitable for the particular model marketed in Britain by all the larger laboratory suppliers.

1. The instrument should be set on a firm table high enough so that the pointer is at eye-level.
2. Adjust the levelling screws at the feet so that the bubble in the spirit level on the base of the instrument is central.
3. Attach the small aluminium pan if this is required.
4. Adjust the main pointer to the zero position using the parallax mirror.
5. Move the locking lever to release the balance arm and adjust the small pointer on the right-hand side to the rest position by turning the knob at the back of the case. Lock the balance arm.
6. Place the material to be weighed on the pan.
7. Release the balance arm and move the handle so that the pointer moves across the main scale until the subsidiary pointer once again points to the rest position.
8. Read the weight of the material from the main scale using the parallax mirror.

Although the torsion balance has not the accuracy of the analytical balance (about $\pm$ 0·2–$\pm$ 0·4 per cent of maximum load is usual) it is cheap, robust and very rapid in use. Models with maximum loads of 5 mg–10g are available. This type of balance is not often used in inorganic chemistry but it is invaluable in the biochemistry laboratory where the rapid weighing of numerous tissue samples, drug doses, etc. may be required.

VOLUMETRIC APPARATUS

The measurement of liquid volumes in volumetric analysis is achieved by four or five main types of graduated glassware. Some of these are graduated 'to contain' others 'to deliver'. Each individual instrument will be labelled with its characteristics according to the B.S.S. The label usually has:

1. Maker's name or trade mark.
2. B.S. number for specification of instrument.
3. Volume contained or delivered in millilitres or cm^3.
4. The letters *Int* meaning 'to contain' or *Ex* meaning 'to deliver'.
5. The temperature (usually 20°C) at which calibration is valid.

The basic measuring instruments required for elementary titration are:

1. The transfer pipette.
2. The burette.
3. The volumetric flask.

Pipettes

Two types of pipette are commonly used in inorganic chemistry (*Figure 4.6*):

1. The common transfer or bulb pipette to deliver.
2. The graduated pipette.

The common transfer pipette is the type used in volumetric analysis on the macro-scale. Its purpose is to deliver a stated volume of liquid and it is available in sizes from 1 cm^3 to 100 cm^3. A glass cylindrical bulb is joined at each end to a narrow tube. The upper (suction) tube has a calibration mark etched around it while the lower (delivery) tube is drawn out to a fine tip.

To transfer a known amount of liquid from one vessel to another:

1. The clean pipette is rinsed three times with small volumes of the liquid to be transferred.
2. The liquid is sucked up by mouth to a level just above the calibration mark.
3. The tip is removed from the liquid and simultaneously the top of the suction tube is closed with the forefinger.
4. The outside of the tip is wiped with Kleenex or filter paper to dry it.
5. The liquid is run out by slightly releasing the pressure of the forefinger until the graduation mark is tangential to the *bottom* of the meniscus when held at eye level. The tip is *just* touched against the surface of the liquid being transferred to remove any drop hanging from the tip.

6. The pipette is moved to the vessel to which the liquid is to be transferred and the contents allowed to discharge by the removal of the forefinger. It is held vertically with the tip just touching the inside of the receiving vessel which is tilted slightly to allow this.
7. After the meniscus has reached the tip of the pipette it is held in position for 15 sec to allow complete drainage. The pipette is then removed and the transfer is complete. The drop of liquid remaining in the tip of the pipette should never be blown or shaken into the receiving vessel.

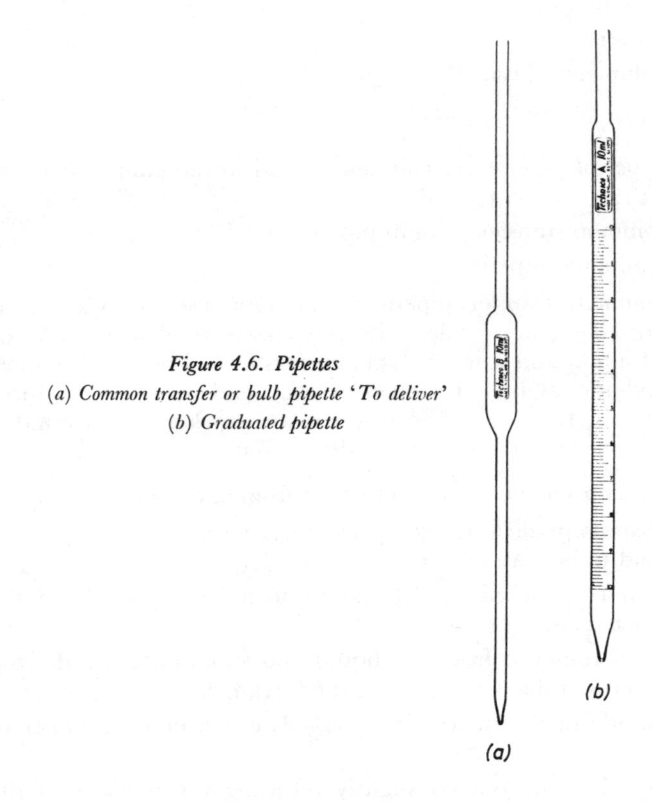

Figure 4.6. Pipettes
(*a*) *Common transfer or bulb pipette 'To deliver'*
(*b*) *Graduated pipette*

Although some authorities recommend slight variations in this technique the method described is that recommended by the N.P.L. in its instructions on the calibration of pipettes, and it is felt that this is sufficient reason for this procedure to be rigidly adopted.

Modifications to the common transfer pipette are mainly safety devices for the measurement of poisonous solutions. The simplest is a large bulb blown in the suction tube above the level of the graduation mark, to act as a reservoir so that about twice the volume of liquid required may be drawn into the tube before the liquid reaches the lips. A popular device is the pipetting bulb (Pro-pipette, Griffin Pipette Filler) a large rubber bulb fitted with three valved tubes. The pipette fits into one of the tubes, and, by squeezing the bulb and manipulating the valves, liquid may be drawn into the instrument, adjusted to the graduation mark, and dispensed without mouth suction.

The graduated pipette is used to deliver any volume of liquid up to the maximum capacity stated. It is by no means as accurate as the bulb pipette and is available in a large number of sizes. Graduated pipettes are used in a similar manner to bulb pipettes. They may, however, be either 'to deliver' or 'to contain'.

Other types of pipette are available for special purposes. These will be described as the appropriate topics occur in later chapters. Occasionally a special technique may be used in the laboratory which requires accurate pipettes of an 'odd' size. The larger manufacturer does not usually consider it profitable to provide a small 'run' of an 'odd' size but some small firms* are prepared to do this, especially if they have graduating equipment.

Burettes

The common burette (*Figure 4.7*) is used to deliver any volume of liquid up to the maximum capacity. Sizes of 10–50 cm^3 are available graduated in 0·1 cm^3.

The recommended method of using the burette is as follows:

1. Check that the tap is properly lubricated to prevent leakage.

2. Rinse the burette three times with the liquid to be used, allowing the rinsing liquid through the tap so that the jet is well wetted. If this causes lubricating grease to clog the tap the latter must be removed and the excess grease extracted with a fine wire.

3. Fill the burette with the liquid to be used to well above the zero mark.

4. If a funnel is used to fill the burette it is removed. Liquid is allowed out of the jet until the zero graduation is tangential to

* One such company is Messrs. Vifleur (Scientific) Ltd., 4 Sutherland Avenue, London, W.9, who the authors have always found helpful in the supply of a wide range of both standard and 'odd' volumetric apparatus.

the meniscus. If a drop of liquid is still hanging from the jet it must be 'touched off' against the receiving vessel. There must be no air bubbles either in the main tube or in the jet under the tap.

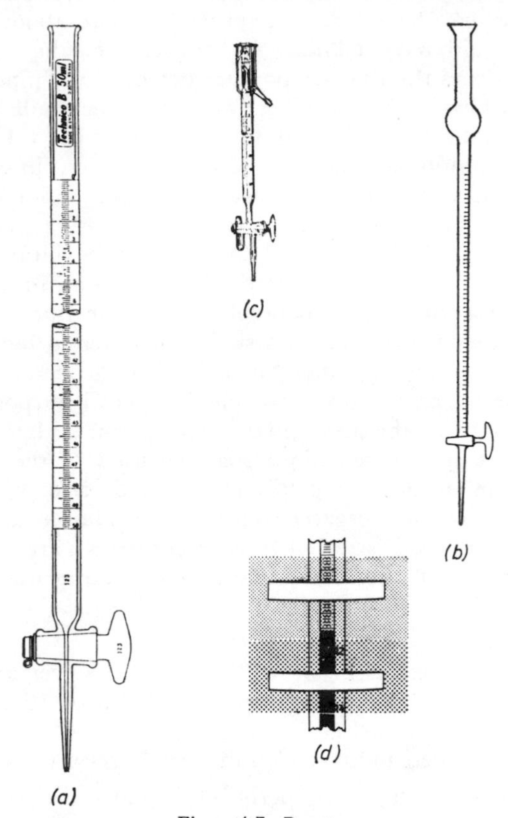

Figure 4.7. Burettes

(*a*) *Common burette* (*c*) *Automatic filling automatic zero burette*
(*b*) *Micro burette* (*d*) *The burette reader*

A useful device for reading the position of the meniscus easily is the burette reader (*Figure 4.7d*). This is a piece of card or dull plastic, half black and half white with four cuts in it so that it can be slipped over the burette tube as shown. The card is adjusted so that the top of the black half is just 0·1 cm^3 below the bottom of the meniscus. The burette must always be read with the meniscus at eye level. Even when the burette is clamped in a stand, its normal position, the meniscus must be brought to eye level.

5. The liquid is run out of the burette as required. The burette tap should be turned with the *left* hand. Although this may seem clumsy at first it prevents the burette tap from falling out and supports the hand at the end of the titration when the liquid is added drop by drop. Before reading, any drop hanging from the jet is 'touched off' as before. The instrument may be read to the nearest $\frac{1}{2}$ division (0·05 cm^3).

Burettes of capacity 1–5 cm^3 are used in semi-micro work. These are usually graduated into 0·01 cm^3 or 0·02 cm^3 divisions. They are similar to those described above but are usually fitted with reservoirs from which the instrument may be refilled as required.

Micro burettes of capacity less than 1 cm^3 may be graduated to fractions of 0·001 cm^3. These are often operated on a micrometer principal. One of the most popular models in this country is the Agla microsyringe (*Figure 4.8*). Right-angled glass adapters are available to use the instrument as a burette. Each revolution of the micrometer head corresponds to 0·01 cm^3 and each graduation on the head to 0·0002 cm^3.

Often a particular series of titrations is regularly performed in a laboratory. It is convenient to set up a *titration bench* as a semi-permanent structure. The burettes used are of the automatic-filling automatic zero type (*Figure 4.7c*). Turning the double stopcock one way fills the burette until it overflows at the top, the excess returning to the reservoir, a large bottle of standard solution. Reversing the tap allows the burette to empty normally. Automatic pipettes of various sizes are made on the same principle.

The Volumetric Flask

Volumetric (or graduated) flasks are made to contain a stated volume of liquid and are obtainable in sizes from 5 cm^3 to 5 dm^3. (*Figure 4.9*). Most are fitted with standard glass or plastic stoppers although occasionally non-interchangeable individually ground glass stoppers will be encountered. In this case stopper and flask must be numbered together to identify the stopper for the flask. For student use, flasks with unground tops are often used with rubber stoppers as closures.

The main use of the volumetric flask is in the preparation of standard solutions. This is usually begun by weighing out the correct amount of material for making up to volume. In elementary courses a small beaker or watch glass is used as a weighing vessel. The weighed material is dissolved in water in the beaker and the procedure for making up to volume in the volumetric flask is as follows:

1. The solution should be allowed to cool if necessary.
2. After cooling it is carefully poured into the flask through a small funnel. Alternatively, the solid is washed off the watch-glass into the flask.

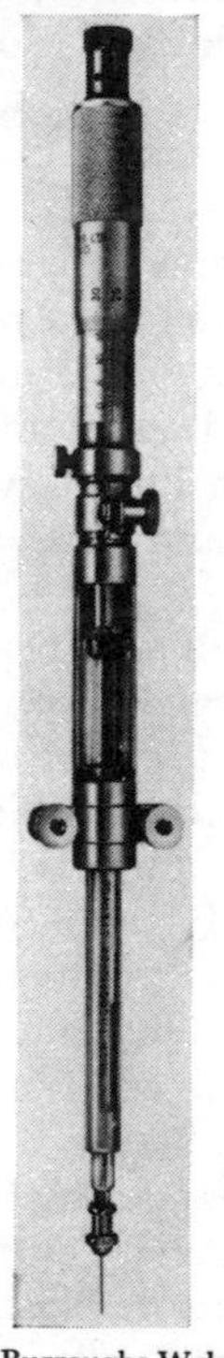

[By courtesy of Burroughs Welcome and Co]

Figure 4.8. Agla micrometer syringe

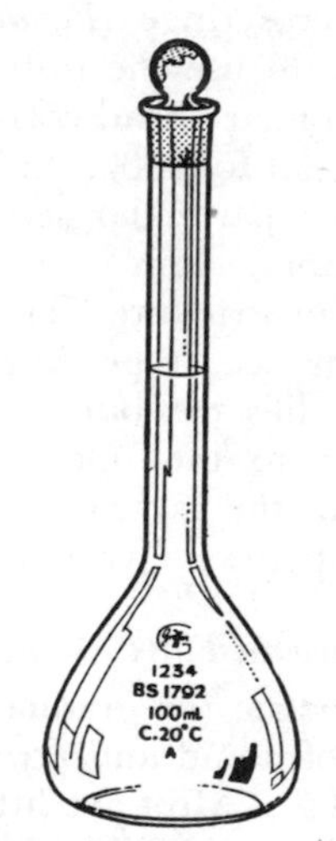

Figure 4.9. Volumetric flask

3. The beaker is rinsed with a little water and the rinsings added.
4. Step 3 is repeated at least three times. This process is called *quantitative transfer*, i.e. the material has been transferred from the beaker to the volumetric flask quantitatively.
5. The funnel is rinsed and removed.
6. Water is added until the meniscus is nearly up to the mark.

7. The final small quantity of water is added with a teat pipette dropwise until the mark is tangential to the bottom of the meniscus.
8. The flask is stoppered and shaken very thoroughly, the stopper being held and the flask inverted and righted several times during shaking to ensure homogeneity.

The Measuring Cylinder

The measuring cylinder (*Figure 4.10*) may be any size from 5 cm^3 to 2 dm^3 and may be stoppered or spouted. It is used to deliver any volume of liquid to a stated maximum. The measurement is only very approximate. Calibration is usually only at the

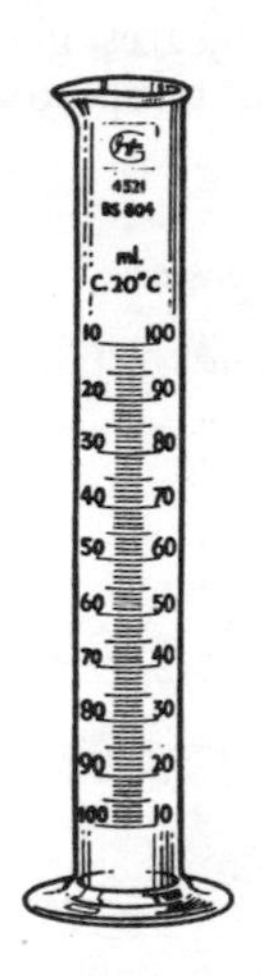

Figure 4.10. Measuring cylinders

maximum graduation mark, the rest of the cylinder being graduated on the assumption that the diameter of the vessel is constant along the whole of its length. This is seldom true and a large measuring cylinder should never be used to deliver a small volume.

The Calibration of Volumetric Apparatus

Most apparatus purchased is guaranteed by the manufacturer to be accurate within the limits laid down by the relevant B.S.S. For very accurate work it may be necessary to calibrate a particular piece of apparatus in the laboratory. Calibration is always a weighing process. The apparatus to be calibrated is cleaned with 'chromic acid' (Chapter 1, page 17) and thoroughly dried.

Pipettes—

1. A vessel of greater capacity than the pipette is weighed.
2. The pipette is filled to the mark with distilled water and the water transferred to the vessel as described above.
3. The vessel is again weighed.
4. The temperature of the water is taken.
5. The process is repeated several times until five separate weights of water are obtained. For a macro pipette the variation should not be greater than about 0·5 per cent between the highest and lowest figure (e.g. 50 mg for a 10 cm³ pipette). A greater variation than this would indicate faulty technique in using the pipette.
6. The mean of the five weights is taken and the actual volume delivered by the pipette calculated from the figures given for the volume of 1·0 g of water at the measured temperature in Table 4.2.

TABLE 4.2. Volume of 1 g Water at Different Temperatures

Temp.°C	*Volume* (cm³)	*Temp.*°C	*Volume* (cm³)
10	1·0016	22	1·0032
11	1·0017	23	1·0034
12	1·0018	24	1·0036
13	1·0018	25	1·0038
14	1·0019	26	1·0041
15	1·0021	27	1·0043
16	1·0022	28	1·0046
17	1·0023	29	1·0048
18	1·0025	30	1·0051
19	1·0026	31	1·0054
20	1·0028	32	1·0057
21	1·0030	33	1·0060

Burettes—To calibrate a 20 or 50 cm³ burette the process is similar.

1. The burette is set up in a stand and filled above the zero mark with distilled water. The temperature of the water is taken and the water is allowed to run out until the meniscus is at the zero mark.
2. The water is allowed to run into a previously weighed vessel until the meniscus is at the 5·0 cm³ mark and the vessel weighed again.
3. The burette is refilled, zeroed, and the water delivered between zero and 10 cm³ weighed.

4. The process is repeated until weights are obtained for sections 0–5 cm^3, 0–10 cm^3, 0–15 cm^3 etc.
5. The volume for each weight is calculated.
6. The process is repeated to give duplicates for each reading. Assuming the temperature to be reasonably constant duplicates should not vary by more than 20 mg.
7. The results should be tabulated to give the correction for each section of the burette. Alternatively a graph may be plotted (burette reading/correction) to be used with the burette.

Another method uses a special pipette device, the burette calibrator. This is a difficult instrument to use accurately, and it is not the technique employed by the N.P.L. Its use is described in the specialist literature.

Volumetric Flasks—For the calibration of volumetric flasks a large balance of about 1 kg capacity and a sensitivity at full load of 5 mg/division is needed.

1. The clean dry flask is weighed.
2. It is filled to just above the mark with distilled water without wetting the neck above the water surface.
3. Water is withdrawn with a capillary pipette until the meniscus is at the mark.
4. The flask is again weighed and the weight of water and hence its volume calculated.
5. The calibration is repeated to give a duplicate. The mean is taken as the volume of the flask.

Micro apparatus is calibrated in a similar manner but using pure mercury instead of water so that the weight/volume ratio is high enough to detect small errors.

The Conduct of a Titration

The handling of the more common apparatus used in volumetric analysis has been described. By summarizing these accounts the technique for conducting a simple titration is obtained.

1. Make up the standard solution required and ensure that both standard and 'unknown' are homogeneous.
2. Check that the pipette, burette, and conical flask in which the titration will be performed are clean. The burette tap should be lighty greased but not clogged.
3. Rinse out both burette and pipette three times with small amounts of the liquids to be used in them. (*N.B.* There is no

general rule about the standard or 'unknown' being in the burette. This usually depends on such factors as the colour change of the indicator.)

4. Rinse the conical flask with distilled water.
5. Fill the burette using a funnel. Clamp it vertically in a burette stand, remove the funnel and adjust the meniscus to the zero mark.
6. Pipette the aliquot of the other solution into the conical flask and add the indicator if necessary. The amount of indicator should be kept to a minimum. Two drops is usually enough although in some special cases a larger amount is specified.
7. Read the burette. Note the reading in a notebook and check again.
8. Put the conical flask under the burette on a white tile or white paper. Run in the solution from the burette swirling the flask gently with the right hand.
9. When the colour change in the flask persists for a short time before vanishing the end-point is near. The inside wall of the conical flask is washed down with a fine jet of distilled water from a wash bottle.
10. The titrant is added one drop at a time swirling between the addition of each drop.
11. When the end-point is reached the burette is read to the nearest $\frac{1}{2}$ graduation. (For a burette graduated in 0·1 cm^3 divisions read to the nearest 0·05 cm^3.)
12. The titration is repeated. If the second titration agrees with the first within 0·1 cm^3 the duplicates are considered satisfactory. If not, the titration must be repeated until two results agreeing within 0·1 cm^3 are obtained.

Standard Solutions

In describing standard solutions terms are used which require definition:

1. The *relative atomic mass* or *atomic weight* of an element is the mass of one atom compared with the mass of one atom of the carbon nuclide ^{12}C. The symbol is A_r.

2. The *relative molecular mass* or *molecular weight* of a compound is the sum of the atomic weights of all the atoms in the molecule. The symbol is M_r.

3. The term *equivalent weight* (and thus the term *normal solution*) are not admissible in the SI system.

4. The *mole* is the amount of a substance which contains as many elementary entities as there are carbon atoms in 0·012 kilogram of ^{12}C. In the case of compounds this would be the molecular weight expressed in grams. The concentration of ions, radicles etc. may also be expressed in molar terms.

Examples:

1 mole of HgCl has a mass of 236·04 grams
1 mole of $Hg_2\ Cl_2$ has a mass of 472·08 grams
1 mole of Ca^{2+} has a mass of 40·08 grams
1 mole of $\frac{1}{2}Ca^{2+}$ has a mass of 20·04 grams

5. A *molar* solution contains one mole of a compound in 1 dm^3 of solution.

6. A *molal* solution contains one mole of the solute in 1 kg of solution.

In the equation

$$H_2SO_4 + 2NaOH = Na_2SO_4 + 2H_2O$$

one molecule of sulphuric acid replaces 16 units of oxygen (one atom). The weight equivalent of sulphuric acid will therefore be

$$\frac{98{\cdot}08}{2} = 49{\cdot}04$$

and a half-molar solution of sulphuric acid will contain 49·04 g/dm^3.

The Preparation of Standard Solutions

Successful volumeteric analysis ultimately depends on pure substances accurately weighed. In very accurate work several samples of the pure substance are weighed separately dissolved in water, and the whole of each weighed sample is titrated with the 'unknown' in the burette. For most volumetric analysis one weighed portion of the pure sample is made up to a known volume and aliquots are pipetted to be titrated with the 'unknown' or the standard is titrated with pipetted aliquots of the 'unknown'.

The ideal substance for the preparation of a standard solution has the following characteristics:

1. Available in a very pure state.
2. Stable in the solid state even at 110–120°C so that it may be

dried in the oven.

3. Neither deliquescent nor hygroscopic.
4. Must speedily react completely with the substance to be standardized.
5. Have a relatively high equivalent weight so that small errors in weighing should have minimum effect.
6. Be readily available and comparatively inexpensive.

Several materials having all or nearly all the above characteristics are used as *primary standards*. This means that a standard solution may be prepared by weighing a known amount of the substance and making it to a known volume with water. Further standardization is unnecessary. Some primary standards are:

1. *Potassium Hydrogen Phthalate* $C_6H_4(COOH)$ COOK. mol. wt. 204·14.

To prepare a 0·1000 M solution 20·414 g of the dry A.R. grade salt is diluted to 1 dm^3 with distilled water. This is an acid standard. Phenolphthalein is the indicator of choice when titrating with this material, as it is, in effect, a very weak acid.

2. *Hydrochloric Acid* HCl, mol. wt. 36·465.

Although concentrated hydrochloric acid is not a suitable substance with which to prepare a primary standard, *constant boiling hydrochloric acid* is an excellent material for this purpose. This is because boiling hydrochloric acid loses water and HCl in such proportions that it eventually attains a constant known composition at a given barometric pressure.

To prepare constant boiling hydrochloric acid A.R. grade concentrated acid is diluted to give a density of about 1·10 (about 500 cm^3 water + 500 cm^3 acid at 20°C and adjust to give density 1·10 by adding water or acid as required).

The diluted acid is distilled. The first three quarters of the distillate is discarded. Three quarters of the remainder is retained as constant boiling acid. The barometric pressure is noted and the weight of liquid required to prepare 1·0 M HCl is found from the composition Table 4.3. (N.B. The barometric pressure is given in mmHg as barometers in SI units are not yet available.)

This is a primary standard of very high accuracy and should not require checking. In some laboratories, however, the acid is standardized gravimetrically with silver nitrate.

3. *Sodium Oxalate* $Na_2C_2O_4$, mol. wt. 133·99.

This is the only common material which may be used as an

TABLE 4.3. The Composition of Constant Boiling Hydrochloric Acid

Atmospheric pressure (mm Hg)	% HCl *by weight*	*Wt. constant boiling* HCl *containing* 1 mole/g
780	20·173	180·621
770	20·197	180·407
760	20·221	180·193
750	20·245	179·979
740	20·269	179·766
730	20·293	179·555

alkaline standard without standardization against one of the acid standards above.

To prepare a 0·1000 M solution 13·399 g of dry A.R. grade salt is made to 1 dm^3 with distilled water.

4. *Silver Nitrate* $AgNO_3$, mol. wt. 169·89.

To prepare a 0·1000 M solution 16·989 g A.R. silver nitrate is dissolved in 500 cm^3 of water. About 100 cm^3 A.R. concentrated nitric acid is added and the mixture made up to 1 dm^3 in a volumetric flask. The solution is stored in the dark in a brown bottle.

Although this is an accurate primary standard some workers check the concentration gravimetrically with hydrochloric acid.

5. *Potassium Iodate* KIO_3, mol. wt. 214·03.

To prepare a 0·1000 M solution 21·403 g A.R. potassium iodate is dissolved in water and made up to 1 dm^3.

6. *Potassium Dichromate* $K_2Cr_2O_7$, mol. wt. 294·22.

To prepare a 0·1000 M solution 29·422 g A.R. potassium dichromate is dissolved in water and made up to 1 dm^3.

The above six solutions are true primary standards and require no further checking. Other substances which are often used in volumetric analysis do not fulfil one or more of the criteria for a primary standard mentioned above. Sodium hydroxide, for example, readily takes up both carbon dioxide and water from the atmosphere and is difficult to weigh accurately. Standard solutions of these substances must be standardized by titration against one of the primary standards described. Some of them are shown in Table 4.4.

The Mathematics of Titration

Students in C.G.L.I. courses often have difficulty in calculating the results of their titration experiments, even though their experimental technique is often immaculate.

To introduce this topic a typical example of a simple titration is given.

TABLE 4.4. Secondary Standard Solutions for Volumetric Analysis

Substance	*Grams for* 1 dm^3 *Molar solution*	*Class of analysis*
Sodium carbonate (Anhyd.)	106	Acid–Base
Potassium hydroxide	56·108	Acid–Base
Sodium hydroxide	40·005	Acid–Base
Oxalic acid ($2H_2O$)	126·046	Acid–Base
Oxalic acid (Anhyd.)	90·016	Acid–Base
Sulphuric acid	98·08	Redox
Potassium permanganate	158·04	Redox
Sodium thiosulphate	248·192	Redox

The Determination of the Molarity of a Solution of Sodium Hydroxide using Standard Hydrochloric Acid Solution

The equation for the reaction is:

$$NaOH + HCl = NaCl + H_2O$$

25·00 cm^3 of NaOH solution is pipetted into the flask for each titration. Standard 0·125 M HCl solution is run in from the burette.

Titration	*1*	*2*	*3*
Burette reading at end	20·90	41·55	20·65
Burette reading at beginning	0·00	20·90	0·00
Titre	20·90	20·65	20·65

Titration 1 is clearly discordant and is ignored.

Taking the average of the two results which agree within the specified limits:

$$25{\cdot}00\ cm^3\ \text{NaOH solution} \equiv 20{\cdot}65\ cm^3\ 0{\cdot}125\ \text{M HCl}$$

$$1\ cm^3\ \text{NaOH solution} \equiv \frac{20{\cdot}65\ cm^3\ 0{\cdot}125\ \text{M HCl}}{25{\cdot}00}$$

$$\therefore\ \text{NaOH is } \frac{20{\cdot}65}{25{\cdot}00} \times 0{\cdot}125\text{M}$$

$$= 0{\cdot}103\ \text{M}$$

If the unknown molarity is always put on the *left* of the equation and the known molarity on the *right*, this method of calculation will be found to be suitable for most titrations.

An alternative method would be:

$$25{\cdot}00\ \text{cm}^3\ x\ \text{M} - \text{NaOH} \equiv 20{\cdot}65\ \text{cm}^3\ 0{\cdot}125\ \text{M HCl}$$

$$25{\cdot}00 \times x \equiv 20{\cdot}65 \times 0{\cdot}125$$

$$x = \frac{20{\cdot}65 \times 0{\cdot}125}{25}$$

$$= 0{\cdot}103\ \text{M}$$

The titration is calculated in molar terms. Normality and equivalent weight are no longer acceptable, and, indeed, practice will show that working in moles rather than equivalents is really easier. Whichever method of calculation is used, it is essential that for any volumetric analysis the equation for the reaction must be known.

Back Titration

It is not always possible to carry out volumetric analysis by the direct titration of two solutions. An example of this problem is the estimation of ammonium salts. If a solution of ammonium salts is boiled with excess sodium hydroxide the ammonia is expelled quantitatively. The sodium hydroxide remaining can then be estimated by titration with acid in the normal way

$$NH_4Cl + NaOH = NH_3\uparrow + NaCl$$

This technique is called *back titration* and is a useful extension of the volumetric method.

It may be summarized as follows: the concentration of a substance A in aqueous solution is required and there is no direct titration method available. Substance A, however, reacts with another substance B to produce a product X which can be removed from the reaction mixture or which is found to take no part in any further reactions. Also substance B is easily estimated by titration with a third substance C.

Then

$$A + B = X$$

$$A + 2B = X + B$$

$$B + C = Z$$

The amount of B present at the start of the reaction is known and the amount of B present at the end of the reaction is known, i.e. the amount of B which did not react with A. The amount of B which reacted with A is easily calculated, and hence the amount of A originally present.

Errors in Volumetric Analysis

Volumetric analysis is not usually considered to be a very accurate method of quantitative analysis. Conway* discusses the sources of error at great length. Errors may be described under two headings:

1. Constant errors
2. Random variations

Constant errors are usually inherent in the apparatus used in a titration or in faults in the quantitative method being used. These may be eliminated or reduced by careful calibration of the apparatus and by standardization of the method used, e.g. by performing indicator blanks by ensuring that solutions of suitable concentration are used.

Random variations are due to human error, usually faulty technique. If, for example, there is difficulty in deciding the exact shade of colour to be taken as the end-point in a titration it is helpful if a separate colour standard is prepared and the titration always taken to this colour.

GRAVIMETRIC ANALYSIS

The technique of gravimetric analysis can be broken-down into a number of individual processes:

1. Precipitation.
2. Filtration of the precipitate.
3. Washing of the precipitate.
4. Drying of the washed precipitate.
5. Ignition of the precipitate.
6. Weighing of the precipitate.

It would be convenient to discuss the technique under the headings listed above.

1. *Precipitation*

This is perhaps the most vital of the processes involved. Unless the conditions for the precipitation are almost ideal the analysis of the component being determined will not succeed. The factors which must be considered are:

(*a*) The insolubility of the precipitate. It must be possible to collect the precipitate with no appreciable loss during filtration.

* *Micro-diffusion Analysis and Volumetric Error* (Crosby Lockwood, 1962).

For all practical purposes the loss, i.e. the amount remaining in the filtrate, should not exceed 0·1 mg, the smallest amount which may be detected by the balance.

(*b*) It must be possible to wash the precipitate free of impurities and of the supernatant fluid from which it is derived.

(*c*) The particles of precipitate must be large enough to be completely held back by the filtering medium and the particle size must not be diminished by the process of washing.

(*d*) The precipitate collected must be convertible into some substance which is both pure and of known composition.

In connection with (*d*) above, it should be noted that the substance weighed is not necessarily the substance precipitated, e.g. in the gravimetric estimation of magnesium the metal ion is precipitated as magnesium ammonium phosphate $Mg(NH_4)PO_4 \cdot 6H_2O$ but on ignition this becomes magnesium pyrophosphate $Mg_2P_2O_7$.

The theoretical basis of the factors which lead to successful precipitation is discussed in several of the books noted in the appendix. The general rules for precipitation in gravimetric analysis are as follows:

(*a*) For complete precipitation hot solutions should be used as solubility usually rises as the temperature rises.

(*b*) Dilute solutions should be used.

(*c*) The precipitant should be added slowly and the mixture thoroughly stirred. This allows the first precipitate particles to act as 'growth nuclei' on which further precipitate collects.

The precipitate is prepared as follows:

(*a*) A borosilicate glass beaker of suitable size should be used.

(*b*) The precipitant is added with a burette, tap funnel, or pipette down the side of the beaker to avoid splashing. The mixture should be thoroughly stirred with a glass rod while the precipitant is being added.

(*c*) Only a small excess of precipitant should be added. Large excesses may lead to excessive contamination of the precipitate, prolonged washing and inaccurate results.

(*d*) After the precipitate has settled, a little of the precipitant is added dropwise to see if any further solid particles are formed in the clear supernatant, i.e. to check if precipitation is complete.

(*e*) The mixture is usually set aside for several hours to *digest*.

Depending upon the particular estimation this may be at an elevated temperature on a steam bath or at room temperature. The beaker should be covered with a watch glass during digestion.

2. *Filtration*

Filtration is the basic separation process of gravimetric analysis in which the substance being determined in the precipitate is quantitatively separated from the mother liquor which contains the impurities.

There are three basic filtration methods in gravimetric analysis:

(*a*) Filter paper.

(*b*) Gooch crucibles using asbestos filter mats.

(*c*) Crucibles with sintered bases in glass, silica, or porcelain.

The method used will depend on a number of factors which are mainly concerned with the character of the precipitate being filtered, although in some laboratories cost is a decisive factor.

(*a*) *Filter paper*—In gravimetric analysis 'ashless' filter paper is used. This term refers to paper in which the weight of the ash is less than 0·1 mg for the circle which is used. The manufacturers of 'Whatman' filter paper, perhaps the most popular of several brands, mark each box of papers with the average ash weight of each paper. If the ash weight exceeds 0·1 mg it must be deducted from the weight of precipitate after ignition.

Filter paper is manufactured in various grades differing in porosity, filtration speed and mechanical strength when wet. Some of the more popular 'Whatman' papers are listed in Table 4.5. The hardened filter papers are treated with nitric acid to give them a higher mechanical strength and a lower ash figure than untreated papers. These are slightly more expensive than untreated papers but should always be used for accurate quantitative work.

The size of paper chosen for any particular determination depends upon the bulk of the precipitate not the volume of liquid to be filtered. The precipitate should occupy less than one third of the volume of the conical 'vessel' formed when the paper is fitted into a 60 degree conical funnel. It is convenient if the amounts of material involved in the determination produce a precipitate suitable for a 9 cm or 11 cm paper. Smaller amounts may be difficult to weigh and larger amounts difficult to wash.

Flow rate	*Qualitative*	*Single acid washed*	*Double acid washed*	*Hardened single acid washed*	*Hardened double acid washed*	*Mean retention index*	*Precipitate retained*
Fast					541	0	Gelatinous or very coarse particles
			54			0	
	4					0	
		31		41		0	Ferric or aluminium hydroxide
	15					0	
Fast/Medium	7					20	Coarse crystalline precipitates Ammonium phosphomolybdate
	1					25	Calcium oxalate
Medium					540	30	Curdy precipitates
			52			30	Medium crystalline
	11					30	Lead sulphate
		30		40		35	Copper oxide
	2					30	Barium sulphate (hot)
	3					80	
Slow				44		90	Fine to very fine precipitates
					544	85	
		32		42		100	Barium sulphate (Cold)
					542	85	
	5					100	
			50			85	

Mean retention index is on an arbitrary scale from **0** (practically no retention) to **100** (practically complete retention).
The figures are placed on the chart in order of comparative flow rate.

The filter paper is folded to fit tightly into a 60 degree conical funnel so that the top edge of the paper is 1–2 cm from the top of the funnel. 'Pyrex' funnels are accurately made and have a particularly long stem which aids filtering considerably. The paper is moistened with distilled water and pressed to the wall of the funnel with a flattened glass rod so that at least the upper half adheres to the wall. On filling with water the stem of the funnel should fill with water and remain filled while liquid passes through it if the paper is properly fitted inside the funnel.

The correct set-up for filtration is shown in *Figure 4.11.* The funnel is placed in a filter stand or clamp with the stem touching the wall of a clean beaker. The liquid is poured on to the side of the filter paper at the thickest part down the stirring rod. If the

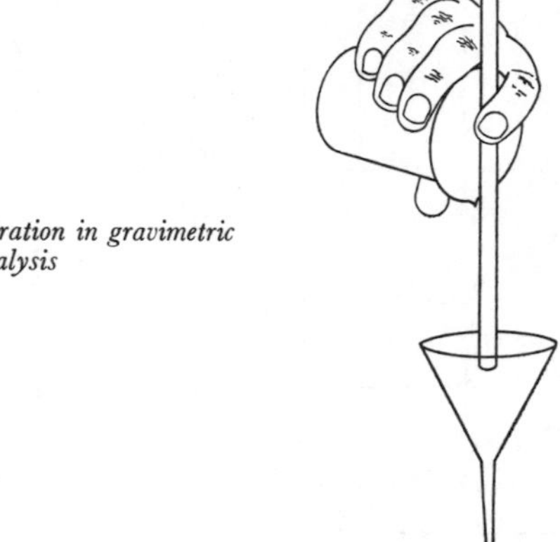

Figure 4.11. Filtration in gravimetric analysis

precipitate tends to stay at the bottom of the beaker it can be washed out of the tilted beaker with a fine jet of water from a wash-bottle. The beaker is held in the left hand the rod being forced against it, resting in the beaker lip with the curled forefinger, and the right hand holding a plastic wash-bottle. If the precipitate sticks to the sides of the beaker it can be scraped together with a rubber 'policeman', a blind-ended rubber tube which slips over the end of the stirring rod and which must be thoroughly rinsed into the funnel to ensure quantitative filtration.

In some cases, e.g. when dealing with particularly gelatinous precipitates, suction may be necessary in quantitative filtration.

The filter paper is then supported in a Whatman filter cone or in a perforated platinum filter cone. Another aid for dealing with slow-filtering gelatinous precipitates is the use of Whatman accelerators which may be in the form of paper clippings, discs, or tablets of 'ashless' paper. Immediately before filtration sufficient of the accelerator is added to the beaker containing the precipitate to double the bulk of solid present.

Finally, the precipitate should not be allowed to dry or cake on the filter. The moist filter paper is transferred to a crucible for subsequent ashing and a dry caked precipitate may be difficult to fit into the crucible and losses may occur.

(*b*) *The Gooch Crucible*—The most common form of Gooch crucible is a 'tall form' porcelain crucible with a perforated base. For very accurate work silica and platinum crucibles are available but it is unlikely that students will come across these.

A filter mat of asbestos is formed on the bottom of the crucible before use. The asbestos used is the long fibred 'amphibole' variety. At one time this had to be pre-treated in the laboratory by boiling with concentrated hydrochloric acid followed by prolonged washing. This is not necessary nowadays as most of the laboratory suppliers market a suitably prepared 'Asbestos for Gooch Crucibles' at a price which makes laboratory preparation uneconomic.

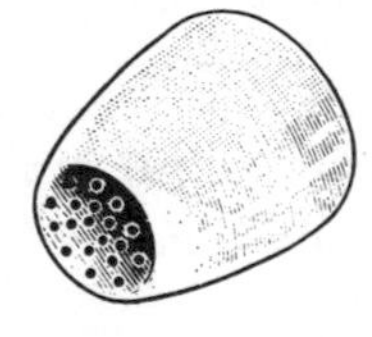

Figure 4.12. The Gooch crucible

The Gooch crucible is supported in a Gooch funnel by means of a shaped rubber gasket. The bottom of the crucible must not touch the glass of the funnel or the gasket, and the tip of the funnel must be below the side arm of the filter flask.

The crucible is half-filled with a well-shaken suspension of asbestos in water and allowed to stand for a few minutes so that the larger particles sink to the bottom. Gentle suction is applied until the water has been removed followed by hard suction to impact the filter mat tightly. The filter mat should have a final thickness of 2–3 mm and when it is held up to the light the outline of the holes should just be visible through the mat. If the latter is too thin the procedure should be repeated until a mat of sufficient thickness has been built up. The mat is thoroughly washed with water under full suction until no more fine asbestos fibres pass into the filtrate. Although the mat is fairly delicate it should not tear if two rules are rigidly applied:

1. No liquid should be poured directly on to the mat but allowed to run on to the centre of the mat from a glass rod.
2. No liquid should be poured into the crucible without suction being applied.

The prepared crucible is then placed on a silica dish and dried in the oven at 250°C to constant weight. This is a common procedure in gravimetric analysis. The crucible is heated for some time in the oven, cooled in a desiccator and weighed. This procedure is repeated until two consecutive weighings are similar.

The precipitate is filtered under suction on to the asbestos mat, the application being similar to that described above.

(*c*) *Crucibles with Sintered Bases*—Sintered glass crucibles are perhaps the most widely used filtration medium in gravimetric analysis. Both 'tall' and 'squat' form Pyrex crucibles are available having sintered glass bases in porosities graded 1, 2, 3 and 4 corresponding to an average pore diameter of 100–200 μm, 40–50 μm, 20–30 μm, and 5–10 μm respectively. No. 4 is used only for the very finest precipitates, e.g. barium sulphate, No. 3 being the most commonly used grade for particles of medium size.

They are used in the same way as Gooch crucibles, but have several advantages over the earlier apparatus:

(i) The glass of which they are made is more inert than the porcelain of the Gooch crucible.

(ii) It is easier to clean than porcelain.

(iii) Glass crucibles may be dried to constant weight at lower temperatures than porcelain crucibles with asbestos mats. A temperature of 110–150°C being adequate for most purposes.

Glass crucibles should not be heated above 200°C. Where a higher temperature is required to dry a particular precipitate silica filtering crucibles should be used. These are manufactured in a similar range of porosities and may be used up to 1000°C. Alternatively, porcelain filtering crucibles with glazed walls and an unglazed porous base are available. These may be heated to a dull red glow, and, if protected by a covering nickel crucible, in the flame of a bunsen or Meker burner. They are relatively inexpensive.

The cleaning of sintered crucibles often presents a problem. This can often be minimized by ensuring that the correct grade of sinter is used so that it does not become clogged with the finer particles of precipitate. The main bulk of solid is shaken out and the filter washed with a suitable solvent or cleaning fluid. A convenient arrangement for back washing sintered glass crucibles is shown in *Figure 4.13*. The cleaning fluid to be used will depend, of course, on the precipitate to be removed. A very useful liquid for many inorganic precipitates is a hot 0·1 M solution of tetrasodium sequestrate (the tetrasodium salt of ethylene-diamine tetracetic acid).

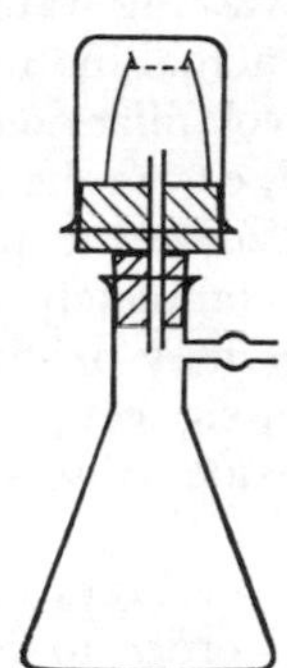

Figure 4.13. Cleaning sintered glass crucibles

3. *Washing the Precipitate*

After the precipitate has been separated from the main bulk of the mother liquors by filtration it must be washed free of the remainder of the soluble impurities formed during precipitation.

For most purposes washing is the only practical method of removing impurities although occasionally impurities sublime off during incineration.

The washing liquid should have the following characteristics:

(*a*) It must not react directly with the precipitate.

(*b*) It must have no appreciable solvent action upon the precipitate.

(*c*) It must have no effect upon the physical size of the precipitate particles. In particular it must not reduce the particle size so that precipitate may be washed through the filter.

(*d*) It must be so volatile as to be completely removed by heating at the temperature at which the precipitate is to be dried.

(*e*) It must not interfere with the testing of the filtrate carried out to ensure that washing is complete.

For most determinations pure water is not a suitable washing liquid as many 'insoluble' precipitates are appreciably soluble in water within the weighing limits of gravimetric analysis. The usual technique is to introduce a common ion to the water thus reducing the solubility of the precipitate, e.g. in the estimation of calcium as calcium oxalate the precipitate is washed with a very dilute solute of ammonium oxalate. The ammonium oxalate remaining is completely volatilized during the subsequent heating of the precipitate. Gelatinous precipitates which tend to form colloidal solutions may pass through the filter if washed with pure water. The addition of an electrolyte to the washing water inhibits the formation of colloidal solutions. The actual nature of the electrolyte is of no importance so long as it volatilizes on heating. For this reason ammonium salts are often used, e.g. in the determination of aluminium as basic aluminium benzoate the precipitate is washed with ammonium benzoate, which completely volatilizes in the subsequent heating. Some precipitates may oxidize during washing and a reducing solution must be used, e.g. copper sulphide precipitate which may oxidize to cupric oxide is washed with acidified H_2S water.

The precipitate should not be over washed or losses may occur by the solution of a detectable amount. In order to find when washing is complete regular qualitative testing of small portions of the filtrate for the impurities is carried out. It will be found that gelatinous precipitates need more prolonged washing than crystalline ones. Washing is best carried out with successive small amounts of washing liquid, a portion being added as the previous portion has just passed through. The precipitate should not be impacted or sucked dry until washing is complete or channelling may result.

Drying the Precipitate

The precipitate has now been separated and washed free of impurities, all that remains on the filtering medium being the precipitate and the remains of the washing liquid. The final stage before weighing is drying or ignition. The distinction between these terms is vague, but in general, *drying* refers to temperatures below 250°C and *ignition* to temperatures higher than this. Precipitates to be dried are usually separated on one or other of the forms of filtering crucible, Gooch, sintered glass, or sintered porcelain. The precipitate is dried at the same temperature as the crucible has been dried before its preliminary weighing after it has been prepared. Drying is once again to constant weight.

Ignition is usually carried out where the precipitate has been separated on filter paper. It this case the precipitate and paper are usually placed together in a previously weighed crucible and the crucible heated until the paper has been reduced to an imponderable ash. The cooled crucible, which now contains the dried precipitate and the almost weightless ash is weighed again to find the weight of the precipitate.

In order to perform this operation great care must be taken to avoid losses through the 'spitting' of the drying precipitate or through the solid being blown out of the crucible.

The procedure is as follows:

1. A crucible and lid, preferably of silica although porcelain will serve, are ignited to constant weight as described above.
2. The filter paper holding the precipitate is removed from the funnel and folded over at the top to form a 'packet' which completely encloses the precipitate. Care must be taken not to tear or squeeze the paper.

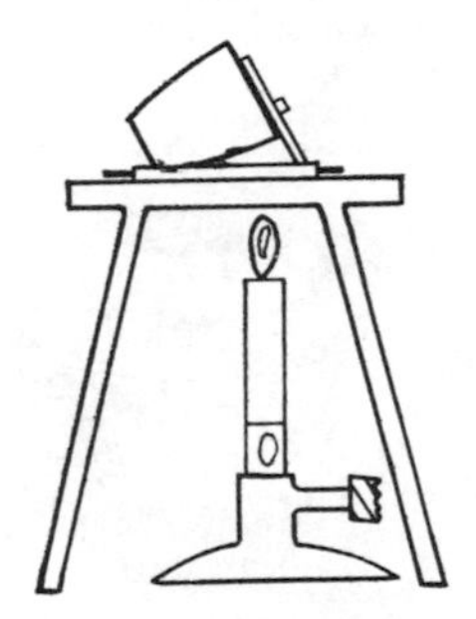

Figure 4.14. Drying a precipitate in a filter paper

3. The paper 'packet' is put into the crucible point downwards. The crucible is rested on a silica triangle as shown in *Figure 4.14*, inclined at an angle and with the lid partially covering the mouth. The crucible *lid* is heated with a small flame from a Bunsen burner. This allows the paper packet to dry slowly.
4. When no more steam is expelled the flame is increased a little. The paper burns slowly without actually catching fire as the flame may cause particles of precipitate to be expelled. If the paper does catch fire the flame may be smothered by moment-

arily covering the crucible with the lid using a pair of crucible tongs.

5. When the carbonization of the paper is complete the flame is moved to the bottom of the crucible and is increased to burn off the carbon.
6. After the carbon has been oxidized the crucible is righted, covered with the lid, and roasted with the flame from a Bunsen or Meker burner until the required temperature has been attained. It normally takes some 20 min to carbonize the paper and about 45 min to roast the crucible to the correct temperature.
7. The flame is removed and the crucible is allowed to cool for a few minutes before being placed in a desiccator and allowed to cool to room temperature. The crucible and lid, together with contents are weighed.
8. The crucible and lid are roasted again for 20 min, cooled as before, and again weighed. This process is repeated until two consecutive weighings are equal.

The incineration of precipitates is made easier by the use of the Main-Smith crucible (*Figure 4.15*). The interior-fitting lid is serrated round the edge. This has several advantages over the traditional form of crucible lid as gases and vapours can escape more easily and the contents of the vessel can be inspected by simply swivelling the lid.

Figure 4.15. The Main-Smith crucible

Some precipitates must be separated from the filter paper before incineration because of the possibility of partial reduction during the oxidation of the filter paper (barium sulphate and cupric oxide precipitates are examples of substances which may react in this manner). The procedure for precipitates of this nature is as follows:

1. The funnel containing precipitate and filter paper is covered by a filter paper larger in diameter than the mouth of the funnel and held by crumpling around the edge.
2. The funnel is placed in an electric oven at 105°C for 2 h so

that the contents may dry completely.

3. The dried filterpaper is carefully extracted from the funnel and as much of the dried precipitate as possible is transferred to a clean clock glass resting on a piece of clean black glazed paper.
4. The filter paper is folded and incinerated in a previously weighed crucible as previously described.
5. The crucible is cooled and the small amount of precipitate contained in it is chemically treated to reverse any reduction from the burning paper.
6. The contents of the clock glass are transferred to the crucible with great care, a camel-hair brush being used to transfer the last few particles.
7. The crucible and contents are roasted to constant weight as previously described.

Micro and semi-micro methods of gravimetric analysis are in everyday use in working laboratories but the special techniques involved do not come within the scope of this book. All that has been given here is a general survey of the main apparatus and methods used. Fuller information together with the materials used for the estimation of specific substances will be found in the specialist books listed in the Appendix.

CHAPTER 5

ELEMENTARY ORGANIC ANALYSIS

DETECTION OF THE ELEMENTS

The analysis of organic substances is a rather more complex procedure than the simple analysis of inorganic salts discussed in Chapter 4. This is because there are now over one million known organic compounds, and a large proportion of these are formed by the combination of two or more of a very small number of elements. When it is considered that there are several thousand separate compounds formed by the combination of the two elements carbon and hydrogen alone, it is clear that qualitative analysis for the elements gives very little information on the identity of an unknown compound. Hydrogen is almost always a constituent of organic substances. The other elements commonly found are oxygen, nitrogen, sulphur, phosphorus and the halogens. In some compounds metals are present, the more common being sodium and potassium from group 1A of the periodic table, and calcium, barium, and magnesium from group IIA.

Before attempting to detect the elements at the beginning of an organic analysis it is customary to carry out two simple and informative preliminary tests.

(*a*) *Ignition on a crucible lid*—About 0·1 g of the substance is placed in the centre of an upturned silica or porcelain crucible lid supported on a pipeclay triangle on a tripod. Heat is applied gently to one side of the lid and is slowly increased until it is red-hot. Any change in appearance is carefully noted, e.g. whether the material is readily inflammable or inflammable only after it has become quite hot or whether it is not inflammable at all; whether the flame produced is clear and blue or yellow and smoky; whether any characteristic odour is associated with ignition; whether a non-volatile residue remains after roasting. The deductions which can be made from this experiment are summarized in Table 5.1.

A black non-volatile residue left after roasting may be merely a residue of carbon which would need prolonged heating to disperse. If it is moistened with a little ammonium nitrate solution and reheated the carbon is oxidized and dispersed and any inorganic residue remains. This is moistened with a little dilute hydrochloric acid and any effervescence or odour noted. A platinum wire flame test often gives sufficient indication of the metal present in the residue.

TABLE 5.1. Heating of Organic Substances on a Crucible Lid

Burns rapidly with clear flame	Aliphatic substance
Burns with smoky luminous flame	Aromatic substance
Substance becomes very hot before igniting with slightly smoky flame	Substance rich in halogen
Burns with difficulty and on strong heating leaves a white infusible residue	Salt of carboxylic acid
Burns with difficulty leaving residue of Na_2S (Moistened with dil. HCl close to filter paper moistened with lead acetate)	Salt of sulphonic acid
Melts, darkens, chars, burns with smell of burnt sugar	Mono- and di-saccharides
Swells up, blackens, odour of burnt sugar but no flame	Tartatic, citric, lactic acid and their salts

(*b*) *Heating with soda lime*—About 0·2 g of the unknown is mixed in a small mortar with about 1 g of soda-lime. An odour of ammonia indicates the presence of an ammonium salt. This mixture is then placed in a hard glass test-tube fitted with a delivery tube bent at a right angle. The tube is held with the mouth slightly lower than the base so that any condensate formed on heating does not run back into the hot tube. The delivery tube is inserted into a second test-tube containing 2–3 cm^3 of water tinted with a little litmus indicator. The mixture is heated, gently at first then more strongly, and any condensable product collected. The deductions which may be made from the results of this experiment are summarized in Table 5.2.

TABLE 5.2. Soda Lime Test

Observation	*Compounds indicated*
Smell of ammonia before heating	Ammonium salt
Smell of chloroform before heating	Chloral hydrate
Ammonia on heating	Acid amide
Methylamine (fishy smell) on heating	Amino acid
Hydrocarbon evolved	Carboxylic acid or salt
Smell of burnt sugar	Carbohydrate or aliphatic hydroxy- acid or salts

It must be emphasized that these preliminary tests can only give the most general indication of the nature of an organic compound, there are exceptions to the deductions which can be made from them in practically every case, and no rigid identification of the class of substance being tested can be made on the results.

Identification of the Elements

(*a*) *Carbon and hydrogen*—These two elements are usually assumed to be present in an organic compound and carbon is usually detected in the ignition test above. The direct test is rarely performed. Cupric oxide is heated strongly in an oven to remove any organic impurity and cooled in a desiccator to prevent any uptake of moisture on cooling. 1–2 g of fine cupric oxide powder is mixed with about 0·1 g of the substance in a dry hard glass test-tube fitted with a cork and delivery tube as in the soda-lime test above. The test-tube is clamped at the upper end at an angle of about 45 degrees with the mouth above the base. The delivery tube is inserted in a second test-tube so that its end is just above a few cm^3 of lime or baryta water. If the delivery tube dips into the lime water uneven heating can cause 'suck back'. The mixture is heated strongly. Droplets of water on the upper part of the reaction tube indicate the presence of hydrogen. A white precipitate in the receiver tube after it has been stoppered and shaken indicates the presence of carbon dioxide formed by the oxidation of carbon.

(*b*) *Nitrogen, sulphur and the halogens*

The Lassaigne sodium fusion—This test converts the elements into ions which can be detected by wet tests similar to those which are used in inorganic analysis. The compound is fused with metallic sodium to form sodium cyanide from nitrogen, sodium sulphide from sulphur, and sodium halides from the halogen elements.

The usual technique for the Lassaigne sodium fusion involves heating the unknown with a small 'seed' of sodium metal in a soft glass ignition tube, and plunging the red-hot tube into an evaporating basin of water. The tube cracks and the inorganic salts formed in the fusion are extracted by the water. The mixture of glass, water, and reaction residue is filtered and tests are carried out in the filtrate. This can be a dangerous procedure especially in inexperienced hands. The violence of the reaction with water often causes the contents of the tube to be blown out, and fires are not infrequently caused by the ignition of unreacted sodium in the water. The following technique is recommended on the grounds of both safety and accuracy. It involves little more trouble than the classical method and gives better results with compounds which yield up their nitrogen with difficulty.

A 150 × 12 mm Pyrex test-tube is wrapped with a few turns of asbestos cloth at its upper end and clamped firmly around the asbestos in an upright position. A piece of freshly cut sodium about the size of a small pea is dropped into the tube which is heated fairly strongly until the metal melts and its vapour rises about 5 cm up the tube. About 50 mg of the substance is added in small portions, by pipette if liquid, with a spatula if solid. Some substances may cause a slight explosion when added, notably chloroform, carbon tetrachloride, azo compounds and other halogen or nitrogen-rich compounds, but if the addition is performed reasonably carefully only a slight 'pop' should result. When the addition is complete the tube is heated to redness for 1 min and cooled. 3–4 cm^3 of pure methanol is added from a dropping pipette to decompose the unreacted sodium. The tube is half filled with water and the contents boiled gently for 2 or 3 min, the solution is filtered and the filtrate tested for anions.

Nitrogen—To about 2 cm^3 of the fusion filtrate 0·1–0·2 g of powdered ferrous sulphate crystals is added. The mixture is boiled gently for a few seconds and hot, dilute sulphuric acid is added dropwise until the precipitate of iron oxides is just dissolved. 1 cm^3 of 5 per cent potassium fluoride solution is added and the mixture filtered through a clean filter paper. Deep blue specks of Prussian blue precipitate indicate the presence of nitrogen. Occasionally the Prussian blue colour is evident in the tube before filtration but where only small proportions of nitrogen are present filtration is the only way of confirming the presence of nitrogen. If a black precipitate is produced before acidification the presence of sulphur is indicated, ferrous sulphide having been formed. The precipitate clears on acidification.

Sulphur—The presence of sulphur will already have been indicated in the test for nitrogen. The element may be confirmed by treating 2 cm^3 of the fusion filtrate with 2 or 3 drops of a fresh 0·1 per cent solution of sodium nitroprusside. A brilliant purple coloration which slowly fades indicates the presence of sulphur. This is a very sensitive test and false results may be obtained unless the apparatus is clean.

Halogens—If nitrogen and/or sulphur have been found, they must be removed by just acidifying about 2 cm^3 of the filtrate with dilute nitric acid and boiling vigorously in an evaporating basin until the volume is halved. The solution is cooled, placed in a clean tube and 1 cm^3 of water is added to replace that lost on evaporation Nitrogen is driven off as hydrogen cyanide, sulphur as hydrogen sulphide.

A few cm^3 of a solution of silver nitrate is added to the acid solution. A white or yellow precipitate indicates the presence of halogen. If the precipitate is readily and completely soluble in ammonia solution, chlorine is present and bromine and iodine absent. If the precipitate is only partially soluble or completely insoluble, or if there is any appreciable delay in solution, any or all of the three elements may be present. The three elements may be differentiated by the standard methods of inorganic analysis.

Fluorine is infrequently found in organic compounds. 2 cm^3 of fusion filtrate is acidified with glacial acetic acid and the mixture boiled vigorously and cooled. A filter paper is soaked in a 5 per cent solution of zirconium nitrate in 5 per cent hydrochloric acid, dried, and dipped into a 2 per cent solution of alizarin-S. The excess dye is washed out with water and the paper again dried. A spot of test solution is applied to the prepared paper with a glass rod. Fluorine is indicated by a bright yellow spot on the scarlet paper.

Phosphorus—If the original fusion is carefully performed any phosphorus present in the compound will be given off as phosphine gas which has a strong smell of rotten fish. The smell is very noticeable even when only small proportions of phosphorus are present. To confirm this about 1 cm^3 of fusion filtrate is acidified with about 3 cm^3 of concentrated nitric acid and the mixture boiled and cooled. 4 cm^3 of ammonium molybdate reagent is added and the mixture warmed to about 40°C. The presence of phosphorus is indicated by a bright yellow crystalline precipitate. Ammonium molybdate reagent is prepared by dissolving about 5 g of pure ammonium molybdate in 10 cm^3 of a 40 per cent aqueous solution of 0·880 ammonia. 12 g of ammonium nitrate is added and the solution made up to 100 cm^3 with water. This reagent 'keeps' better than the more common acid ammonium molybdate and is quicker to prepare requiring no decantation. It must be remembered, however, that the yellow precipitate of ammonium phosphomolybdate will form only if the acid in the test solution is sufficient to completely neutralize the alkalinity of the reagent.

The Middleton zinc-sodium carbonate fusion

In 1935 Middleton introduced an improved method for the detection of nitrogen, sulphur, and the halogens which has distinct advantages over the classical sodium fusion. A fusion mixture of pure zinc powder and A.R. anhydrous sodium carbonate is prepared by grinding the two substances together in a clean, dry mortar in the proportion 2:1 by weight. About 0·2 g of the solid or 3 drops of the liquid unknown is thoroughly mixed with enough of the fusion powder to give a column about 12 mm high in a small hard glass test-tube. More fusion powder is added to make the column about 38 mm high. The tube is held horizontally in a test-tube holder and the mixture heated gently at the open end. The heating is gradually moved along the tube until the whole is red hot, when the tube is turned to the vertical and roasted for a short time. It is immediately plunged into about 20 cm^3 of cold distilled water in an evaporating basin, a wire gauze being held about 15 cm above the basin in the other hand to protect the face from any flying glass which may arise from the disintegration of the hot tube. The contents of the basin are boiled gently for a few minutes and filtered while hot. The residue is retained for the sulphur test.

Nitrogen—2 cm^3 of the filtrate is treated with 2 cm^3 of 5 per cent sodium hydroxide solution and 0·1 g powdered ferrous sulphate added. The mixture is boiled for a short time and acidified with dilute sulphuric acid very carefully as the carbon dioxide formed causes effervescence. A Prussian blue precipitate indicates the presence of nitrogen.

Halogens—The procedure is exactly as described for the Lassaigne test including the elimination of cyanide if nitrogen is present.

Sulphur—A filter paper is moistened with sodium plumbite. About 10 cm^3 dilute hydrochloric acid is added to the residue in the dish. The prepared filter paper is immediately held over the dish. A brown or black stain of lead sulphide, often with a strong smell of hydrogen sulphide indicates the presence of sulphur.

The main advantages of the Middleton method are that nitrogen and sulphur are detected quite independently, and that compounds like carbon tetrachloride and nitro-compounds do not explode on fusion. A disadvantage is that zinc powder is rarely pure and blanks must be performed on each batch of reagent. The method takes rather longer than the Lassaigne technique which is still the method used by most working organic chemists.

QUANTITATIVE ESTIMATION OF THE ELEMENTS

After the elements in an unknown compound have been found the proportion of each element present must be determined as a percentage of the total in order to construct an empirical formula.

The principals involved in quantitative organic analysis were developed in the middle years of the nineteenth century on a macro scale. In 1911 the Austrian Pregl introduced the first of a long series of papers describing modifications of macro procedures using no more than 5–10 mg of material. These methods were based upon the use of a microbalance produced by a team of workers led by Pregl which could weigh such amounts with a precision of about 1 µg. The methods require a great deal of manipulative skill and semi-micro methods are more commonly used. It may be noted, however, that with only 15 mg of pure androsterone Butenandt performed two complete analyses, prepared a derivative and analysed it.

Methods using even smaller amounts of unknown have been reported, samples of about 50 µg being used. A small crystal of cane sugar weighs about 200 µg and sub-micro analysis uses a sample size as small as can be handled without optical magnification and mechanical manipulation. Sub-micro methods have followed the development of balances on which samples of up to 1 g can be weighed with a precision of $\pm 0{\cdot}05$ µg.

A further recent innovation is the production of apparatus in which samples can be analysed automatically. A discussion of 'automation' in the laboratory will be found in Chapter 8.

Estimation of Carbon and Hydrogen

Liebig described a method of estimating carbon and hydrogen together in 1831. A weighed sample of the substance is ignited in a silica or glass tube in a slow stream of oxygen. Cupric oxide in the tube is kept at red heat to aid the combustion of the sample. The carbon in the sample is oxidized to carbon dioxide which is absorbed in a weighed tube of soda lime; the hydrogen is oxidized to water vapour which is absorbed in a weighed tube of magnesium perchlorate. From the increase in weight in the absorption tubes the weight of carbon and hydrogen in the sample can be calculated.

A modern combustion train for the semi-micro determination of carbon and hydrogen is shown in *Figure 5.1*. Oxygen from a cylinder is passed in a slow stream through a transparent silica tube (R_1) partially packed with M.A.R. grade wire form cupric oxide heated

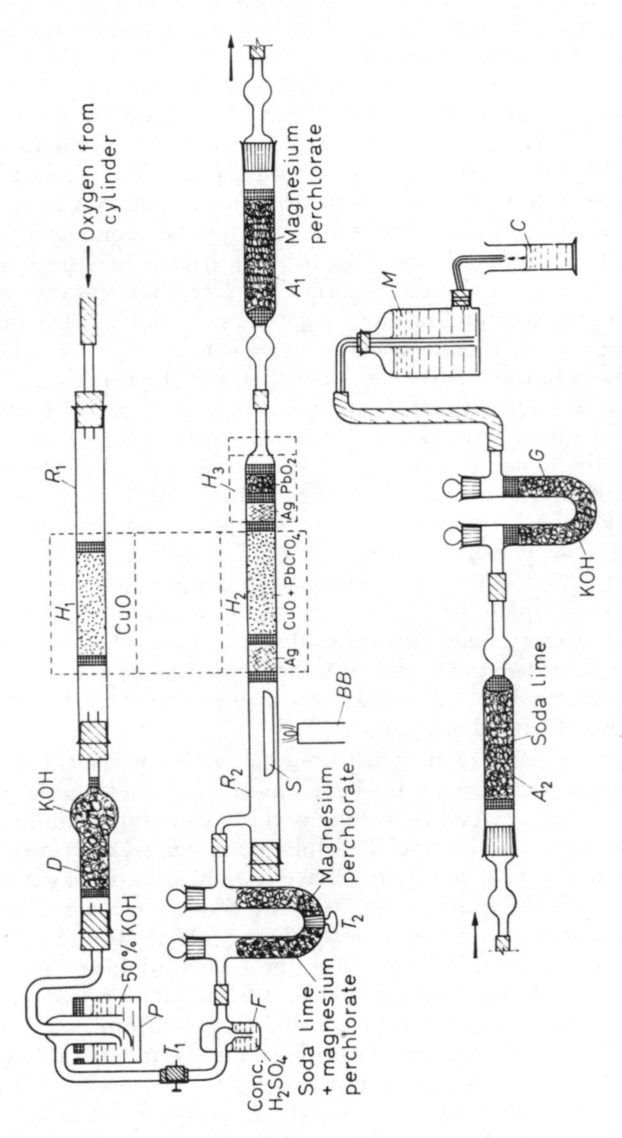

Figure 5.1. Assembly for the estimation of carbon and hydrogen

to 680° by a tube heater H_1. The oxygen leaves the tube heated to the correct temperature with any trace of organic impurity already ignited. The gas then passes through a large tube of potassium hydroxide pellets (D) in which any water vapour present is absorbed. The pressure regulator (P) the precision pinchcock (T_1) and the bubbler (F) regulate the gas flow to a U-tube. The near arm of the U-tube contains soda lime (M.A.R. Carbasorb obtainable from B.D.H.) to which a little magnesium perchlorate has been added (M.A.R. Anhydrone from B.D.H.). The further arm contains anhydrone. There is a tap between the two arms (T_2) to ease the filling of the U-tube with the two compounds. Oxygen enters the combustion tube (R_2) already heated and free of any organic impurity, moisture, or carbon dioxide. The gas thus enters and leaves the combustion tube in the same condition. The sample is weighed out in a platinum or porcelain boat (S), and is heated by a bunsen burner (BB). The 'beak' end of the combustion tube is packed with four separate materials. Silver wool removes halogen as non-volatile silver halide. Copper oxide mixed with lead chromate in equal proportions oxidizes any partially combusted organic matter, the lead chromate holding back any oxides of sulphur formed in the combustion. The silver and cupric oxide-lead chromate packings are heated by a tube heater (H_2 similar to H_1) to 680°. The second silver packing which follows takes up any halogen which has passed through the earlier silver wool. The final packing, lead peroxide, absorbs any oxides of nitrogen formed. These two packings are heated by a small electric thermostatic mortar to 180°. The packings in the tube are separated by small plugs of ignited asbestos.

The stream of oxygen which now contains the carbon dioxide and water vapour formed in the combustion of the sample passes first into the weighed absorption tube containing anhydrone (A_1) then into a similar weighed tube containing Carbasorb (A_2). After the products of combustion have been absorbed the gas passes through a KOH guard tube (G) and finally into an aspirator bottle of a special type, a Mariotte bottle (M). By the time the oxygen reaches the absorption tubes, the small initial excess pressure it had initially has been dissipated, and the slight negative pressure applied by the fall of the water level in the Mariotte bottle helps to draw the gas through the absorption train. The rate of flow into the cylinder (C) is about 5 cm^3/min to give a satisfactory flow through the apparatus. The apparatus is capable of giving good results if used with reasonable manipulative skill. The weight of moisture and air absorbed in the tubes is very small in relation to the total weight of the tubes, and weighings must be carried out carefully, preferably in a balance room, where temperature and humidity conditions are reasonably constant, so that, if the tubes are allowed to stand in the balance for some time before weighing, any effect from the change in gas content of the tube or from the condensation of moisture on the outside surface of the tubes may be cancelled out. All rubber tubing and rubber stoppers must be matured before use by steeping in 40 per cent KOH solution for several hours and then passing steam through them for 2 h. Glass joints and stoppers are held firmly in place with Krönig cement composed of one part white wax melted with four parts colophony.

The results are calculated as follows:

$$(1)\ \%\text{Hydrogen} = \frac{\text{Wt } H_2O \times 2{\cdot}016 \times 100}{\text{Wt of sample} \times 18{\cdot}016}$$

$$(2)\ \%\text{Carbon} = \frac{\text{Wt } CO_2 \times 12 \times 100}{\text{Wt of sample} \times 44{\cdot}000}$$

Estimation of Oxygen

No method for the estimation of oxygen has gained much favour with organic chemists, and the element is usually determined by difference. Several methods have been suggested, among the best are those based upon the technique produced by Schütze in 1939. Nitrogen is passed over a weighed sample mixed with carbon and heated to 1000°. The oxygen present is converted to carbon monoxide which is passed through a tube of iodine pentoxide, the CO being estimated by determination of liberated iodine. Although accurate results may be obtained by this method it has never become as generally used as the techniques for estimating the other elements.

Estimation of Nitrogen

There are three main methods for the estimation of nitrogen in organic compounds. The technique having the most general application is based upon the method introduced by Dumas in 1830. The Kjeldahl method, published in 1883 is unsuitable for many compounds, especially those containing nitro or nitroso groups, but is a particularly rapid and accurate method for the estimation of nitrogen in biological material and is mostly used in biochemical laboratories. The third technique is the very elegant diffusion method developed by Conway.

(*a*) *The Dumas method*—The apparatus for the semi-micro determination of nitrogen in organic compounds is shown in *Figure 5.2.* The source of carbon dioxide is very important. It may be a cylinder of gas or a Dewar flask filled with solid carbon dioxide. Alternatively, a modified Kipp generator charged with marble chips and sulphuric acid may be used. It is important that no air should enter the combustion tube with the carbon dioxide. The gas passes into the combustion tube (T) over a spiral of oxidized copper

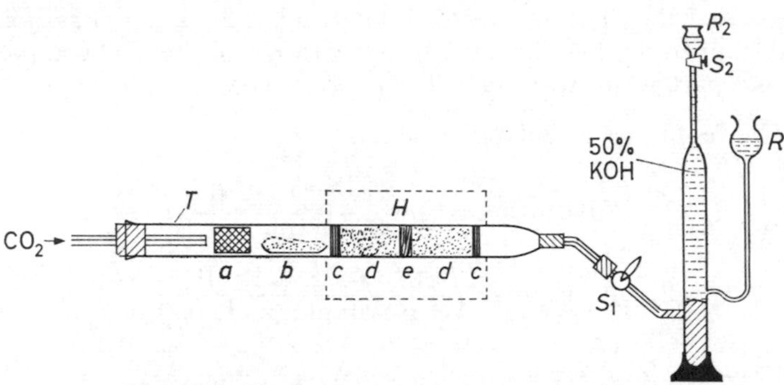

Figure 5.2. The Dumas apparatus for nitrogen estimation

gauze (a) and a porcelain boat containing about 25 mg of the sample mixed with an equal quantity of freshly ignited copper oxide (b). The further end of the combustion tube, surrounded by a tube furnace at 750°, (H), contains the main packing. This consists of a layer of reduced copper oxide (e) between two layers of copper oxide in short 'wire form' lengths (d), the packing being held in place by short lengths of copper gauze wound spirally (c). The beak end of the combustion tube is connected to a nitrometer tube filled with 50 per cent potassium hydroxide through a tap (S_1). The tap at the upper end of the nitrometer (S_1) is for filling the graduated stem into the reservoir (R_2) to prevent splashing of the concentrated alkali. The movable reservoir (R) is used as a levelling bulb.

The gas is passed through the tube for 10 min with the heater on. A bunsen burner flame is applied to the tube so as to heat the copper spiral until it is red hot, and the flame is gradually moved toward the sample until the latter is heated strongly. The carbon dioxide oxidizes the heated sample to carbon dioxide, nitrogen and oxides of nitrogen. The heated packing reduces the oxides of nitrogen to nitrogen. The potassium hydroxide absorbs the CO_2 and any other acidic oxides formed in the combustion and the nitrogen passes into the nitrometer to be measured. The volume of nitrogen is corrected to S.T.P.

Calculation

$$\text{Weight of nitrogen at S.T.P.} = \frac{V \times P \times 28{\cdot}016}{(1 + T/273) \times 760 \times 1{,}000 \times 22{\cdot}415}$$

$$\% \text{ of nitrogen} = \frac{V \times P \times 1{\cdot}2502 \times 100}{(1 + T/273) \times 760 \times 1{,}000 \times W}$$

Where V = nitrogen volume (corrected and in cm^3)

P = corrected barometric pressure

T = nitrometer temperature (°C)

W = Weight of sample (*g*)

The correction for nitrogen volume is to allow for the vapour pressure of the alkali in the nitrometer and for its viscosity which prevents complete drainage in the calibrated tube. Pregl stated that a reduction of the observed volume by 2 per cent was an adequate correction for these sources of error.

(*b*) *The Kjeldahl method*—Both macro and micro versions of the Kjeldahl method have been described. A sample of the unknown substance is weighed into a Kjeldahl flask. The sample is digested with concentrated sulphuric acid, the nitrogen in the sample being converted quantitatively to ammonium sulphate. The digested acid mixture is made strongly alkaline to liberate ammonia, and steam distilled, the distillate being collected in boric acid solution. The ammonia is determined by titration with standard acid, the nitrogen being calculated from the ammonia found.

The micro method is universally employed in biochemistry for the determination of protein. Other methods for the determination of protein are always related to the nitrogen content. (See biochemical methods—Chapter 11.)

Reagents—

(1) Concentrated sulphuric acid A.R.

(2) Potassium sulphate crystals A.R.

(3) Selenium dioxide A.R.

(N.B. Many workers prefer to use 'Kjeldahl catalyst tablets' marketed by B.D.H. Each tablet contains 1 g potassium sulphate and 0·1 g selenium dioxide. Alternatively, copper sulphate may be used instead of selenium dioxide, similar tablets being available. Tablets containing 1 g potassium sulphate only are also supplied for those who wish to use a mercury catalyst.)

(4) 40 per cent w/v sodium hydroxide A.R. solution.

(5) Saturated boric acid solution.

(6) Mixed methyl red/methylene blue indicator.

(*a*) 1 per cent methylene blue in distilled water.

(*b*) Saturated solution of recrystallized methyl red in 95 per cent ethanol.

A mixture of $1\frac{1}{2}$ cm^3 of (*a*) with $12\frac{1}{2}$ cm^3 of (*b*) is added to $2\frac{1}{2}$ dm^3 saturated boric acid to give a stock reagent solution.

(7) Standard M/70 hydrochloric acid. This is prepared by direct weighing of constant boiling acid. (See preparation of reagents, Chapter 4.)

Digestion—About 5 mg of the sample is weighed into a 100 cm^3 Kjeldahl flask (*Figure 5.3a*), 2 cm^3 A.R. grade concentrated sulphuric acid is added. 1 g potassium sulphate is added to raise the boiling point of the digestion mixture and 0·1 g of selenium dioxide is added to act as an oxidation catalyst. Alternatively, an appropriate catalyst tablet is used. The flask is placed on the digestion stand (*Figure 5.3b*) and warmed until the contents are boiling gently. After the excess water has boiled off the flame is raised slightly until the digestion mixture begins to char. The digestion must be carefully carried out and foaming or the deposition of black residues in the neck of the flask must be avoided. Excessive foaming may be minimized by the addition of three or four drops of octyl alcohol. If residues collect in the neck of the flask it must be cooled, and the deposits rinsed down with water. Digestion may then be recommenced. As the digestion proceeds the mixture clears and becomes colourless. Boiling is allowed to continue for at least half an hour after the solution clears. The flame is turned off, the flask allowed to cool to room temperature, and about 10 cm^3 of distilled water added. During the digestion of the sample a 'blank' is set up consisting of the digestion reagents alone. This is carried through the digestion, distillation, and titration processes in order to give a blank figure for any traces of nitrogen present in the reagents.

Several patterns of steam distillation unit have been described. A convenient and common type, the Markham still, is shown in *Figure 5.3c*. Before a series of distillations is carried out the apparatus

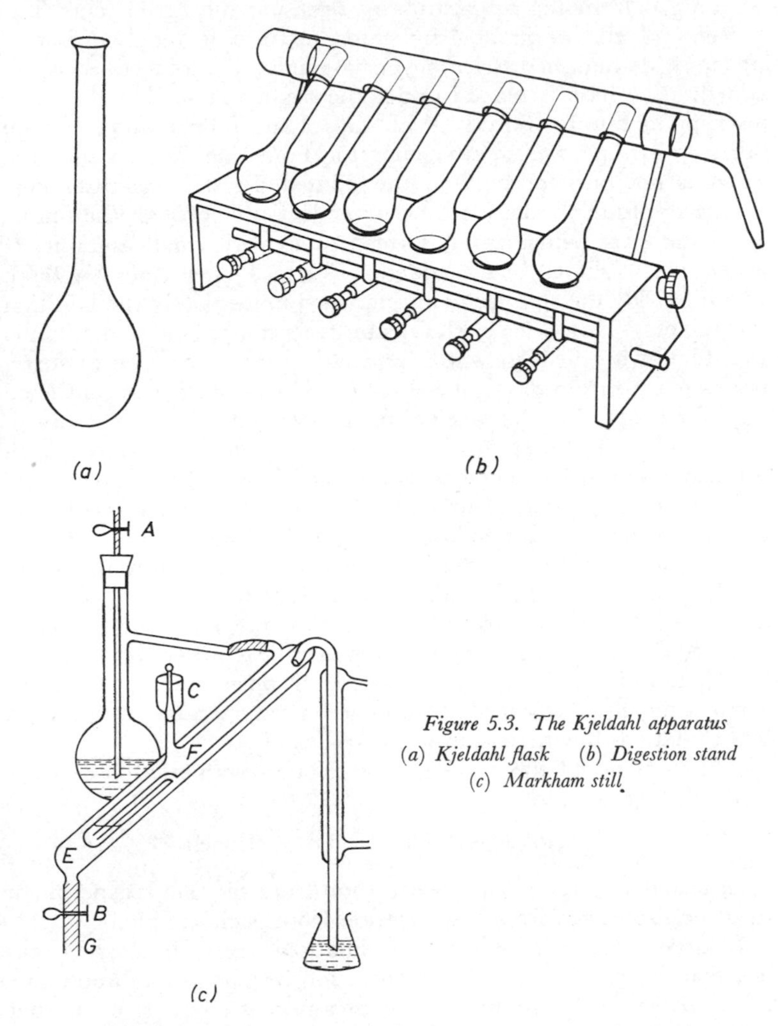

Figure 5.3. The Kjeldahl apparatus
(*a*) *Kjeldahl flask* (*b*) *Digestion stand*
(*c*) *Markham still*

must be steamed out to remove any traces of alkali left from earlier use. The spring clips *A* and *B* are applied and the water in the distillation flask boiled vigorously for at least 15 min, the condensed steam being collected in a 50 cm³ wide-mouthed Erlenmeyer flask.

After this time the flame is removed from the steam generator, the cooling of the water causing any liquid condensed in the inner jacket, *F*, to be sucked back into the outer jacket, *E*, where it can be drained from the apparatus by releasing the spring clip *B*.

5 cm³ of the boric acid/indicator mixture is pipetted into a 50 cm³ wide-mouthed Erlenmeyer flask and 5 cm³ of distilled water added. The flask is placed under the condenser so that the tip is just dipping into the liquid. With clips *A* and *B* open and a bunsen burner warming the steam generator from one side so that the water is hot but not boiling, the diluted digest is quantitatively transferred from the flask to the inner jacket *F* through the funnel at *C*, the flask being rinsed several times with small amounts of water the washings being added. 10 cm³ 40 per cent NaOH is added quickly the stopper being replaced immediately the addition is complete. The water in the steam generator is boiled, the funnel at *C* filled with distilled water, and clip *B* closed as soon as steam begins to issue from the rubber tube *G*. The distillation is continued until the contents of the receiver flask have approximately doubled. The receiver is lowered, the tip of the condenser rinsed with distilled water into the receiver and the bunsen burner removed from under the steam generator. As the boiling water cools, the rubber tube above the steam generator is pinched between the fingers and the residue in the jacket *F* is sucked back into the jacket *E*. The clip *B* is opened to allow the residue to drain out, replaced and the stopper at *C* is pulled out to allow the water in the funnel to enter *F*. When the stopper is replaced this water too is sucked into *E* and drained out. The clip *B* is left open and the apparatus is ready for the next sample. An experienced operator can perform 20 distillations/h with this apparatus.

The contents of the receiver are titrated with M/70 HCl from a 5 cm³ microburette.

$$1{\cdot}0 \text{ cm}^3 \text{ M}/70 \text{ HCl} = 0{\cdot}2 \text{ mg nitrogen}$$

In calculating the result the ttiration from the blank experiment must be subtracted from the titration from each sample.

If large numbers of Kjeldahl estimations are to be carried out, it is convenient to set up a small titration bench with an automatic filling reservoir burette and a 10 cm³ automatic pipette connected to a 1 dm³ reservoir of ready diluted boric acid/indicator mixture.

Although the Kjeldahl technique is not usually considered satisfactory for the estimation of nitrogen in nitro and nitroso compounds, Niederl and Niederl have reported a modification of the method in which the sample is reduced by phosphorus and hydriodic acid before digestion with sulphuric acid and potassium sulphate using mercuric acetate as a catalyst.

(*c*) *The Conway microdiffusion technique*—Although the use of the Conway unit is given as a microtechnique for the estimation of nitrogen in organic compounds this is only one example of the large range of quantitative measurements which can be performed by the various modifications of this method. Brȧdy lists about 50 different compounds or groups which may be estimated on a micro or submicro scale by the Conway microdiffusion technique.

The Conway unit consists of a circular dish with a thick outer wall and with a central chamber having a wall concentric with the outer boundary and rising to just under half the height. (*Figure 5.4.*) Into the outer chamber is pipetted an aliquot of the

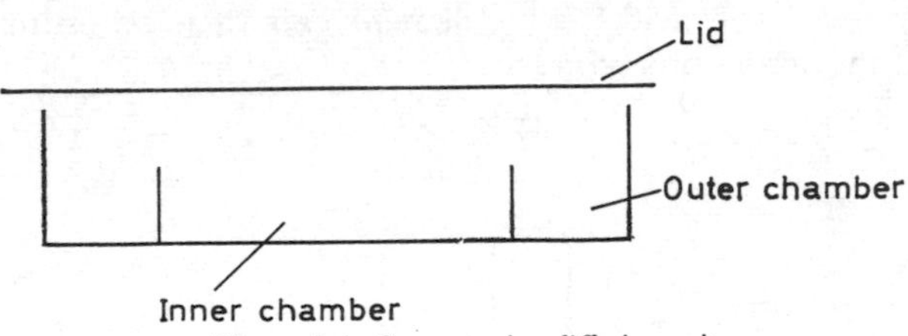

Figure 5.4. Conway microdiffusion unit

substance to be analysed and a reagent (liberator) which liberates the group to be estimated in gaseous form. The inner chamber contains a reagent (absorbent) which absorbs the vapour given off from the outer chamber. The unit is sealed with a glass plate coated with an appropriate fixative and left for some time at room temperature. When diffusion from the outer to the inner chamber is complete the estimation can be completed by titration or by colorimetric examination of the contents of the inner chamber.

The estimation of nitrogen is a modification of the Kjeldahl technique already discussed. The sample is digested in an exactly similar manner and, after cooling and diluting, the digest is transferred quantitatively to a 25 cm^3 volumetric flask and made up to the mark. 1·0 cm^3 of the solution is pipetted accurately into the outer chamber of the Conway unit and 1·0 cm^3 of the Kjeldahl boric acid/indicator reagent into the inner chamber. 1·0 cm^3 40 per cent potassium hydroxide solution is pipetted quickly into the outer chamber and a glass cover coated with a fixative of paraffin wax-liquid paraffin mixture is placed on top of the unit. After standing at room temperature for 2 h the cover is removed and the contents of the inner chamber titrated against M/70 hydrochloric acid.

This elegant method is very convenient where large numbers of nitrogen estimations must be carried out, the average time spent on each estimation being very small. The use of special indicator mixtures allows quantities of nitrogen as low as 0·1 μg to be measured with a reasonable degree of accuracy. Only a brief survey of the technique can be given here; the standard work on the use of the Conway unit may be consulted for exhaustive details of the numerous analytical procedures in which it may be employed.

Estimation of Sulphur and the Halogens

Sulphur and the halogens are estimated in organic compounds by variations of the Carius technique. This involves the conversion of the elements to ionic forms by heating in a sealed tube with concentrated nitric acid. The element can then be estimated by standard inorganic procedures.

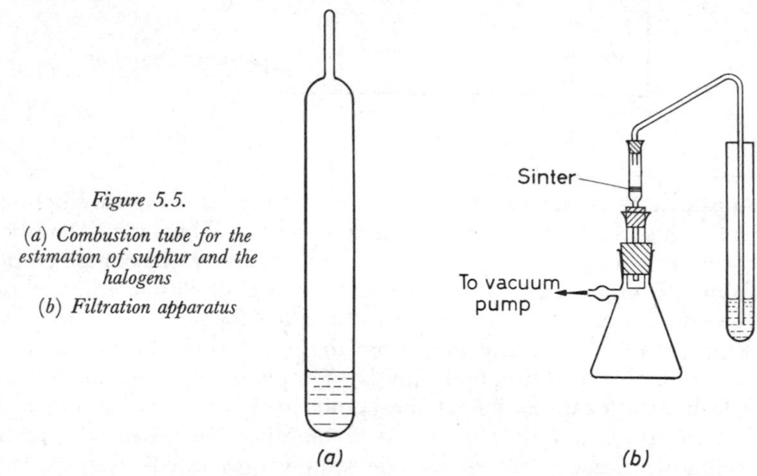

Figure 5.5.

(*a*) *Combustion tube for the estimation of sulphur and the halogens*

(*b*) *Filtration apparatus*

Estimation of halogens—Heavy walled soda-glass tubing is used to form a combustion tube about 350 mm long × 9 mm i.d. × 13 mm o.d. (*Figure 5.5a*) rounded at one end. About 1 g of A.R. silver nitrate and about 25 mg of the substance is carefully weighed into the tube using a long stemmed weighing funnel so that the sample does not stick to the walls of the tube. Using a long stemmed, graduated tap funnel 0·4–0·5 cm³ A.R. grade fuming nitric acid is poured over the sample, the funnel being carefully withdrawn to avoid droplets of acid being deposited on the walls. The tube is sealed and placed in a tube furnace in a cast-iron jacket and the temperature raised to 280° over half an hour. This temperature is maintained for 4 h, the heat is turned off, and the tube allowed to cool overnight. The pressure inside the cooled tube is relieved by playing a fine flame on the tip of the tube which is retained in the furnace. As the glass softens, the gases inside the tube blow a small hole in the tip and escape. The top of the tube is cut off care being taken that the cut is clean and that no glass falls into the tube. The contents of the tube are washed down and diluted with 3 cm^3 of distilled water and the base of the tube gently warmed until the liquid begins to boil. This breaks up the silver halide precipitate so that any occluded excess silver nitrate is dissolved out.

The apparatus for the collection of the silver halide is shown in *Figure 5.5b*. The sintered glass funnel is carefully washed with water alcohol and acetone and dried for 15 min at 140° and then weighed. The long tube is inserted in the reaction vessel and gentle suction applied at the Buchner flask so that the precipitate is drawn over and filtered off. The reaction tube is washed four times with 3 ml of distilled water the washings being similarly drawn through the filter each time. Two washings each of 5 cm^3 of A.R. ethanol are drawn over. The funnel is removed, dried for 15 min at 140° and again weighed, the gain in weight representing the weight of silver halide formed. The weight of halogen present may be calculated from the factors:

1·000 g silver halide = 0·2474 g chlorine or 0·4256 g bromine or 0·5406 g iodine.

In those cases where two or more halogen elements are present in the molecule a special separation procedure has been described by Niederl and Niederl.

Where both bromine and chlorine are present in the same compound the weighed silver halide is treated with ammonium bromide and strongly heated. The mixture of silver bromide and silver chloride is all converted to the bromide, the ammonium salts being volatile disappear during heating. The silver bromide is weighed and the chloride and bromide in the original sample can be calculated.

Estimation of sulphur—The procedure for the estimation of sulphur is precisely the same as that for the estimation of halogens except that about 50 mg of barium chloride is substituted for the silver nitrate. The sulphur in the compound is oxidized to sulphuric acid which is precipitated as barium sulphate. The precipitate must be washed thoroughly to remove the excess barium chloride which may be occluded.

1·000 g $BaSO_4$ = 0·1374 g sulphur in the original compound

The Parr Bomb

Bomb methods have superseded the Carius technique in many laboratories. The Parr bomb consists of a pure nickel thimble with very heavy walls fitted with a heavy screw-on lid which is made to fit tightly on the thimble by a lead washer (*Figure 5.6*). The whole bomb fits into a heavy screw casing made of iron. The sample is weighed directly into the thimble and about 4 g of A.R. sodium peroxide added together with about 0·3 g of a

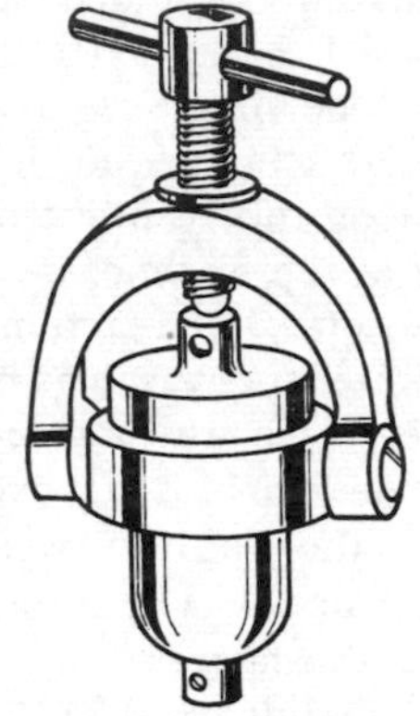

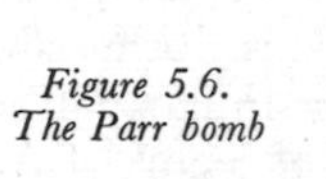

Figure 5.6. The Parr bomb

mixture of equal parts of potassium nitrate and lactose. The solids are carefully mixed and the bomb fitted together in its case. The casing is filled with water and the suspended bomb is heated strongly in a pointed flame from a blowpipe of the type used for glass-blowing. When the water in the casing begins to boil, the heat is removed, and the bomb immersed in water until it is quite cold. The material inside the thimble is carefully washed into a beaker and the element to be estimated is analysed by standard inorganic procedures. Chlorine is estimated by the Volhard method. Bromine and iodine, which form the bromate and iodate respectively, are reduced by hydrazine sulphate to bromide and iodide. Sulphur is estimated as sulphate with barium chloride; phosphorus is estimated as the phosphate with ammonium molybdate.

Analytical Services

Except for the Markham technique for nitrogen the methods described above have been given in outline only. The Kjeldahl method is used in most biochemistry laboratories and may reasonably be expected to come within the scope of the worker who is not primarily an analyst. Detailed accounts of the other methods can be found in the literature listed in the appendix. It must be emphasized, however, that organic analysis is a complicated and highly skilled procedure and a great deal of experience is necessary before reliable results are obtained. Many laboratories find that quantitative estimation of this type is rarely required, and then usually as a check on the synthesis of a new or uncommon compound. Under these circumstances it is usual to employ one of the specialist analytical services which are expert in this type of analysis. Rather than spending large sums of money on analytical equipment which is rarely used and which few workers have any real experience of using, a few milligrammes of the material, purified and dried may be sent to a microanalytical laboratory and a complete or partial analysis performed, in duplicate if necessary, inexpensively and quickly. The author has had dealings extending over several years with one of these firms.*

THE EMPIRICAL FORMULA

When the proportion of each of the elements in an organic compound has been determined, the empirical formula can be calculated. This shows the relative number of different atoms in the compound. The calculation is carried out by dividing the percentage of each element present by its atomic weight. An example is perhaps the best method of explaining this procedure.

An organic compound on analysis gave 20·02 per cent carbon, 46·64 per cent nitrogen and 6·67 per cent hydrogen. Find its empirical formula.

(C = 12·00, H = 1·008, O = 16·00, N = 14·01)

Element	*% At. wt.*	*Proportion in whole numbers*
C	$\frac{20{\cdot}02}{12{\cdot}00} = 1{\cdot}67$	1
H	$\frac{6{\cdot}67}{1{\cdot}008} = 6{\cdot}67$	4
N	$\frac{46{\cdot}64}{14{\cdot}01} = 3{\cdot}33$	2
O	$\frac{26{\cdot}67}{16} = 1{\cdot}67$	1

Empirical Formula = CH_4N_2O

* Drs Weiler and Strauss, Banbury Road, Oxford.

THE DETERMINATION OF MOLECULAR WEIGHT

The molecular formula of the 'unknown' compound is obtained by multiplying the empirical formula by a whole number, e.g. the compound above would have a molecular formula $(CON_2H_4)_n$ where n is a whole number. To find the factor n the molecular weight of the compound must be determined. Very high accuracy is not required for compounds of low and medium molecular size as a reasonable approximation will show clearly the value of the factor n.

Broadly speaking the methods for molecular weight determination may be divided into two groups, those depending upon the physical (colligative) properties of compounds being more commonly used for organic compounds, while those depending upon gravimetric techniques and chemical methods are more often used in inorganic chemistry. There is a vast literature dealing with the various methods available and only some of the simpler and more common techniques suitable for class experiments are described.

Molecular Weight by Depression of the Freezing Point

The method depends upon the observed fact that the freezing point of a pure solvent is depressed by having another substance dissolved in it. The depression of the freezing point is proportional to the amount, expressed in molar terms, of solute used. The cryoscopic constant K of a solvent is the depression in freezing point caused by dissolving 1 g molecular weight of a pure solute in 100 g of the solvent. This may be expressed mathematically as a formula for the calculation of molecular weight by this method.

$$M = \frac{w \times 100 \times K}{W \times \Delta T}$$

where M = molecular weight of solute

w = weight of solute

W = weight of solvent

ΔT = difference in freezing point between solution and pure solvent

K = cryoscopic constant of solvent.

Although the expression refers to the solution of 1 g molecular weight of solute in 100 g of solution, this high concentration would never be used in experimental practice, indeed it could rarely be achieved. If suitable solvents with high K values are employed, very dilute solutions of up to 5 per cent concentration are adequate for a sufficient ΔT to be observed.

Rast's melting point method—Perhaps the simplest experimental determination of molecular weight is the method devised by Rast based upon the depression of the melting point of camphor. This compound has an unusually high cryoscopic constant and results of sufficient accuracy can be obtained quickly and conveniently with the simpler apparatus.

A little pure resublimed camphor is powdered with a spatula on an unglazed tile until it is throughly pulverized. The melting point (about 180°) is carefully determined as described in Chapter 2 (page 74). About 200 mg pure naphthalene is weighed into a clean hard glass test-tube. 2 g of the powdered camphor is added and the tube is again weighed. The contents of the tube are melted in an oil bath heated to about 190° the chemicals being thoroughly mixed by swirling. After a homogeneous solution has been formed it is allowed to solidify until it is quite cold, the solid, extracted by carefully cracking the tube, being ground to powder in a clean mortar. The melting point of the solution is determined and the cryoscopic constant of camphor (K) calculated from the difference between the two melting points (ΔT), the molecular weight of pure naphthalene being 128.

$$K = \frac{\Delta T \times 128 \times W}{100 \times w}$$

where W and w are the weights of camphor and naphthalene respectively. If the process is repeated substituting the unknown substance for naphthalene the molecular weight of the unknown can be calculated from the equation

$$M = \frac{K(100 \times w)}{\Delta T \times W}$$

The Rast technique is very useful as a speedy method of determining the approximate molecular weight of a compound in order to calculate a molecular formula, but even if the low accuracy normally required is to be achieved certain criteria must be followed. It is obvious, for example, that solute and solvent must be completely miscible in the concentration used, and neither may react with the other or catalyse any reaction in the other.

Although Rast used camphor in his original method, several other solids have been found to be useful solvents. A list of these together with cryoscopic constants is given in Table 5.3.

TABLE 5.3. Solvents with Rast's Molecular Weight Determination

Solvent	M.P. (°C)	K (°C)
Camphor	176–180	37–40
Camphene	49	31
Bornylamine	164	41
Cyclohexanol	25	43
Cyclopentadecanone	66	21
Dicyclopentadiene	32	46

Other cryoscopic methods—The Rast method uses solvents which are solid at room temperature and to which standard melting point techniques can be applied, but the method cannot be employed if no suitable solid is available. In these circumstances common organic solvents may be used. There are several of these with freezing points between 3 and 10° which have a high cryoscopic constant. This freezing point range is convenient because ice-water can be used as a cooling agent.

In these methods a differential thermometer is employed. The Beckmann thermometer (*Figure 5.7*) has a scale extending over 6° (S_1) and can be read accurately to 0·01. Absolute readings of freezing point are not required but only the difference in freezing point between pure solvent and solution. The thermometer can be set to cover a temperature range around the freezing point of the solvent. Setting the older type of Beckmann thermometer was a tiresome procedure but the modern instrument has an auxiliary scale (S_2) which speeds up the process considerably. It must be noted that the upper bulb is reversed and that, therefore, the graduations above the zero mark are in fact 'minus' degrees and those below the zero point are 'plus' degrees.

The procedure for setting is as follows. Suppose the solvent to be used is cyclohexane. The thermometer must be set to give an upper reading of about 7·5°, i.e. about 1° above the freezing point of the solvent should be equivalent to the zero mark:

1. The bulb of the thermometer is placed in a beaker of water adjusted to 8–9° with ice, using a separate thermometer.

2. When the mercury level is constant the instrument is carefully inverted and gently tapped so that the mercury in the upper bulb *B* falls to *A*. The thermometer is restored to its original position, the mercury remaining at *A*.
3. The bulb of the thermometer is warmed in water at about 40° so that the capillary from the bulb to *A* is filled with mercury.

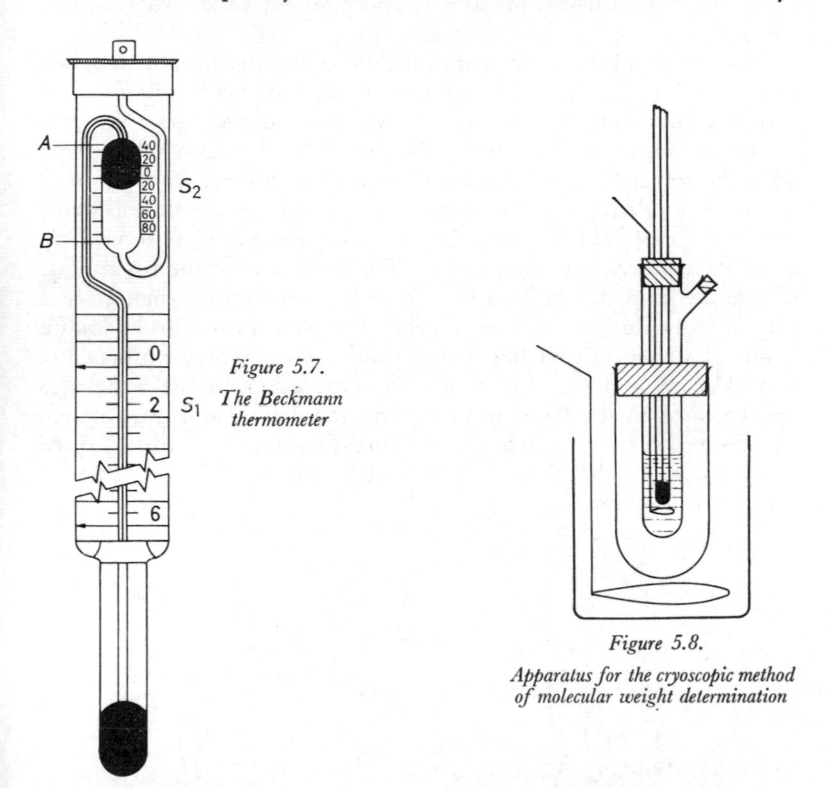

Figure 5.7. The Beckmann thermometer

Figure 5.8. Apparatus for the cryoscopic method of molecular weight determination

4. The thermometer is cooled. This causes the meniscus in *B* to rise. When the meniscus is at about 6·5 on the scale S_2 the metal cap on the thermometer is tapped sharply. This causes the mercury thread at *A* to break. The remainder of the mercury in *B* falls to the bottom.
5. The thermometer is now set. If the bulb is placed in melting solid cyclohexane the thermometer will record at the zero mark on the scale S_1.

The apparatus for determination of molecular weight by the cryoscopic method is shown in *Figure 5.8.* A tube about 2·5 cm in diameter and 18 cm long with a slanting side arm at its upper end is fitted with a cork through which pass a Beckmann thermometer appropriately set and a glass stirring loop. This tube fits through a cork in a short wide tube which acts as an air-jacket. The jacket fits through an asbestos lid to a round glass jar of ice water which has a heavy copper wire stirring loop.

The side arm tube is cleaned and thoroughly dried. It is weighed empty and again after 25 cm³ of solvent has been added. The cork holding thermometer and stirrer is fitted and the tube pre-cooled by placing it directly into ice water with occasional stirring until the solvent solidifies, then allowing the solvent to thaw until the last crystals have just melted. The outside of the tube is dried and it is fitted into the air-jacket. The glass stirrer is moved up and down slowly and uniformly. The mercury on the main scale of the thermometer falls until crystallization begins when it rises very slightly due to the latent heat of crystallization and remains steady until the solvent has frozen solid. The highest temperature at this stage is noted. The tube is removed from the apparatus and warmed in the hand until the crystals have again just melted and the process is repeated. Three consistent readings are obtained to 0·01° using a hand lens to read the scale, care being taken to avoid parallax errors.

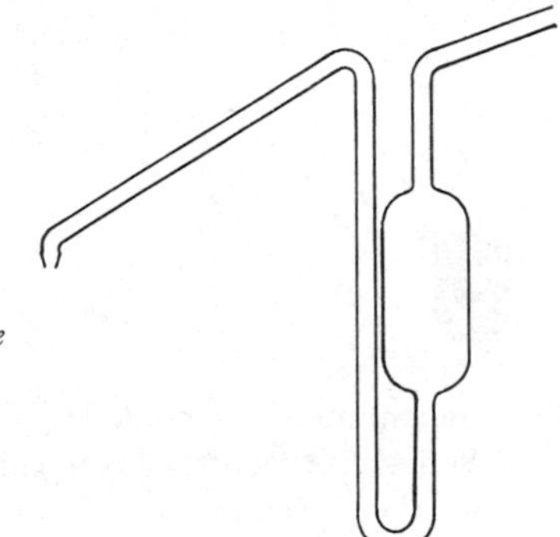

Figure 5.9. Pyknometer or weighing pipette

The sample of solute can now be added. If this is a solid, two tablets are prepared with a screw tablet press and weighed. The first is dropped into the solvent through the side arm and stirred until completely dissolved. The freezing point is determined as described. A liquid solute can be added from a pyknometer or weighing pipette with a long bent tip (*Figure 5.9*) so that the liquid is injected directly into the solvent. After the freezing point has been determined following the addition of one of the weighed tablets, the second tablet is added and the process repeated. For a liquid solute, of course, a second weighed addition is made from the weighing pipette.

Three freezing points have now been determined. The molecular weight is calculated from the formula given for the Rast method, e.g. for the determination of the molecular weight of anthracene in cyclohexane the following results might be obtained.

(*a*) Weight of cyclohexane = 21·74 g
Cryoscopic constant of cyclohexane = 201
Freezing point of cyclohexane (on 'set' Beckmann thermometer) = 0·76°, 0·77°, 0·76° (Mean = 0·76)

(*b*) Weight of first tablet of anthracene = 186 mg (0·186 g)
Freezing point of solution = 1·73, 1·74, 1·73
(N.B. Thermometer has reverse Scale) = (Mean = 1·73)

$$\Delta T = 1{\cdot}73 - 0{\cdot}76 = 0{\cdot}97°$$

$$\text{mol. wt. anthracene} = \frac{0{\cdot}186 \times 100 \times 201}{21{\cdot}74 \times 0{\cdot}97} = 178$$

(*c*) Weight of second tablet of anthracene = 192 mg (0·192 g)
Freezing point of solution = 2·74°, 2·73°, 2·74°, (Mean = 2·74°)

ΔT = 1·98°
Total weight of anthracene = 378 mg (0·378 g)

$$\text{mol. wt. anthracene} = \frac{0{\cdot}378 \times 100 \times 201}{21{\cdot}74 \times 1{\cdot}98} = 176$$

Molecular Weight by Elevation of the Boiling Point

The theory of molecular weight determination by elevation of the boiling point is exactly parallel to that for the cryoscopic techniques.

A simple apparatus for class experiments is shown in *Figure 5.10*. The apparatus is cleaned, dried and weighed. 20–25 cm³ of the solvent is added and the apparatus again weighed. The Beckmann thermometer (adjusted as previously described) and the condenser are fitted and the apparatus heated with a small flame from a micro-burner. The burner should be adjusted by a screw clip on the gas tubing so that the solvent refluxes from the condenser at the rate of about 1 drop/sec. Readings are taken on the thermometer every 2–3 min. When three consistent readings have been obtained

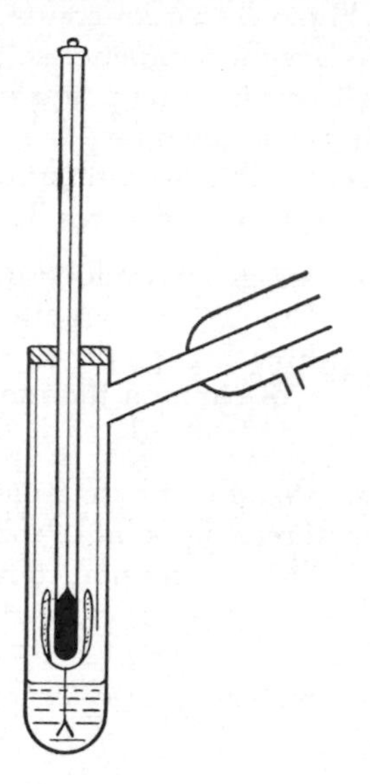

Figure 5.10. Apparatus for the ebullioscopic method of molecular weight determination

a weighed pellet of sample may be added by momentarily removing the condenser and inserting the pellet with a pair of forceps. If the condenser is replaced immediately there will be no loss of solvent. A few minutes after the pellet has dissolved readings are taken every few minutes until three successive readings are consistent. A second weighed pellet is added and the process repeated. The formula for the calculation of the molecular weight of the sample is similar to that described above for the cryoscopic method.

It is more difficult to obtain consistent results by boiling point methods than it is by depression of the freezing point and the apparatus must be used very carefully to eliminate inconsistencies.

(*a*) The solvent must boil smoothly and steadily so that superheating is kept to a minimum.

(*b*) Variation in the rate of heating can greatly affect the thermometer reading. This is especially noticeable when the pellet of sample is added as the new boiling point may barely be achieved.

The heat should therefore be increased very slightly after the addition of sample, but not so much as to cause bumping or superheating.

A number of micro and semi-micro methods have been developed and are described in the specialist literature. One of these may be briefly mentioned, the semi-micro method of Sucharda-Bobranski. The apparatus is shown in *Figure 5.11.* It is constructed entirely

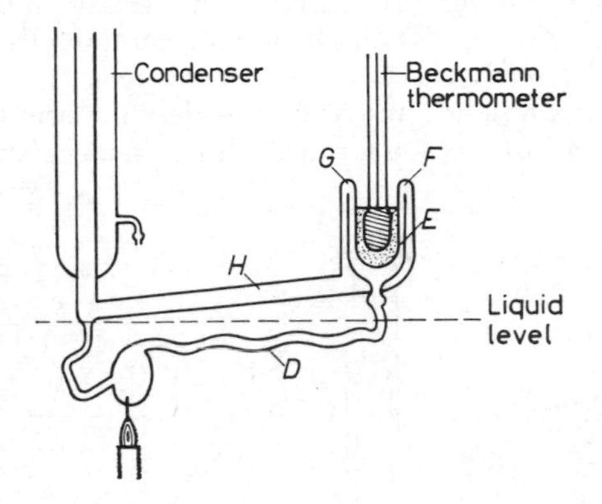

Figure 5.11. Sucharda Bobranski apparatus

of glass. A measured volume of solvent is run down the condenser, enough being used to fill the bulb and lower horizontal tube to the level shown. When the platinum wire is heated the solvent boils and circulates along *D*, over the inner cup *E* into *F* and *G* and finally returns to the bulb via the sloping tube *H*. After the boiling point of pure solvent has been found on the Beckmann thermometer the weighed pellet (25–50 mg) is dropped through the condenser and the process repeated to give the elevation of the boiling point. The weight of solvent is calculated from its volume and density.

TABLE 5.4. Solvents for Cryoscopic Molecular Weight Determination

Solvent	M.P. (°C)	*K* (°C)
Benzene	6	4·9
Dioxan	12	4·6
Nitrobenzene	6	6·8
Water	0	1·9
Acetic acid	17	3·8

This apparatus gives much more consistent readings than the simple macro method previously described, the determination is quickly completed, and good results are obtained with very small samples.

A list of suitable solvents for the ebullioscopic method is given in Table 5.5.

Molecular Weight by Determination of Vapour Density

The Victor Meyer method is almost universally employed for the determination of the vapour density of volatile organic compounds. It can be shown that molecular weight = 2 × vapour density at S.T.P. (from Avogadro's hypothesis) and thus the determination of vapour density is an important method for determining molecular

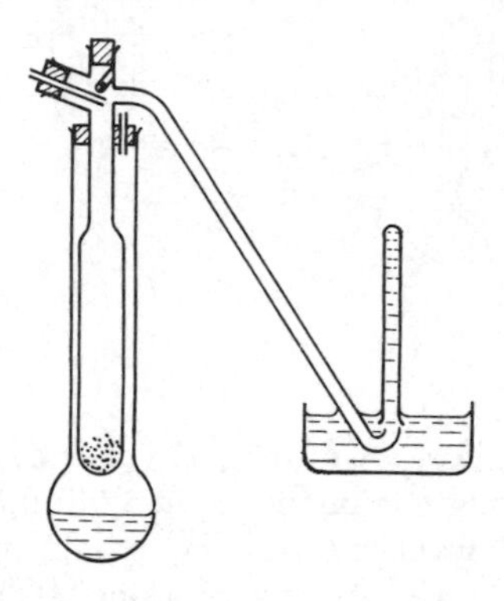

Figure 5.12. The Victor-Meyer apparatus

weight. The apparatus is shown in *Figure 5.12*. The inner vaporization tube is about 76 cm long, the main tube is about half an inch in diameter, the bulb at the end having a volume of about 100 cm³. The outer jacket contains a liquid having a boiling point 50–70° higher than that of the sample. The sample bottle which has a capacity of about 0·5 cm³ is cleaned, dried and weighed. About

TABLE 5.5. Solvents for the Ebullioscopic Molecular Weight Determination

Solvent	B.P. (°C)	*K* (°C)
Ethyl ether	35	2·2
Acetone	56	1·7
Carbon tetrachloride	77	5·0
Ethanol	78	1·2
Benzene	80	2·6
Water	100	0·52

0·2 cm^3 of the sample is pipetted into the bottle which is stoppered and again weighed. It is then fixed into the top of the vaporization tube as shown. The liquid in the outer jacket is heated until it is boiling steadily and no more bubbles appear from the delivery tube. The calibrated receiving tube is filled with water and placed over the delivery tube. The sample bottle is allowed to fall to the bottom of the vapour tube by partially removing the upper rod and replacing it. The sample rapidly volatilizes, blowing out the stopper, and bubbles of air leave the delivery tube and enter the receiver. When no more air escapes into the delivery tube the receiver is stoppered with a finger and transferred to a gas jar full of water so that the level of water inside and outside can be equalized. The volume of air, room temperature and barometric pressure are noted. If the vapour density of chloroform is measured using chlorobenzene as the heating liquid the following figures give an example of the calculation.

Weight of chloroform taken = 0·2360 g
Volume of air collected = 46·3 cm^3
Temperature = 15·5°C Barometric pressure = 766·0 mm
Vapour pressure of water at 15·5°C. (from standard tables)
= 13·1 mm

$$\therefore \text{ Volume of air collected at S.T.P.} = 46{\cdot}3 \times \frac{273}{288{\cdot}5} \times \frac{766 - 13{\cdot}1}{760}$$

$$= 43{\cdot}4 \text{ cm}^3$$

43·4 cm^3 of vapour weighs 0·2360 g
11 210 cm^3 of vapour weighs 61·1 g
∴ Vapour density = 61
Molecular weight of chloroform = 2 × 61 = 122
(Theoretical for $CHCl_3$ = 119·5)

Other Methods for the Determination of Molecular Weight

The methods which have been described use fairly simple apparatus and the techniques are suitable for class practical work. In working laboratories many other ways of measuring molecular weight are found. The elementary procedures which have been discussed are only suitable for substances of fairly low molecular weight. For proteins, plastic polymers, etc. where the molecular weight may be measured in millions special techniques must be employed, e.g. the tobacco mosaic virus has a molecular weight of 50 million, and serum albumin, a comparatively simple protein, has a value of 150 000.

The two factors most often used for molecular weights of this order are osmotic pressure and sedimentation value. The results obtained from the sedimentation value, measured by ultracentrifugal analysis, are usually considered the most accurate.

Other procedures which have been described depend on the measurement of rates of diffusion or vapour pressure of liquids.

All these methods require very elaborate apparatus as the physical measurements involved are of a very small order. Because of the high cost of the apparatus it is only found in specialist laboratories. The techniques are described in the specialist literature.

THE CLASSIFICATION AND IDENTIFICATION OF ORGANIC COMPOUNDS

The procedures so far described lead to a molecular formula for an unknown compound. This is only part of the way to actually identifying the compound. The next stage is the *classification* of the compound by chemical methods, followed by the *identification* of the compound by physical methods.

The reactions of the substituent radicals in an organic compound, are, in general, independent of the influence of the rest of the molecule. By chemical tests it is possible to determine the substituent radical or radicals present in the molecule and to say to what class of compound (alcohol, fatty acid, ketone, etc.) it belongs. The results of these chemical tests are rarely as definite as those of inorganic chemistry, neither is the method of analysis so systematic. The argument has been described by Klyne as running on the following lines: 'The compound has the same general reactions as ethyl alcohol. It therefore contains a hydroxyl group.'

Although compounds in the same class show little chemical difference from each other, they show great differences in their physical properties. After the compound has been classified by chemical tests it may be identified (i.e. the determination of which *member* of its class it is) by the measurement of the physical properties of the compound or one of its derivatives.

The measurement of melting-point is of the very greatest importance in this respect as by the use of the mixed melting-point method the identity or non-identity of two apparently identical substances may be decided. Tables of melting points have been compiled so that a classified compound, or one of its derivatives, may be identified (see mixed melting points, page 76).

The full process of classification and identification is beyond the scope of this book. The syllabus on which it is based requires

TABLE 5.6. Classification of Substances Containing C H (O)

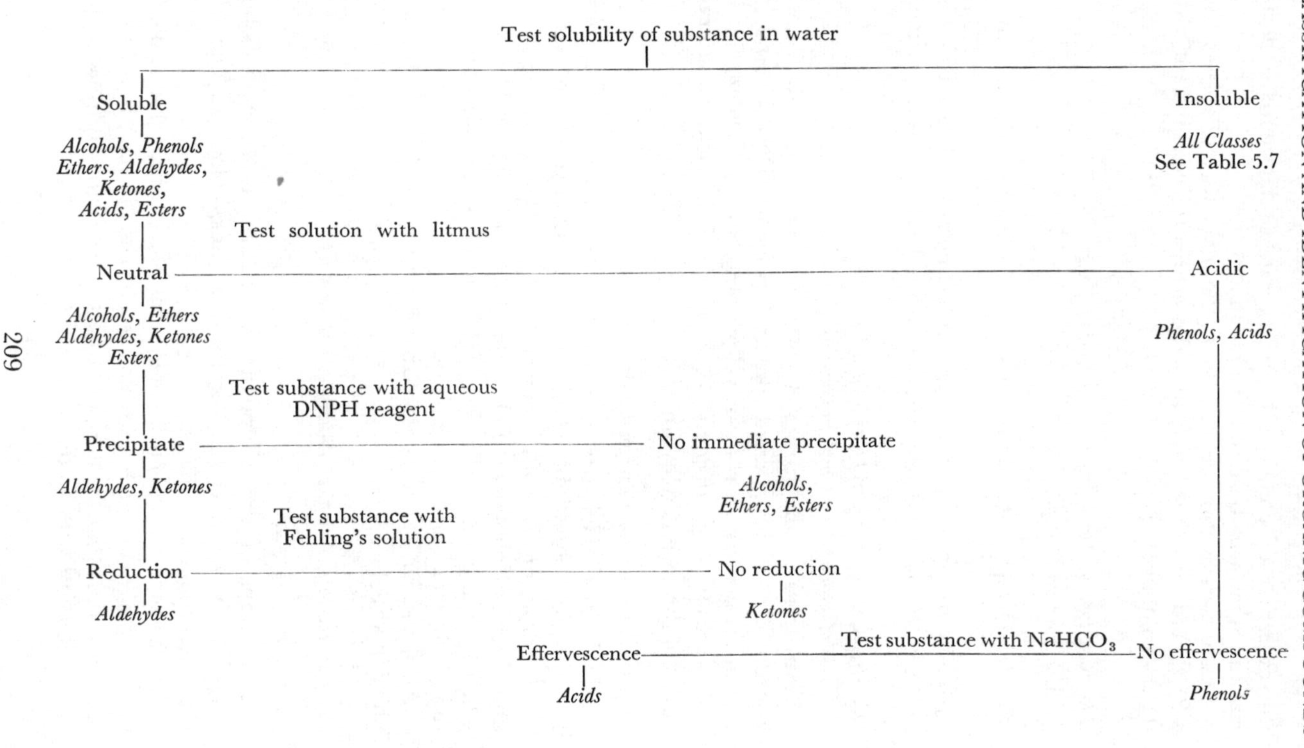

the detection of compounds in several of the more important classes. This will be discussed hereunder.

Before classification begins, the preliminary tests (Tables 5.1 and 5.2) should be carried out, and the elements present identified as described above.

The Reactions of Some Important Groups

1. Aldehydes and ketones

Aldehydes $\left(\begin{matrix}R\\H\end{matrix}\!\!>\!CO\right)$ and ketones $\left(\begin{matrix}R\\R^1\end{matrix}\!\!>\!CO\right)$ both contain the carbonyl group ($= C = O$). Acetaldehyde (CH_3CHO) and acetone ($(CH_3)_2CO$) are suitable examples for class experiments.

(*a*) *Preparation of dinitrophenylhydrazone*

(*i*) *Water soluble carbonyl compounds*

Five drops of the carbonyl compound are added to 5 cm^3 of aqueous 2:4-dinitrophenylhydrazine reagent (DNPH) and the mixture is shaken. A copious yellow precipitate is formed.

Aqueous DNPH reagent: 2 M hydrochloric acid is saturated with 2:4 dinitrophenylhydrazine.

(*ii*) *Water insoluble carbonyl compounds*

The higher aldehydes and ketones may not be water soluble but are probably soluble in alcohol. Benzaldehyde is an example.

Five drops or 0·1 g of the compound is dissolved in 5 cm^3 ethanol and 5 cm^3 alcoholic DNPH added. The mixture is boiled in a water bath and 3 drops of conc. HCl added. On cooling the yellow diphenylhydrazone crystallizes out.

Alcoholic DNPH reagent: Ethanol is saturated with 2:4-dinitrophenylhydrazine.

(*b*) *Differentiation between aldehydes and ketones*

(*i*) *Schiff's reagent*

Two drops of the compound are added to 2 cm^3 of Schiff's reagent. Most aldehydes produce a pink colour in the cold. Ketones produce no colour.

Schiff's reagent. 20 cm^3 water is saturated with sulphur dioxide and 0·2 g *p*-rosaniline hydrochloride dissolved in it. The solution is kept in a dark tightly stoppered bottle overnight and made up to about 200 cm^3 with water.

TABLE 5.7. Classification of Substances Containing C H (O)

Substances insoluble in Water

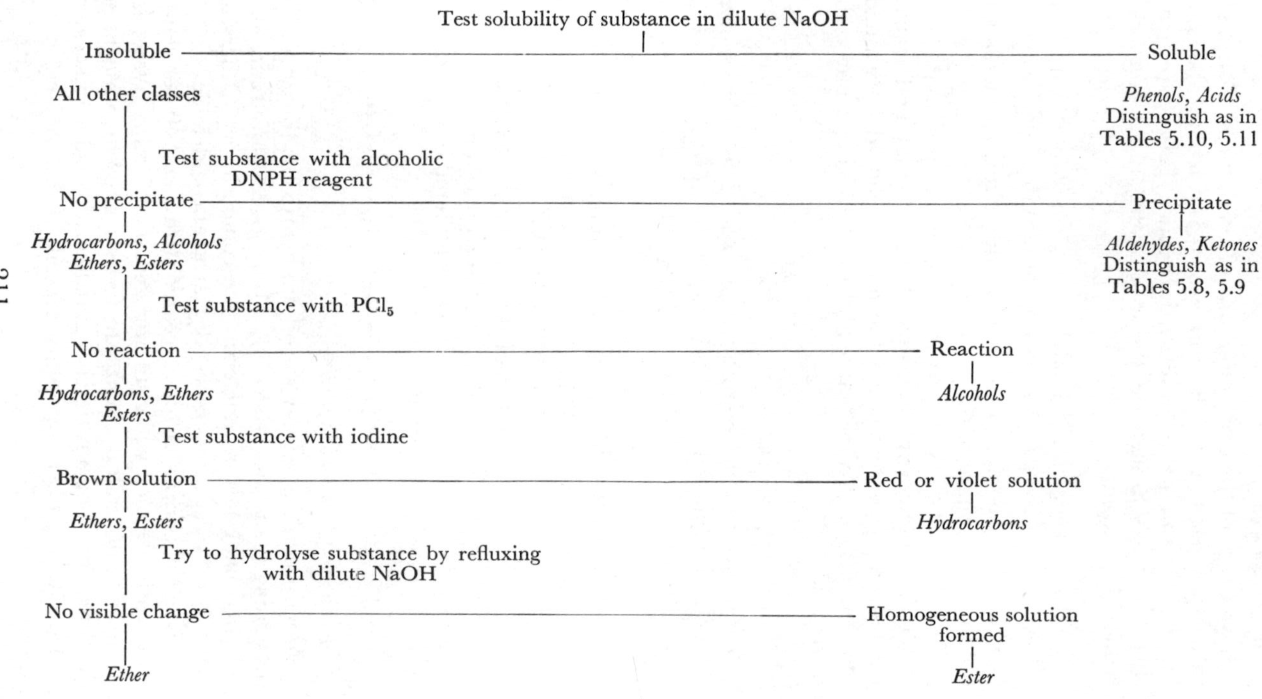

(*ii*) *Tollen's reagent*

Two drops of the compound are added to 2 cm³ Tollen's reagent in a clean test-tube. The mixture is warmed to about 40° in a beaker of warm water. Aldehydes reduce Tollen's reagent and a silver mirror is formed on the wall of the tube. Ketones do not reduce the reagent.

Tollen's reagent: *A*. 3 g silver nitrate in 30 cm³ water.
B. 3 g sodium nitrate in 30 cm³ water.

Mix 1 cm³ each of *A* and *B* and add dilute ammonia dropwise until the precipitate of silver oxide just dissolves.

Care should be taken with this reagent. It should not be allowed to stand or highly explosive silver fulminate may be formed. After the test has been completed the mixture should be washed down the sink with plenty of water and the tube rinsed out with dilute nitric acid.

(*iii*) *Fehling's solution*

Three drops of the compound are added to 4 cm³ of Fehling's solution and the mixture boiled. A red precipitate of cuprous oxide indicates the presence of aldehydes. Ketones do not reduce Fehling's solution.

Fehling's solution: *A*. 35 g of A.R. cupric sulphate are dissolved in water, slightly acidified with sulphuric acid, and made up to 500 cm³.

B. 60 g of A.R. sodium hydroxide and 173 g of A.R. sodium potassium tartrate are dissolved in water, filtered and made up to 500 cm³.

The reagent consists of equal parts of *A* and *B*.

(*iv*) *Action of sodium hydroxide*

Aldehydes often react with sodium hydroxide, the type of product depending upon the aldehyde. Many give yellow resins or yellow coloured materials of indeterminate composition. Others undergo Cannizzaro's reaction, especially aromatic aldehydes.

0·5 cm³ of benzaldehyde is treated with 2 cm³ 40 per cent NaOH and gently warmed while stirring for about 5 min. Water is added to dissolve the sodium benzoate formed and the aqueous solution decanted off from any unchanged benzaldehyde. Excess concentrated HCl is added to produce crystals of benzoic acid on cooling.

TABLE 5.8. Tests for Aldehydes

Test	*Formaldehyde* (40% *soln.*)	*Acetaldehyde* (*soln.*)	*Paraldehyde*	*Benzaldehyde*	*Salicylaldehyde*
Schiff's	+	+	+	+	−
Tollen's	+	+	+	+	−
Fehlings	+	+	+	Very slight	−
Iodoform	−	+	−	−	−
Nitroprusside	−	+	+	−	−
NaOH	Cannizzaro	Yellow resin	Yellow resin	Cannizzaro	Yellow colour ppt. on standing
$NaHSO_3$	−	−	−	Immediate ppt.	

TABLE 5.9. Tests for Ketones

Test	*Acetone*	*Acetophenone*	*Benzophenone*
Iodoform	+	+	−
Nitroprusside	+	+	−
$NaHSO_3$	+	−	−

TABLE 5.10. Tests for Phenols

Test	*Phenol*	*Resorcinol*	*Catechol*	*Hydroquinone*	*O-Cresol*	*m-Cresol*	*p-Cresol*	1-*Naphthol*	2-*Naphthol*
$FeCl_3$ colour	Purple	Purple	Green	Red-green crystals	Purple	Purple	Purple	—	—
Phthalein	Red	Green	Blue	Purple	Red	Blue purple	—	Green	Faint green fluorescence

The equation for the reaction is

$$2C_6H_5CHO \longrightarrow \underset{\text{benzyl benzoate}}{C_6H_5COOCH_2C_6H_5}$$

$$C_6H_5COOCH_2C_6H_5 + NaOH \longrightarrow \underset{\substack{\text{Benzyl}\\\text{alcohol}}}{C_6H_5CH_2OH} + \underset{\substack{\text{Sodium}\\\text{benzoate}}}{C_6H_5COONa}$$

$$C_6H_5COONa + HCl \longrightarrow \underset{\substack{\text{Benzoic}\\\text{acid}}}{C_6H_5COOH} + NaCl$$

(*v*) *Action of Sodium Bisulphite*

Aldehydes and ketones often give addition compounds with sodium bisulphite. 1 cm^3 of benzaldehyde or 1 cm^3 of acetone is treated with about 0·5 cm^3 of saturated $NaHSO_3$ solution. A white addition compound is precipitated.

(*vi*) *Iodoform test*

Iodoform (CHI_3) is readily prepared by the action of potassium iodide and sodium hypochlorite on any compound which contains the $CH_3CH(OH)$ group, e.g. ethanol (q.v.) acetaldehyde (which in aqueous solution exists partly as the hydrated form $CH_3CH(OH)_2$) acetone, iso-propanol.

3 cm^3 10 per cent KI solution and 10 cm^3 of fresh NaOCl solution are added to 0·5 ml acetone. Yellow crystals of iodoform separate immediately in the cold.

(*vii*) *Nitroprusside reaction*

1 cm^3 of freshly prepared sodium nitroprusside solution is added to 0·5 cm^3 acetone. Excess 10 per cent NaOH solution is added. A deep red colour is produced.

2. Alcohols and phenols

Alcohols and phenols both contain the hydroxyl group (— OH). In phenols the group is directly attached to the aromatic nucleus. Alcohols may be classified as primary, secondary and tertiary.

$R.CH_2OH$	$\begin{matrix}R\\R'\end{matrix}\Big\rangle CHOH$	$\begin{matrix}R\\R'\\R''\end{matrix}\Big\rangle COH$
primary	secondary	tertiary

These can be distinguished by their response to oxidation.

(*a*) *Esterification*

Hydroxy compounds all react with acid anhydrides or acid chlorides. Esters are formed.

(*i*) 1 cm^3 ethanol and 1 cm^3 acetic anhydride are heated together in a boiling water bath for a few minutes. The mixture is cooled and made just alkaline with sodium hydroxide solution. Ethyl acetate floats to the surface as an oil. It has a pleasant fruity smell.

(*ii*) The above experiment is repeated using 0·1 g of phenol. Phenyl acetate is formed.

(*b*) *Reaction with sodium*

All hydroxy compounds react with sodium to form compounds RONa (alkoxides or phenoxides) with the evolution of hydrogen.

(*i*) A tiny piece of sodium is added to 2 cm^3 of dry ethanol in an evaporating basin. A vigorous reaction ensues. When the reaction has subsided the solution is evaporated to dryness. A white residue, sodium ethoxide, remains in the basin.

(*ii*) The experiment is repeated using a solution of phenol in dry ether. The ether may be carefully removed by distillation. The white residue in this case is sodium phenoxide.

(*c*) *Differentiation between alcohols and phenols*

(*i*) *pH of aqueous solutions*

An aqueous solution of phenol in water is acid to litmus. Aqueous solutions of alcohols are neutral.

(*ii*) *Reaction with ferric chloride*

Phenols give characteristic colours with neutral ferric chloride solution. One drop of neutral ferric chloride is added to 2 cm^3 of a dilute solution of phenol. A vivid violet colour results.

Alcohols give no colour with ferric chloride.

Ferric chloride reagent: The solid is almost always contaminated with excess hydrochloric acid which might interfere with the reaction. The reagent is prepared, from a 5 per cent solution of ferric chloride. Dilute sodium hydroxide solution is added dropwise until a faint permanent precipitate appears. The mixture is filtered and the filtrate used as the reagent.

(*iii*) *Millon's reagent*

2 cm^3 of Millon's reagent is added to 2 cm^3 dilute phenol solution. The mixture is heated in a boiling water bath for 1 min. A deep red colour is produced. Alcohols give no colour with Millon's reagent.

Millon's reagent: 1 part mercury is dissolved in 1 part conc. HNO_3 with gentle heating. 2 parts water and a few crystals of KNO_3 are added.

(*iv*) *Diazotization*

Three drops of aniline are dissolved in 1 cm^3 conc. HCl and 3 cm^3 water added. The mixture is cooled in ice and 5 drops 20 per cent sodium nitrite solution is added. This forms a diazonium salt. A little phenol is dissolved in excess NaOH solution and cooled. The diazonium salt is added. An azo dye, red in colour is precipitated. Other phenols form azo dyes of different colours.

(*v*) *Phthalein test*

0·2 g phenol and 0·2 g phthalic anhydride are mixed in a hard glass test-tube and moistened with 2 drops (only) of concentrated H_2SO_4. The mixture is fused for 1 min. After cooling excess 10 per cent NaOH solution is added. A red solution of phenol phthalein is produced. Other phenols produce other characteristic colours.

(*vi*) *Oxidation*

When alcohols are oxidized by potassium dichromate-conc. H_2SO_4 mixture the red dichromate ion is reduced to the green chromic ion. ($Cr_2O_7^{2-} \rightarrow Cr^{3+}$).

Five drops of ethanol are treated with 2 cm^3 10 per cent potassium dichromate and 5 drops concentrated H_2SO_4. On gentle heating the red coloration changes to green.

Phenols are oxidized to a brown complex mixture.

(*vii*) *The iodoform reaction*

Many alcohols give the iodoform reaction (see aldehydes and ketones) because they are oxidized in the process to the corresponding carbonyl compounds which then undergo halogenation and hydrolysis.

Five drops of ethanol are treated with 5 cm^3 of iodine solution. 10 per cent NaOH is added dropwise until the solution is pale yellow in colour. The mixture is heated for 1 min in a boiling water bath and cooled. Iodoform crystallizes out.

N.B. Heating is necessary for this reaction but is not required for the iodoform test with aldehydes and ketones.

3. Carboxylic acids

Carboxylic acids all have the group — COOH.

(*a*) *Solubility in* NaOH

All carboxylic acids are soluble in 10 per cent NaOH solution.

TABLE 5.11. Tests for Carboxylic Acids

Test	*Formic*	*Acetic*	*Oxalic*	*Succinic*	*Lactic*	*Tartaric*	*Citric*	*Benzoic*	*Salycilic*	*Phthalic*	*Cinnamic*
Soda lime	H_2	CH_4	H_2	C_2H_4	Smell of burnt sugar			Benzene	Phenol	Benzene	Benzene
$FeCl_3$	Red	Red	Pale yellow	Pale brown	Yellow	Yellow	Yellow	Pale brown	Purple	Pale brown	Pale brown
Tollen's	+	—	—	—	+	+	—	—	—	—	—

TABLE 5.12. Tests for Aromatic Amines

Test	*Aniline*	*o-Toluidine*	*m-Toluidine*	*p-Toluidine*	*1-Naphthyl-amine*	*2-Naphthyl-amine*	*Diphenyl-amine*	*Monomethyl-aniline*	*Dimethyl-aniline*	*Triphenyl-amine*
Isocyanide	+	+	+	+	+	+	—	—	—	—
Nitrous acid	← Diazotizes (Red dye with alkaline β-Naphthol) →						← Yellow nitrosamine →		Green nitroso compound with alkali	—
Peroxidase	Purple turning brown	Green turning deep blue	Deep red-purple	Orange turning blood red	Opal-escent deep blue	Brown	Purple	Blue	Yellow turning green	Pink on prolonged standing

(b) Solubility in Na_2CO_3

Carboxylic acids dissolve in Na_2CO_3 solution with the evolution of CO_2. Although some phenols are soluble in Na_2CO_3 there is no evolution of gas.

(c) Ester formation

1 cm³ of ethanol is heated with 0·5 cm³ of the acid and 1 drop of concentrated H_2SO_4 for 1 min. After cooling the mixture is poured into a small beaker of water. Many of the esters formed in this way have characteristic odours, e.g. ethyl acetate smells of apples, ethyl salicylate smells of oil of wintergreen.

(d) Ferric chloride reaction

For this test the acid must be neutralized. 0·5 cm³ of acetic acid is treated with ammonia solution until just alkaline to litmus, and boiled until no more ammonia is evolved. The mixture is cooled and a few drops of neutral $FeCl_3$ added (see Phenols above). A deep red colour is produced which turns to a reddish brown precipitate on boiling. Other acids give other colours.

4. Amines

Amines are the only group of nitrogen containing substances described here. Like alcohols they may be primary, secondary or tertiary and may be considered to be based on ammonia:

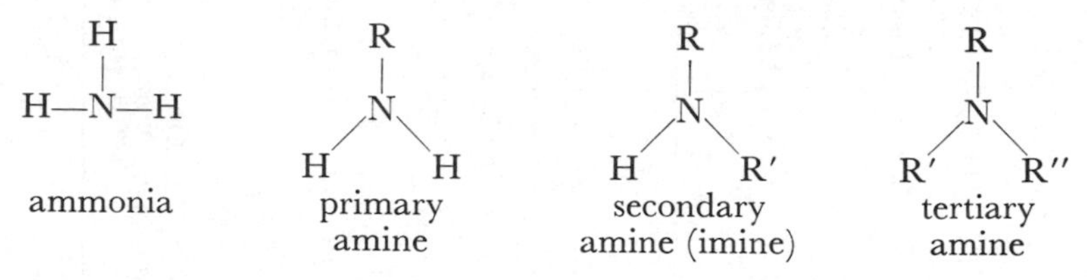

Only the reactions of the simple primary amines will be considered.

(a) The isocyanide (carbylamine) reaction

A few drops of chloroform are added to a few drops of aniline and 2 cm³ alcoholic NaOH solution added. After mixing the tube is warmed gently. Isocyanide is produced which has a particularly foul odour. As soon as the smell is detected the tube must be cooled and excess concentrated HCl added—isocyanide is extremely poisonous. Most primary amines give the isocyanide reaction.

(b) Diazotization

0·2 cm³ aniline is dissolved in 1 cm³ concentrated HCl, diluted with 3 cm³ water, cooled in ice and 5 drops sodium nitrite solution added. This mixture is added to a chilled solution of 2-naphthol in excess of 10 per cent NaOH. A red dye is produced. Diazotization is characteristic of aromatic primary amines.

(*c*) *Action of bleaching powder*

Two drops of aniline are shaken with 5 cm^3 water and a few drops of bleaching powder solution ($CaOCl_2$). A purple colour is produced which rapidly becomes brown. Other aromatic primary amines give different reactions.

(*d*) *Reaction with hydrogen peroxide and peroxidase*

This is a beautifully delicate and elegant test which may be used to distinguish between a large number of aromatic primary amines in very low concentration. The enzyme peroxidase which is one of the reagents used must be active, i.e. the commercial preparation must be dry and in fresh condition. It may be possible to use a suspension of freshly ground turnip or horseradish which are potent sources of the enzyme.

About 0·1 cm^3 aniline is dissolved in 0·1 cm^3 glacial acetic acid and about 5 cm^3 water added. About 0·5 g peroxidase is ground with about 25 cm^3 water in a mortar and filtered. Four drops of the filtrate is added to the aniline acetate solution followed by 1 drop of '20 volume' hydrogen peroxide. A purple colour turning brown on standing is observed. Other amines give a range of colours by which they may easily be identified.

The Systematic Classification of Organic Compounds

The classification of the simple organic compounds discussed containing only carbon, hydrogen, and oxygen may be easily accomplished by the use of Tables 5.6 and 5.7. The former should be followed for water soluble compounds, the latter for water insoluble materials. Amines may be detected by the preliminary tests given in Tables 5.1 and 5.2 and by the simple tests given under the appropriate heading.

Identification of Some Simple Substances

After classification the correct procedure for the certain identification of a compound is to prepare a derivative on which a melting point determination can be performed. This technique is beyond the scope of this volume. It is possible to identify some of the more simple compounds in each group by spot-tests. Some of these are given in Tables 5.8 to 5.12. Others are noted below.

1. *Alcohols*

(*a*) *Methanol* (CH_3OH)

1 cm^3 of methanol is heated gently with 0·5 g sodium salicylate and 1 drop of conc. H_2SO_4. After 1 min the mixture is poured into a boiling-tube containing 5 cm^3 of water. A strong smell of methyl salicylate (oil of wintergreen) is given off.

(*b*) *Ethanol* (C_2H_5OH)

The above experiment is repeated using ethanol and glacial acetic acid. Ethyl acetate has a fruity odour.
N.B. Methyl and ethyl salicylate have a similar odour as do methyl and ethyl acetate. These tests cannot be used to distinguish between the two alcohols.

(*c*) *Benzyl alcohol* $C_6H_5CH_2OH$)

0·5 cm^3 of the alcohol is treated with 0·5 cm^3 concentrated H_2SO_4. A white gelatinous polymer forms with the evolution of heat.

(*d*) *Glycerol* ($CH_2OH{\cdot}CHOH{\cdot}CH_2OH$)

0·5 cm^3 glycerol is gently heated with 1 g of $KHSO_4$ powder. Acrolein is evolved which has a characteristically sickening and pungent odour.

2. *Urea* ($CO(NH_2)_2$)

Urea (carbamide) is of great biological importance. Physically urea is a white crystalline solid, m.p. 132° very soluble in water and alcohol and insoluble in most other organic solvents. Its reactions are mostly those of simple amides ($RCONH_2$), a group which has not been discussed above. There are, however, special tests for urea which are quite specific for the compound.

(*a*) *Biuret reaction*

On heating, urea undergoes the reaction

$$2CO(NH_2)_2 \rightarrow \underset{\text{Biuret}}{NH_2CO{\cdot}NH{\cdot}CO{\cdot}NH_2} + NH_3\uparrow$$

With alkaline copper sulphate biuret gives a purple colour (cf. proteins page 453).

A little urea is heated gently in a test-tube. It liquifies at first and then re-solidifies. The smell of ammonia may be noted and the presence of gas tested with a damp red litmus paper held over the mouth of the tube. The mixture is allowed to cool and the solid dissolved in 3 cm^3 10 per cent NaOH solution. One drop of copper sulphate solution is added to give a purple colour. Too much copper sulphate gives a blue precipitate of copper hydroxide which masks the biuret reaction.

(b) Urease test

The most specific test for urea is the action of the enzyme urease which hydrolyses the compound to ammonium carbonate. Urease in tablet form may be crushed to a powder, or jack bean meal, which is high in urease, may be used.

Five drops of phenol red are added to about 0·2 g urea in 5 cm^3 water. A similar amount of phenol red is added to 0·2 g jack bean meal in 5 cm^3 water. On mixing the contents of the two tubes the indicator turns red as the mixture becomes alkaline.

(c) Hypobromite reaction

This reaction is noted as it is often used in the quantitative estimation of urea.

Alkaline sodium hypobromite solution is added to a solution of urea. A brisk effervescence results, the reaction being

$$NH_2 \cdot CO \cdot NH_2 + 3NaOBr \rightarrow 3NaBr + H_2O + CO_2 \uparrow + N_2 \uparrow$$

Other Compounds of Biological Interest

The analysis of carbohydrates, proteins, lipids etc. are discussed in Chapter 11.

CHAPTER 6

ABSORPTIOMETRY

The most popular methods of analysis are those in which concentrations of substances in solution are measured by the amount of colour they produce or absorb. This popularity is because of the speed at which results are obtained, that in the more complex machines judgement is made instrumentally, and because of the greater sensitivity that can be obtained. It is the latter virtue that makes the absorptiometric method important to the biochemist who often requires to analyse micro-quantities when working with samples from living subjects, animals and plants. It may also effect great economy when working with rare and expensive compounds and elements. Indeed, many of the long accepted volumetric methods have been adapted for use with an absorptiometer. In fact, most substances if not coloured themselves form coloured derivatives, or by reaction with specific reagents may be made to form coloured solutions, the intensity of colour bearing a direct relationship to the concentration of the substance.

NATURE OF LIGHT AND COLOUR

Newton discovered that white light (solar light) was compound and when we view a mixture of all colours in the right proportions we see white light. He called the image of a beam of white light that has traversed a prism the *spectrum* and we now know that this is but a small section of the electromagnetic spectrum (*Figure 6.1*). The nature of the electromagnetic spectrum varies with wavelength from cosmic rays below 0·005 nm to Hertzian waves above $2{\cdot}2 \times 10^5$ nm, the very long waves. Wavelength is measured in nanometres (nm). 1 nm $= 10^{-9}$ metres. The obsolescent units are the Ångstrom (Å) and the millimicron (mμ), 1 nm = 1 mμ = 10 Å. Visible light has wavelengths from 380 nm of the blue to 780 nm of the red.

The colours of the objects about us are due to two causes, reflection and absorption. An image viewed in a mirror is the same colour as the object, all the wavelengths of light being reflected. Paper appears white for the same reason but the lack of an image is due to random reflection. Red paper receives all the wavelengths but reflects and transmits only the red ones, the rest being absorbed. Black paper absorbs all the visible wavelengths. A colourless solution allows all light to pass through it, very little being reflected or absorbed so the intensity of light that emerges is only slightly less than that of the incident light. If a crystal of dye is dissolved

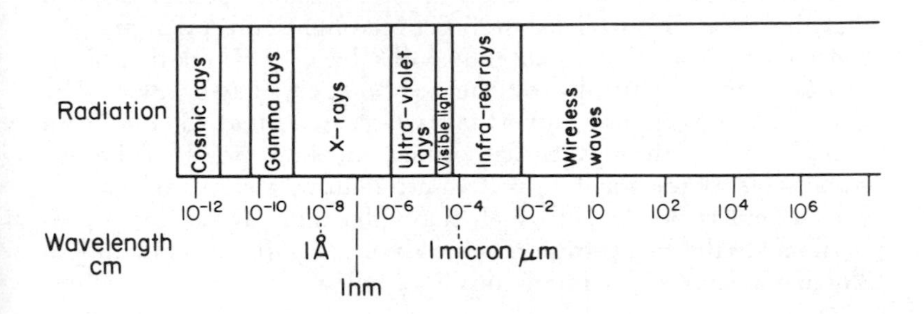

Figure 6.1. The electromagnetic spectrum

in water, the solution appears coloured after white light has passed into it. In other words, it 'filters out' some wavelengths by absorption. Such filters are extremely useful in absorptiometry and optical instruments. They are sometimes in the form of solutions but generally are made of coloured glass or thin films of gelatin sandwiched between two pieces of glass. In absorptiometry the intensity of light is often measured, that is, the more of the absorbing substance present, the less light passes through. Filters can improve the sensitivity of this effect. If a substance giving a yellow solution is analysed by yellow light, the emergent beam will only be a little less intense than the incident beam, even when the concentration of the yellow substance is considerably increased. If blue light is used, a much higher proportion of the light is absorbed. This means that a smaller portion of the yellow substance is detectable and measurable by blue light than by yellow light.

For greatest sensitivity the complementary colour should be used, this is all the wavelengths of light of the spectrum excluding those transmitted by the substance under test. If these two colours were to be projected on to a screen and focused at the same spot, white light would be seen. If two filters that were complementary to each other were superimposed in a beam of white light, they would appear black.

Theory

A beam of light passing through a coloured solution is partly absorbed, the amount of absorption will depend on the number of coloured ions or molecules in its path. If the number of coloured 'particles' in the solution is doubled, all other things being equal, twice as much light will be absorbed. If this increase in colour is due to double the concentration of the substance it will be seen that there is a direct relationship between light absorption and the concentration of coloured substance dissolved. This relationship is expressed mathematically by the combination of two laws. The first is the Bouguer or Lambert law which states that when a beam of light passes through an absorbing medium (solid or liquid) equal stages of this medium will absorb light by a constant amount. The second is Beer's law which states that light absorption is proportional to the concentration of the absorbing substance in solution. These are expressed mathematically as follows:

$$\log_{10} \frac{I_0}{I} = Ecl$$

Where I_0 = intensity of incident light
I = intensity of transmitted light
E = absorption coefficient
c = concentration
l = depth of absorbing layer.

The law is truly valid only for monochromatic light.

It is usually possible to find a concentration range in most methods over which this law holds, but there are exceptions, due to the nature of the chemical reaction. Other variations are due to limitations of apparatus, method and technique and these should be examined when there is apparent disagreement with this law.

Colour Matching methods

Many methods have been devised over the past 100 years, some are still in use, but the methods to be described will be confined to those having practical use today. Colour matching methods are broadly divided into three techniques, colour comparators employing reagent standards for comparison, comparators employing artificial standards and the balancing method.

The first technique requires a set of Nessler tubes. These are made of glass, have flat bottoms and are of uniform bore (*Figure 6.2*). Better ones have optically flat discs fused to the bottom.

They have either one or two graduations generally at 50 and 100 cm³. The tubes are best mounted in a rack which has a white tile or an illuminated strip of flashed opal for the tubes to stand on. The colour of the unknown solution is compared with that of a series of standards.

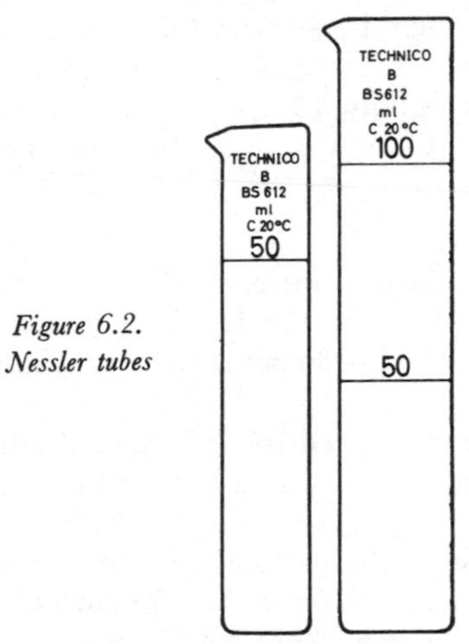

Figure 6.2. Nessler tubes

Method

First a set of standards covering a wide range is produced. The colour intensity of the unknown will be seen to come between two of these. A further set of standards covering the range between these two previous standards are made and a careful comparison with the unknown will then ascertain its concentration more accurately.

The following experiment will illustrate the technique: It is possible to estimate inorganic phosphorus by the method of Fiske and Subbarow* using Nessler tubes. A series of tubes are prepared with the solutions listed below, making a set of phosphate standards ranging from 1–5 mg.

Nessler tubes	1	2	3	4	5
Phosphate standard solution	2·5	5	7·5	10	12·5
Milligrammes of phosphorus	0·1	0·2	0·3	0·4	0·5
Water to 50 cm³ mark					
Molybdate solution	10	10	10	10	10
Aminonapthol sulphonic acid solution	4	4	4	4	4
Water to 100 cm³ mark. Mix					

* *J. biol. Chem. 66* (1925) 375.

A blue colour develops in each tube and the intensity is proportional to the quantity of phosphorus present. A phosphorus solution of unknown concentration prepared as above in a Nessler tube may be compared with the standards. Using even illumination the tubes are viewed from above through their depth. It will quickly be seen between which concentrations of standards the unknown falls. A second set of standards is then produced using the same technique as before rising by 0·02 mg of phosphorus. A careful examination will then show the accurate concentration of the sample.

Reagents

Phosphate standard solution: Dissolve 0·1755 g of potassium dihydrogen phosphate in 0·05 M sulphuric acid, make up to 1 dm^3 in a volumetric flask. 10 cm^3 of this solution contains 0·4 mg of phosphorus.

Molybdate solution: Dissolve 25 g of ammonium molybdate in 200 cm^3 of water. Add this to 300 cm^3 of 5 M sulphuric acid and dilute to 1 dm^3.

Aminonaphtholsulphonic acid reagent: Dissolve 0·5 g of 1,2,4-aminonaphtholsulphonic acid in 195 cm^3 of 15 per cent sodium bisulphite solution (clear) then add 5 cm^3 of 20 per cent sodium sulphite. Mix, if not fully dissolved add more of the sodium sulphite solution. Store in a dark bottle.

With some reactions it is possible to find the approximate concentration by having the standard solution in a burette and slowly adding it to the colour reagents until, when viewed through the depth the colour matches that of the unknown.

For routine purposes the making up of a series of standards is time consuming. This objection may be overcome by using an instrument employing sets of artificial standards such as the B.D.H. Lovibond Nesslerimeter (*Figure 6.3*). Two Nessler tubes, one a reagent blank and the other the unknown, are placed in the black bakelite box. The requisite disc containing a set of Lovibond colour standards is placed above the tubes, positioned over the reagent blank and revolved until, when viewing from above, a colour match is obtained. The figures beside this standard are the concentration of the unknown. A similar instrument is the Lovibond comparator. In this the colour match is made by viewing through the sides of the tube. The latter are matched test-tubes, or cuvettes when greater accuracy is required. The illumination for both instruments must be an even diffuse light from a north window or from a lamp shielded by a piece of flashed opal glass.

Assay methods are supplied by the makers and these must be strictly adhered to. They cover a wide range of chemical and biochemical techniques.

The most popular colour matching method is the balancing technique. The method is more easily explained by the following experiment using the unknown and one standard from the phosphorus experiment and by constructing the simple apparatus in *Figure 6.4*. The exact volumes are now not important as the

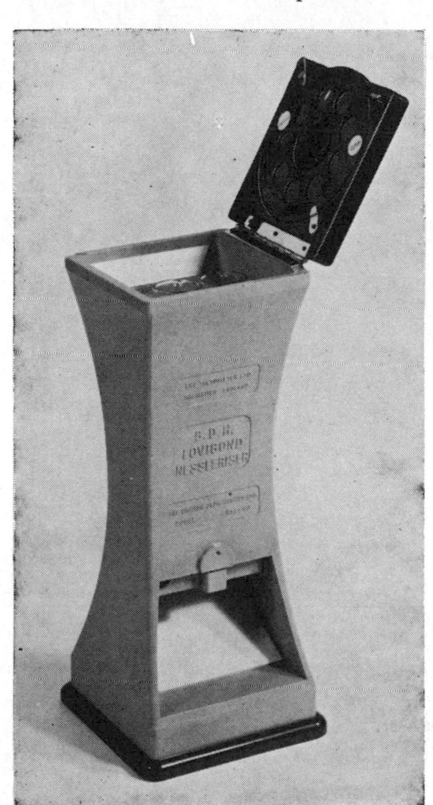

Figure 6.3. The Lovibond Nesslerimeter

[By courtesy of Lovibond Ltd]

colour match is going to be effected by varying the depths of the solutions. Tube 2 should contain the solution with the deeper colour, it does not matter if it is the standard or the unknown. It will be convenient if the solution in Tube 1 is adjusted to a depth of 100 mm, but this is not essential. Apply a mild vacuum to the separating funnel and with a finger placed over Tube *A* open the tap and draw the liquid up until, when viewing from above, there is a colour match. If too much liquid is removed, remove the finger from A and allow the solution to run back. When the colours are the same, close the tap. Using a millimetre rule measure

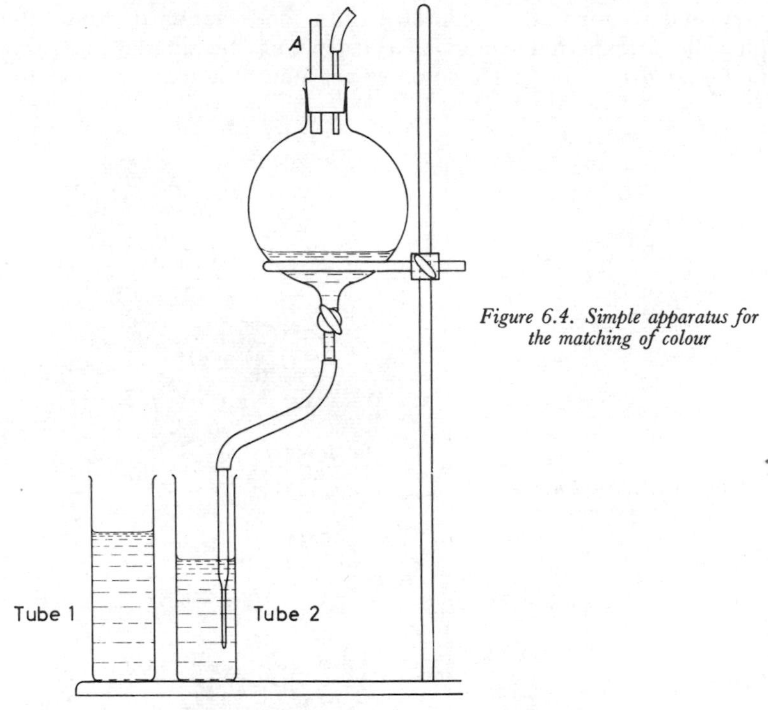

Figure 6.4. Simple apparatus for the matching of colour

the depth of the solutions. We know from Beer's law that log I/I_0 or absorbance is Ecl, therefore for Tube 1 it is Ec_1l_1, and Tube 2 it is Ec_2l_2. As the tubes are adjusted to give the same absorbance and as E is the same for both, it will be seen that $c_1l_1 = c_2l_2$. If Tube 1 is the unknown we arrive at the equation

$$c_1 = \frac{c_2 l_2}{l_1}$$

If, as suggested, l_1 was 100 mm and if the standard c_2 was 0·4 mg/100 cm³ and if by measurement l_2 was found to be 69 mm, then the unknown

$$c_1 = \frac{0{\cdot}4 \times 69}{100} = \frac{27{\cdot}6}{100} = 0{\cdot}276 \text{ mg/100 cm}^3$$

This type of experiment may be more easily performed using Hehner's cylinders. These are cylinders graduated in millimetres that have a tap at the side. The balancing of two solutions is performed by opening the tap of the cylinder containing the denser solution until a match is obtained.

Dubosq designed an absorptiometer (a colorimeter) that used this relationship between depth of solution and concentration. There are several instruments now on the market of similar design, one being illustrated (*Figure 6.5*).

[By courtesy of James Swift and Son Ltd]

Figure 6.5. The Dubosq colorimeter

The solutions are contained in two glass cells that are dished at the top, these are placed on two movable brackets fixed to pieces of racking held in guides. The brackets are raised and lowered by turning a pinion wheel in mesh with the racking. The wheel is calibrated in millimetres with a vernier for greater accuracy and representing vertical movement of the cells. Positioned centrally over these are two fixed glass plungers, above which is a prism and lens optical system that directs light from both plungers into the eyepiece, illuminating respectively two halves of a circular field. Light is supplied from a filament lamp made diffuse by a flashed opal or mottled glass screen; some instruments have provision for coloured filters to be superimposed over the light source.

The bottoms of the cells and plungers are of optically flat glass and are cemented in position.

To prevent stray light entering, the cells may be of opaque glass, or a cowling covering cells and plungers may be supplied.

Before using the instrument the optical parts must be examined for cleanliness. A lens tissue usually suffices to remove any smears.

Operation of the Dubosq Colorimeter

Place both cups into position and raise them carefully until the plungers touch the bottoms of the cells. The scales should now read zero. If not, loosen the set screw on the scales, make the adjustment to zero, then tighten them again. Rinse both cups with a portion of the standard solution, then fill them both with this solution to the top of the narrow part of the cup. Raise the cups until the ends of the plungers are immersed and adjust the scales to the same reading. Look through the eyepiece and if the illumination is uneven, adjust the lamp by revolving it or moving it horizontally in the lamp housing, then tighten the locking screw.

The instrument is now ready for use. Replace the standard solution in one cup with a solution of an *unknown* and set this at a convenient depth, say 15 or 20 mm. If a cowling is supplied, this is now put into position. Raise or lower the *standard* until the field is evenly matched and note the depth indicated on the scale. This is repeated several times from both directions and an average figure taken. The final result may be obtained very rapidly, the necessary readings being taken within one minute.

The calculation is the same as before $c = \dfrac{c_2 l_2}{l_1}$. Using the standard as the variable makes the calculation simpler l_1 being a round figure. The colour developed must obey Beer's law. If this is not known or a new assay method is being developed a test must be made. This is done by taking some of the coloured solution and making a series of dilutions. Regard one solution as a standard and estimate the concentration of the others. Draw a graph, plotting the curve of the actual concentration against a curve estimated by use of the Dubosq. For example, make a series of dilutions of a dye, e.g. methylene blue having the following concentrations:

$$0{\cdot}0025,\ 0{\cdot}005,\ 0{\cdot}01,\ 0{\cdot}02,\ 0{\cdot}04\ \mathrm{g/dm^3}$$

Take the 0·01 g/dm³ dilution as a standard, put it in one of the cups and this time keep it at a constant depth. Place the other solutions

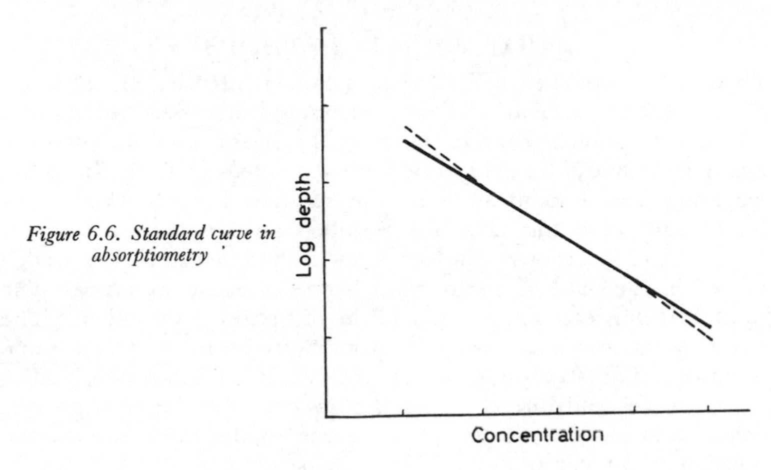

Figure 6.6. Standard curve in absorptiometry

in the other cup in turn and note the depth at balance. Draw a graph of depth against concentration (*Figure 6.6*). Calculate what the depth should be providing Beer's law holds.

$$c_1 = \frac{c_2 l_2}{l_1}$$

$$\therefore \qquad 0{\cdot}02 = \frac{0{\cdot}01 \times 15}{l_1}$$

$$\therefore \qquad l_1 = \frac{0{\cdot}01 \times 15}{0{\cdot}02} = 7{\cdot}5\ \mathrm{mm}$$

Draw a second curve using these calculated depths. If they overlap, obviously Beer's law is obeyed throughout that range of concentration and the Dubosq may be used without fear of error.

If the curves diverge at any point as in *Figure 6.6*, the analytical method must be modified so that the colour density of the unknown falls on the part of the curve that does obey Beer's law.

When using the Dubosq, the unknown should not have an optical density of less than half that of the standard or more than double. In all visual colour matching methods it is unwise for more than one person to take readings during an experiment, as each individual will have his own idea of what is an exact colour match. This is one of the disadvantages of visual comparison methods, the possibility of human error is overcome in instruments that employ photoelectric cells as an 'eye'.

PHOTOMETER METHODS

Photometers are instruments that measure intensity of light. If the intensity of a beam of light is measured, a vessel containing a coloured solution is superimposed in the beam, and the intensity again measured; by subtraction of one intensity from the other we know the amount of light the solution has absorbed. An instrument using this principle is called an *absorptiometer*.

The *visual photometer* employs a calibrated neutral glass wedge placed in the path of one of twin beams of equal intensity. The solution being examined is placed in the path of the other. The wedge is moved until the light from both beams is of the same intensity. The absorption of the wedge and the solution is similar and may be read directly from the wedge. The latter is grey in colour and steps must be taken to ensure the same hue as the solution under comparison. This can be done by light filters.

If light intensity is measured with the use of a photoelectric cell, the instrument is called a photoelectric absorptiometer.

Photoelectric Photometer Methods

These instruments are also called photoelectric filter photometers as filters are superimposed in the light path for greater sensitivity as described earlier, and a photoelectric cell is used to measure light intensity.

Before describing the instruments an expansion of the theory and terms used is necessary.

The Beer-Lambert law is expressed as follows:

$$\text{Log}\,\frac{I_0}{I} = Ecl$$

This is called the absorbance (A)

If a substance in solution obeys Beer's law, and the absorbance is determined for a series of standards, they will produce a straight line graph.

Transmission is the ratio $\dfrac{I}{I_0} = T$

This will be less than 1 and is often expressed as a percentage. It will be seen that if

$$A = \log \frac{I_0}{I} \text{ and } T = \frac{I}{I_0}$$

$$A = \log \frac{1}{T} \text{ or } - \log T$$

When transmission is 100 per cent, absorbance is 0.

Transmission is measured in terms of the incident light and transmitted or emergent light, but in actual fact, no instrument actually measures the incident light. If it is measured and used in the calculation, the absorbing solution will appear more concentrated than it really is, due to the absorption of the glass and reagents.

It is more practical to measure I_b or the transmitted light of a reagent blank. Then $T = \frac{I}{I_b}$.

The concentration of an unknown may be found if Beer's law is applicable by estimating the absorbance (A) of a standard and the unknown using the same light and light path. As A is directly proportional to the concentration, then:

$$\frac{A_1}{A_2} = \frac{C_1}{C_2}$$

Therefore, if the unknown concentration is C_1

$$C_1 = \frac{A_1 \times C_2}{A_2}$$

Most absorptiometers are calibrated in absorbance and often in percentage transmission as well.

Light Filters

Two types of light filters are found. First is the chemical filter already mentioned on page 223. These act by absorbing all the unwanted light and heat allowing only a small proportion of radiation to pass. As most light sources emit wavelengths well into the infra-red this has the effect of heating the filter and can cause fading and chemical breakdown.

Another type of filter that does not suffer this disadvantage is the interference filter. This comprises a thin film of magnesium fluoride whose thickness is one half the wavelength of the light required. This film is sandwiched between two semi-transparent layers of silver by the technique of vacuum deposition. Most of the incident light suffers 'interference' and is totally reflected, and for this reason does not readily absorb heat. These filters can be made to pass a bandwidth of only 6 nm.

Two groups of filters are generally supplied by the manufacturers. The first is made by Ilford called Spectrum Filters, and they cover the visible range providing near monochromatic light. A combination of filters may be used, for example, 602 with a transmission peak of 470 nm and 603 of 490 nm when combined provide a transmission peak at 480 nm, but this will reduce the percentage transmission considerably. In fact, it may be found that the spectrum filters absorb too much light with certain analytical procedures, not allowing a full scale deflection of the meter. In this case, the Chance filters may be more useful as they have a higher percentage transmission, but also transmit over a wider portion of the spectrum.

Photoelectric Cells

Several types of photoelectric or photo-emissive cells are used in absorptiometers. Filter photometers use the barrier layer cell which only requires a simple electrical circuit and is robust. It consists of a steel disc (*Figure 6.7*) (*a*) on which is applied a layer of pure

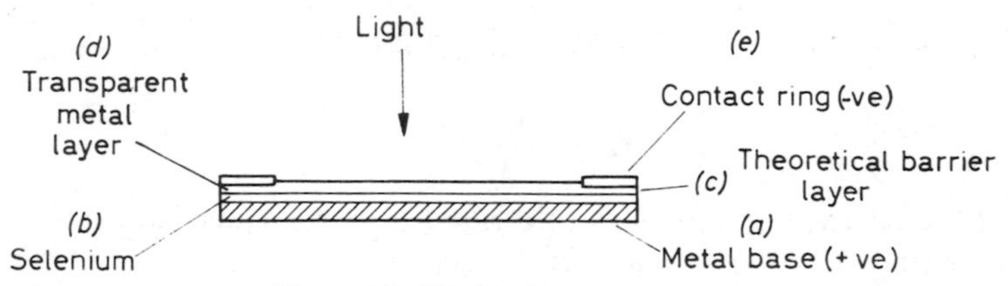

Figure 6.7. The barrier layer cell

molten selenium (*b*) this is heat and pressure treated to give a thin even coating. Over this is a layer of metal (*d*) that is so thin that it is transparent. Between these two coatings is a 'theoretical barrier layer' (*c*) through which electrons flow in one direction when light falls upon the cell. A ring of low melting point metal is sprayed on forming a negative contact (*e*) and finally a protective lacquer is applied. Enough current is produced to operate a microammeter in series.

Another type of photocell is photoresistive, that is, its resistance varies with the intensity of light falling on it. The material used is cadmium sulphide. This type is found in simple densitometers especially those used in photography. They are very sensitive to heat and are used in some spectrophotometers (see page 242) that operate in the near infra-red region. Temperature changes may cause errors and some instruments use a second cell to compensate for background radiations. Both barrier layer and photoresistive types have a slow recovery time and are not suitable for use with a recorder where a fast response is needed.

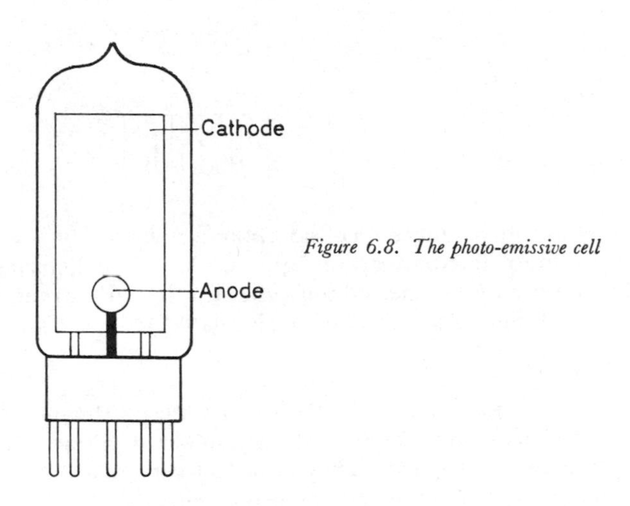

Figure 6.8. The photo-emissive cell

The more sensitive instruments use the photo-emissive cell. This looks like a simple radio valve and has two electrodes (*Figure 6.8*). The cathode is usually a metallized coating on the inside, an area being left untreated as a window for light to enter. The anode is a centre rod of wire. The glass envelope is either evacuated or filled with inert gas. If light falls on the light-sensitive cathode, electrons are emitted which flow to the anode being held at a positive potential. In a vacuum tube the current produced is proportional to the intensity of light. This current is then amplified.

In some modern instruments a light detector called a photomultiplier is used. This is based on the fact that electrons attracted to an electrode at a higher potential liberate an increased number of electrons from it. If this phenomenon is allowed to happen in

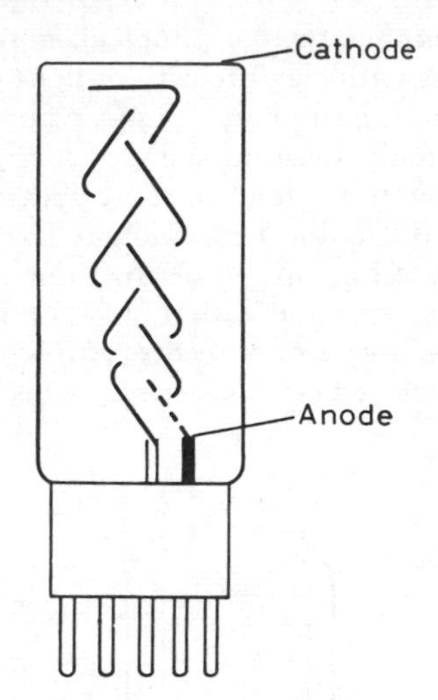

Figure 6.9. The photo multiplier

several consecutive stages a considerable amplification is achieved, thereby minute fluctuations in light intensity are detected. The amplification may be one million times the initial current, but the output is still only a few milliamps (*Figure 6.9*).

Filter Photometer

There are a number of reliable instruments on the market, but it is proposed only to give details and method of use of one of these due to the limitations of space. They all employ a small tungsten lamp of the torch bulb type, a barrier layer cell and a microammeter. The main difference between the types of instruments is the method of control of illumination, types of cell and cell housing and in the positioning of the filters.

The E.E.L. 105 model A portable colorimeter is a robust compact instrument that has given good service to the author over many years (*Figure 6.10*).

The schematic layout and the wiring diagram is shown in *Figure 6.11*. The components of the instrument are exposed by loosening two knurled screws and removing the cover. The light source is a 2·5 V, 0·3 A, M.E.S. bulb which is easily replaced and two spares are provided in clips mounted on the frame. The life of the bulb is very long as it is underrun. The electric supply is 2 V provided by a constant voltage transformer, or a battery or accumulator (see *Figure 6.11*).

The photocell is housed in the right-hand side and is 22 × 40 × 2 mm. It is connected directly to the meter, no power supply being necessary. The meter has an absorbance scale. Special matched test-tubes are supplied 80 × 12 mm and 90 × 6 mm. These are

[By courtesy of Evans Electroselenium Ltd]

Figure 6.10. The E.E.L. portable colorimeter

supplied in matched sets and when re-ordering the code number must be quoted so that identical tubes will be supplied.

The tubes are marked and when inserted into the light path this mark must coincide with another on the colorimeter. A hinged cover is provided over the tubes to prevent stray light entering.

Some instruments also use cells or cuvettes of rectangular section, the glass sides being fused together preserving the optically flat surfaces. These are more accurate but have the disadvantage that reactions cannot be carried out in them, thereby eliminating a stage in the analytical method.

Method of Use

The correct filter must be selected and if the Spectrum filters

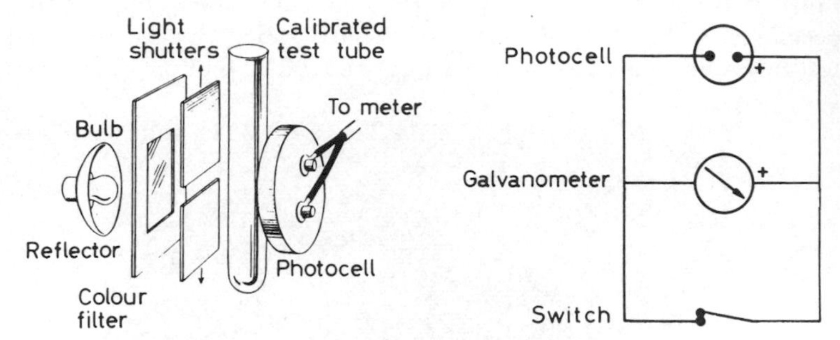

Figure 6.11. Layout of the E.E.L. portable colorimeter

are used the most suitable may be found by trial and error. First prepare the solutions, including the standard and a reagent blank and develop the colour. A reagent blank is a solution that has all the reagents except the substance under test and has been put through the analytical procedure. For example, if ammonia is being estimated by Nessler's method the reagent blank would contain Nessler's solution which is the colour producing reagent, and water instead of ammonia. Turn on the instrument and allow it to stabilize for a few minutes before use. Place the tube containing a reagent blank into the colorimeter and set a filter into the space provided. Revolve the wheel that operates the light shutter until the meter reads zero. Replace the blank with one of the prepared standards and note the absorbance. Replace the reagent blank, and the meter should return to zero. If not, repeat the operation until it does. Replace the filter with another and again balance the meter to zero with the blank in position. Change the blank for the standard previously used, and again note the absorbance. Repeat this procedure with the other filters. The filter that produces the highest absorbance, that is the maximum absorption for the given concentration, will provide the most sensitive analysis.

With the chosen filter in position place the tube containing the unknown in the absorptiometer and note its absorbance. If Beer's law is operative its concentration may be worked out as follows:

Concentration of unknown =

$$\frac{A \text{ of unknown} \times \text{concentration of standard}}{A \text{ of standard}}$$

If many estimations are to be made it may be worth plotting a graph of the concentration of a series of standards against A. This may be used indefinitely provided the same technique and the same instrument is used, and the graph is checked with the occasional standard. If the reaction does not follow Beer's law a graph must be plotted.

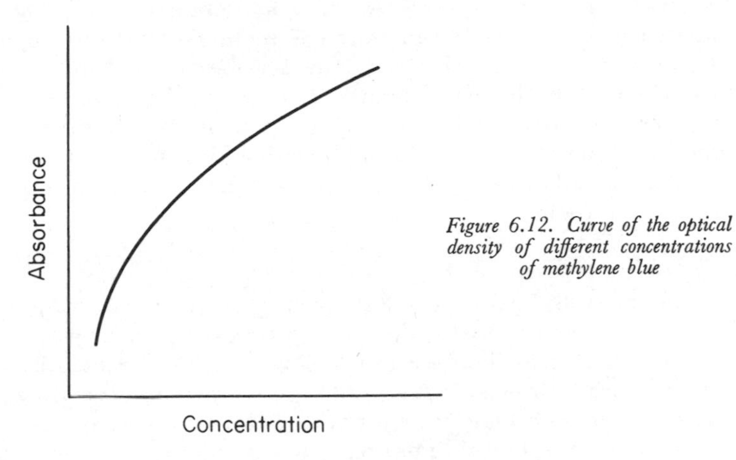

Figure 6.12. Curve of the optical density of different concentrations of methylene blue

For example, if you read the A of methylene blue concentrations as described earlier you will have the curve illustrated in *Figure 6.12.* This again would not have to be repeated when using the same instrument.

MEASUREMENT OF TURBIDITY

In certain instances a precipitate is more readily obtained than a coloured solution when seeking a means of analysis. Many gravimetric methods are used simply because no accurate volumetric or absorptiometric method is available. It is possible to circumvent many of the tedious gravimetric techniques of weighing and preparation of crucibles by estimating the concentration of a suspension by the amount of light it absorbs or by the amount of light it scatters.

The former method, *turbidimetry*, requires only an absorptiometer. The E.E.L. may be used as described, care being taken to ensure that light absorption is due to the suspension and not to variations in colour density of the standard and test solutions. This may be checked by centrifuging the solutions to remove the precipitate and re-reading them in the absorptiometer. Only *very dilute* solutions may be used otherwise error will arise due to light being scattered, or refracted, giving a higher result.

When the precipitates are too dense for this method the scattered light may be used as a measure instead. This is called *nephelometry*. The solutions must still be very dilute and capable of transmitting at least 10–20 per cent of the light.

The instrument used is a nephelometer. One type is similar in construction to the photoelectric absorptiometer, the essential difference being that the photocell is mounted at right angles to the incident ray. The Dubosq may also be adapted for use as a nephelometer by having a light source beside the cells, and using plungers with opaque barrels, and cells with opaque discs at the bottom. Care must be taken in all types of instruments to exclude stray light and with the Dubosq the readings should be taken in a darkened room.

General Technique

The most difficult procedure in nephelometry is to prepare reproducible suspensions. The particles must be uniform throughout the analysis as light scattering power varies considerably with particle size. In this respect amorphous precipitates are to be preferred to crystalline suspensions, the size and form of the latter being difficult to control. The great disadvantage of most amorphous precipitates is the rapidity of sedimentation. This is overcome in some instances by using gums such as gum ghatti and gum arabic or gelatin.

The quality of the light is important and generally the shorter wavelengths are more scattered than the long. Filters should be used such as the Chance OR types to remove wavelengths that are absorbed readily by the solution.

The standard or series of standards must be prepared at the same time and undergo the same procedure as the test solutions. In some absorptiometric methods it is possible to omit, for example, the heating stage when preparing standards, but this would invalidate results in nephelometry as the suspension would certainly be of a different type.

Care must be taken to ensure no variation in pH, temperature, concentration of reagents and developing time.

Silver chloride and barium sulphate precipitates may be measured and some protein suspensions. It is especially useful in following the growth of micro-organisms.

FLUORIMETRY

The phenomenon of fluorescence may be employed for analytical purposes. Fluorescent substances have the ability to absorb light, usually of the lower blue wavelengths, and emit light of a higher wavelength. The fluorescent light emitted is maximal with the wavelength of light that is most highly absorbed, the absorption spectrum and the fluorescent spectrum being mirror images of each other.

The intensity of fluorescence will be found to be proportional to the concentration of fluorescent substance when in very dilute solution, and when absorption of the incident light obeys Beer's law. At high concentrations 'quenching' occurs due to the lower transformation of radiant energy, and to the absorption of the incident light as it passes through the first layer of solution.

Fluorimeters

These instruments are similar in design to nephelometers, the fluorescence being observed at right angles to the beam of light. The main difference is the light source, which is usually a high-pressure mercury vapour lamp. The Duboscq may again be adapted, but if this is done by the technician care should be taken to shield the lamp efficiently as the u.v. emission may damage body tissue.

Most absorptiometers may also be adapted. Modern fluorimeters usually employ photomultiplier tubes as detectors as they are more sensitive. Two sets of filters are used. The first narrows the band of the incident light, and Wood's glass is often used transmitting at 365 nm. The second set is placed in front of the photocell to remove stray reflected u.v. and, in some instances, fluorescence from other substances in solution. Ilford Spectrum filters may be used.

Evans Electroselenium Ltd. market a fluorimeter having two photomultiplier tubes in opposition, balancing a sensitive galvanometer which is adjusted by a calibrated potentiometer. Two equal beams of light emerge from the lamp house at right angles to each other. One of these beams passes through a standard solution and the resultant fluorescent light is collected by one photomultiplier tube. The other beam passes through the test solution and the radiation from it is collected by the second photomultiplier tube.

General Technique

It is possible to detect some fluorescent substances in very low concentrations 0·1 $\mu g/cm^3$ or less. It is important to purify the substance under examination as traces of fluorescent contaminants may cause errors. Other contaminants that are colourless may absorb u.v. and others if in colloidal suspension may absorb the test substance itself. Turbid solutions should not be used as emitted fluorescent radiation will be enhanced by scattered light. Apart from the above special precautions normal absorptiometric techniques may be employed.

SPECTROPHOTOMETERS

In the instruments so far described the width of spectrum band used has depended on the type of filter used. For a single filter this is usually not less than 80 nm. This disadvantage is overcome in the spectrophotometer by using the ability of a prism or diffraction grating to produce a continuous spectrum. A narrow part of this spectrum can then be allowed to pass through the test solution on to the photocell. In effect, the spectrophotometer is a refined filter photometer, but this is not its only use as the absorption spectra for pure substances may be obtained, and also other physical and chemical properties.

Spectrophotometers consist of a light source, a monochromator (that is, a device to produce a monochromatic beam of light from any point on the spectrum), a vessel containing the solution and positioning it in the light beam, and a detector with its amplifier. The light source used for the visible spectrum is the tungsten lamp, usually a single filament car headlamp bulb. This gives a peak energy output at about 950 nm and covers the range 320–3000 nm. The electric supply must be constant as the radiant energy must not vary. This is provided by an accumulator or car battery. A constant voltage transformer or stabilized power pack is used in most instruments. The source of u.v. light is the hydrogen or deuterium discharge tube which supplies light of 180–350 nm. A stabilized power supply is essential for these lamps. The tube has a quartz window and when adjustments are made this should not be touched. If a more intense source is required the mercury arc or xenon arc lamp is used.

Infra-red radiation is supplied by a Nernst filament, a rod of oxides of zirconium, yttrium and thorium, which requires a heater winding for starting, the operating temperature being about 1350°C. A stabilized voltage is required.

A description of the infra-red spectrophotometer will be given later as there are basic differences in design.

The visible and u.v. type employ either a glass or quartz prism or a diffraction grating in the monochromator. Some instruments use both prism and grating to improve resolving power.

In general, the light is focused on a slit and thence to a collimating mirror which produces a parallel beam. This is directed into a prism and is diffracted. The back surface of the prism is mirrored and the diffracted beam is therefore further diffracted as it passes back (through the prism) to the collimating mirror. This in turn focuses the spectrum on to another slit, or the upper part of the first. The selected light then passes through the cells and on to the detector. The slit opening is carefully controlled and is usually calibrated up to 2 mm. At the narrowest

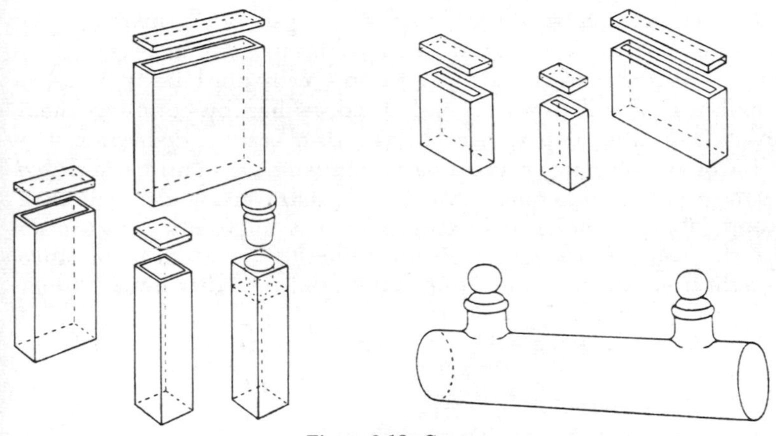

Figure 6.13. Cuvettes

slit, width adjustments of 0·001 mm may be made. The rectangular sectioned cells or cuvettes are of glass or silica, the sides being fused together. This method of construction preserves the optically flat surfaces, and allows the use of most solvents and cleaning fluids. A choice of cells allows the path length of light through the liquid to be from 1 to 100 mm. Micro-cells are available making possible maximum path length with the minimum use of precious solution. It is possible with these to use only 0·5 cm^3 and still have an effective path of 10 mm (*Figure 6.13*). Modifications have been described (Lowry Bessey, 1946) that make it possible to examine the absorption on as little as 0·05 cm^3 of solution. When working in the u.v. region, silica cells must be used as glass is not transparent to u.v. light.

Care of Cuvettes

The accuracy of measurements will depend to a large extent on the condition of the cells. The optical faces must be scrupulously clean and not damaged in any way. When handling them, the optical faces should not be touched by the fingers, as a slight film of grease and dirt may be deposited that could cause considerable error by absorption of light. Care should be taken to avoid scratching, a good rule being to lay the cells down on the semi-opaque faces and, when not in use, to put them in the manufacturer's cases after careful cleaning and drying.

Cells should not be cleaned with chromic acid mixture as traces of chromate will cause enhanced absorption in the u.v. region. Nitric acid or aqua regia (HCl/HNO_3, 3:1) should be used if strong measures are required. On no account may abrasive soaps or soap powders be used. After cleaning, cells must be thoroughly rinsed with distilled water and dried, either by standing them inverted on filter paper in a covered dish or in a desiccator.

In use the cells are placed in a holder which fits on to a movable carriage in the cell compartment. Light traversing the cells then passes into the photocell housing which is light-tight apart from the entry slit. Two photocells are generally used, one is sensitive to light from about 180 nm to 600 nm, and the other from 500 nm to 1000 nm.

The photocells are quickly changed by operating a slide or lever.

Use of a Spectrophotometer

In filter photometry light undergoing maximum absorption was used and the filters giving this light were found by trial and error.

In spectrophotometry if the wavelength giving greatest absorption is unknown an *absorption curve* must be obtained, using a standard solution. This is done by finding the absorbance at various wavelengths and plotting a graph of A against wavelength. It will be apparent from the diagram (*Figure 6.14*) that the best wavelength to use for maleic acid is 210 nm. Many substances have more than one peak. This may be an advantage if it is known that a contaminating substance present absorbs light at the wavelength usually used in the analysis. Another absorption peak could then be used instead. In fact, it is often possible to analyse two compounds in the same solution providing absorption peaks are different.

A pure substance will have its own characteristic absorption curve and this may be used in identification. Reactions may be followed

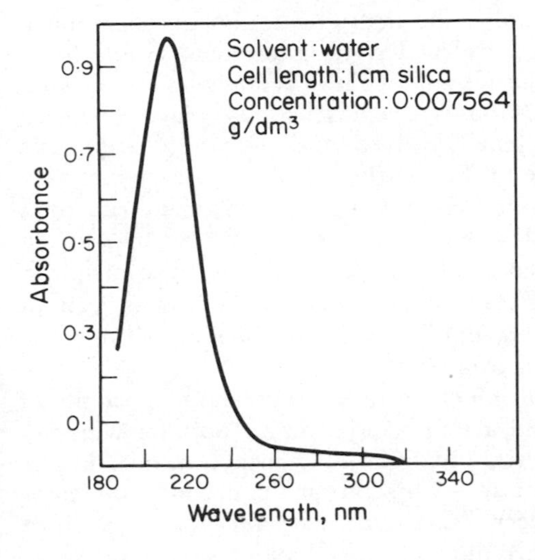

Figure 6.14.
Absorption curve for Maleic acid

in the spectrophotometer using an absorption peak of the product, and the rate of reaction found, provided it is slow enough. This technique is especially useful in following enzymic reactions and some instruments have a constant temperature housing attachment to make this possible.

Operation

The instrument is turned on and allowed to warm up for a few minutes. The illumination must be stable before starting. Most instruments have a 'dark current' control which allows the balancing of the electronics with no light entering the photocell. This is done after the wavelength has been selected. With a reagent blank in the light path the galvanometer or potentiometer is set at zero. The blank is then replaced by the test solution and the A or transmission is noted. The instrument should be zeroed with the blank in position with each successive wavelength. This is a general outline, but the maker's instructions should be strictly followed. Most manufacturers supply a servicing manual with detailed directions for fault finding.

Infra-red Spectrophotometry

It has been seen that pure substances have their own individual absorption spectra in visible and ultra-violet light. This absorption phenomenon carries on into the invisible longer wavelengths of the infra-red. Unfortunately, the instrumentation becomes much more complicated as glass and water, transparent to visible light, have strong absorption bands in the infra-red region, in fact all liquids have absorption bands. The last twenty years have seen the solving of the problems involved and now fully automatic recording instruments are on the market.

Infra-red light is obtained from the Nernst filament run at a temperature of about 1300°C which is usually protected by a water jacket. Other sources used are the *Globar*, which is a rod of silicon carbide, and a nichrome coil on a ceramic core. As glass cannot be used the beam is directed by a series of mirrors through the sample and into the monochromator.

Several transparent substances transmit over large portions of the infra-red spectrum and are therefore suitable both for windows of sample cells and as prism material. Quartz transmits in the near infra-red from 0·8 μm to 3·5 μm but as the productive area of absorption bands is between 2·5–50 μm it is rarely used. Sodium chloride or potassium bromide crystals have been found most useful as between them they transmit up to 25 μm. It will be obvious that no moisture must be allowed near the cells, windows or prisms. Storage of these components must be in a desiccated compartment, and to illustrate the necessity for care it is recommended that these components be held underneath the hand to prevent moisture rising on to them.

Cell holders are constructed of metal having two windows of potassium bromide spaced apart by lead, copper or gold washers. These are usually amalgamated. If this is done just before assembly of the cell, a solvent tight fit is assured as the washers swell. The light path length through the sample is small as the majority of substances absorb strongly, the usual size being from 0·01–0·5 mm. Some instruments are supplied with an adjustable cell fitted with PTFE gaskets and a gas cell, the latter having a length of from 1–20 cm.

Solid samples may be examined either by suspension as a fine powder in a mineral oil called Nujol, that has only a few narrow absorption bands, or by milling with KBr powder and then by means of a high pressure producing a solid semi-transparent disc.

The choice of solvents in which to dissolve samples is limited, carbon tetrachloride being the most useful having only one strong absorption area from 12–14 μm. Other solvents may be needed due to poor solubility in CCl_4, these include chloroform, cyclohexane and pyridine, but these all have several absorption bands and the specialized literature must be consulted for these and other solvents.

The prism is usually made of NaCl or KBr, other materials used are LiF, CaF_2, CsBr and SiO_2, these are capable of good resolution in defined areas of the spectrum. The prism is Littrow mounted, that is, the back surface of the prism is mirrored so that the beam is doubly refracted. The beam is then directed by mirrors on to the detector. This is usually a thermocouple (cf. page 397), other devices used are thermistors and bolometers, these exhibit changes in resistance with temperature and are connected in a Wheatstone bridge circuit. As resistance and voltage change is slight, amplification is difficult. The beam falling on the detector is 'chopped' or interrupted by a revolving slotted disc which produces a pulsating signal, and this can then be amplified by an a.c. amplifier. Another ingenious detector is the Golay cell. This is a small cell filled with xenon. Infra-red rays heat the gas which expands and moves a small mirror. Light from a separate source focused on the mirror is reflected into a photocell.

Most instruments are provided with a pen recorder which plots the spectrum automatically over a period of time ranging from a few minutes to several hours. This is almost essential as absorption peaks are very sharp and numerous, so manual scanning would be laborious.

Infra-red spectroscopy has its greatest application in qualitative analysis, the absorption spectra of pure compounds having been likened to fingerprints because of their individuality. It also gives an indication of the structure of the molecule, different groups and radicals having their own absorption peaks. Impurities are readily shown and it has, therefore, a useful application in quality and production control in industry. Substances may be analysed quantitatively as in visual absorptiometry, the Beer-Lambert law being operational.

Many instruments are calibrated in wave number v instead of wavelength λ . $v = \frac{1}{\lambda}$ and is expressed as cm^{-1}. λ is measured in microns, $1\ \mu m = 0{\cdot}0001\ cm = 1000\ nm$.

$$\text{Therefore at } 2\ \mu m \quad v = \frac{1}{0{\cdot}0002} = 5000\ cm^{-1}$$

$$\text{at } 20\ \mu m \quad v = \frac{1}{0{\cdot}002} = 500\ cm^{-1}$$

Spectrofluorimeters

Spectrofluorimeters are able to give more information than the fluorimeter which is most suitable for routine assays. The light source is either a mercury or xenon arc lamp. The beam of light passes into an excitation monochromator that directs a narrow waveband into the sample cuvette. A second monochromator receives fluorescent light from the sample at right angles to the incident beam and passes a selective wave band into the photomultiplier. Unlike absorption spectrophotometry in fluorescent analysis two spectra can be determined.

1. The excitation spectrum is found by setting the emission monochromator at between 400–500 nm then varying the excitation wavelength over a range 200–400 nm and recording the intensity of the fluorescent light. As mentioned earlier (page 241) this spectrum should be a mirror of the absorption spectrum. However, slight variations will occur due to the instruments.
2. The emission spectrum is found by choosing a strong peak from the excitation spectrum of the sample and varying the emission monochromator over a waveband of 350–550 nm.

Fluorometry is considerably more sensitive than absorption spectrophotometry as the fluorescent light is measured directly and if at the right dilution is proportional to the concentration. Unlike absorptiometry sensitivity can be improved by increasing the intensity of the incident light.

The calculation is similar to that on page 233.

$$C_1 = \frac{F_1 \times C_2}{F_2}$$

Where F_1 = fluorescent reading for the sample and F_2 = fluorescent reading for the standard.

Flame Photometry

This form of analysis utilizes the fact that many elements when heated to a high temperature transmit light, each having a distinct spectrum. It will be remembered from the flame tests of qualitative inorganic analysis that sodium gives a bright yellow colour, potassium a lilac colour, etc. Each of these colours is made up of several narrow wavebands of light. Potassium, for instance, has one blue and two red bands. As each element has several wavebands this makes it possible to quantitatively determine the concentration of several metals in a mixture. Under constant conditions the intensity of light emitted is proportional to the metal present as in previous absorptiometric methods. The problems involved in obtaining these conditions have been solved in the main, and flame photometry is now a routine method of analysis.

Principle

The liquid is introduced into a combustible gas mixture as a fine spray, and is burnt with it. Light from the flame passes through heat and light filters on to a photoelectric cell connected to a galvanometer, which measures the intensity of light produced.

The Apparatus

Two classes of instruments are available, the flame filter photometer and the flame spectrophotometer. The latter instrument is more versatile, capable of measuring concentrations of a greater variety of elements, due to its ability to isolate a narrow waveband of light. All instruments must have a steadily burning flame and to this end the gas and compressed air or oxygen are regulated. The temperature of the flame is important; for greatest sensitivity this must be high to excite the atoms of the element to emit light of high intensity. This is not so important with sodium and potassium as they emit sufficient light with a well-designed Bunsen burner, and for their analysis a mixture of coal gas or calor gas with compressed air is suitable. For high temperatures compressed oxygen is used, sometimes in combination with acetylene. The introduction of the sample must be considered with the design of the burner. The most popular type is a modified Meker burner, the sample is atomized by the compressed air or oxygen and both are then mixed with gas in a chamber (*Figure 6.15*). This removes large droplets of the sample before the mixture is burnt. The sample is drawn up a capillary as in a throat spray, the amount actually burnt depends on the efficiency of the atomizer, even so the majority of the sample (at least 80 per cent) goes to waste. The burner is positioned centrally in a chimney which protects both the instrument from heat and the flame from draughts.

In the simplest instruments the light emitted falls on a barrier layer cell that is connected directly to a sensitive galvanometer. To increase the light falling on the photocell and in order to improve sensitivity a concave reflector is positioned opposite the photocell. This doubles the sensitivity if the flame is at the centre of curvature. Other instruments use two photocells, one to measure the element

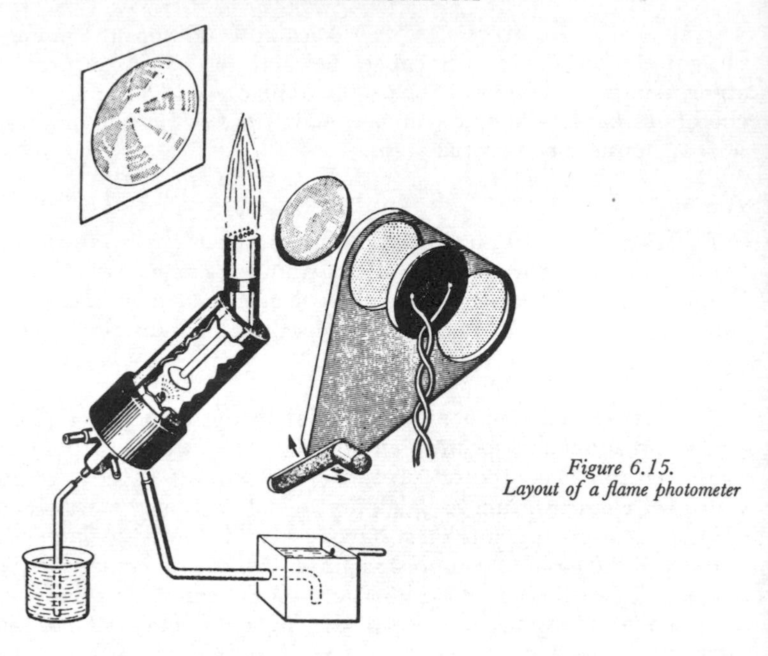

Figure 6.15. Layout of a flame photometer

under test, the other, a standard quantity of lithium incorporated in the sample. The unknown can therefore be referred to the lithium concentration, accurate results can be obtained although there are variations in flame conditions. These affect the analysis of both elements but their intensities are in a ratio with each other. In the flame spectrophotometer the waveband is selected by a monochromator optical system of a prism or grating and an adjustable slit. The light detector is a photo-emissive cell or photo-multiplier tube.

Heat from the burner is prevented from reaching the photocell by the insertion of glass heat filters or a cell of water in the light path. Selected light filters are then placed in the beam depending on the wavelength being measured. Both the Ilford and Chance filters are suitable, the former are more liable to heat damage as they are made of gelatin. Some instruments use interference filters which allow only a very narrow waveband to pass through, these have the added advantage that they transmit a high proportion of the incident light. They are made of ultra-thin layers of metals and other materials produced by a vacuum coating process. These filters are expensive but their use makes a filter flame photometer more versatile.

Flame spectrophotometers use a quartz 60 degree prism and a variable slit to isolate a near monochromatic band. The intensity of the light emerging is very low, and a sensitive detector is needed. This disadvantage is off-set by the fact that bands in the ultraviolet can be utilized, for example, magnesium that transmits at 285·2 nm.

Outline of Method of Use

This will, of course, vary with each instrument and maker's instructions should be followed. First the flame should be lit. This is done by turning on the gases in the correct sequence which is compressed air or oxygen and then fuel gas, and immediately applying a flame to the burner. When analysis is completed the fuel gas is turned off first, this prevents soot being deposited on the optical surfaces. The instrument is left to stabilize, this will be for 30 min with a spectrophotometer, distilled water being run in at the sample port during this time. Standard solutions are then introduced into the flame and a calibration curve plotted, and then the unknown samples determined. The test sample may be preceded by a standard solution and followed by another standard, and the concentration found as in absorptiometry. Approximately 2 cm^3 of sample is required, this allows time for the galvanometer to become steady and a reading to be taken. Samples usually have to be diluted with distilled water, for example, for serum potassium the dilution is 1:50, and for serum sodium 1:200. Great care must be exercised in performing dilutions or considerable error may be introduced. A high degree of purity of both reagents and distilled water is required for the standards. Unicam recommend storing them in polythene bottles and that dilutions be made in Pyrex volumetric flasks.

ATOMIC ABSORPTION SPECTROPHOTOMETRY

The flame photometer and flame spectrophotometer are instruments which assay substances by the emission of light by a flame. Superficially the atomic absorption spectrophotometer is similar because the sample is dissociated by a flame in a very similar manner. In this case, however, the concentration of an element is determined by its capacity to absorb light of a defined wavelength while in the vapour state, i.e. in a flame. In other words an apparatus very similar to a standard visible u.v. spectrophotometer is used, except that instead of a light absorbent solution in a cuvette being placed in the light path, a flame containing vaporized sample is used.

When the technique was first described by Walsh in 1955 the main problem was the provision of a suitable light source. For atomic absorption spectrophotometry an intense stable light of a very narrow waveband is needed. High output light sources used in ordinary spectrophotometers were not suitable as the radiation over the required bandwidth, which may be as narrow as 1 nm, was too feeble. The solution came with the development of the *hollow cathode lamp* which is capable of emitting, at high intensity, spectral lines for most of the elements. Each hollow cathode lamp is specific for a particular element and individual lamps are available for the estimation of over seventy elements. Most of these are for single elements although some multi-electrode lamps for the estimation of a group of elements are manufactured. The hollow cathode lamp is shown diagrammatically in *Figure 6.16a.* It consists of a quartz envelope with a flat end window. The atmosphere within the envelope is an inert gas, usually neon or argon at low pressure.

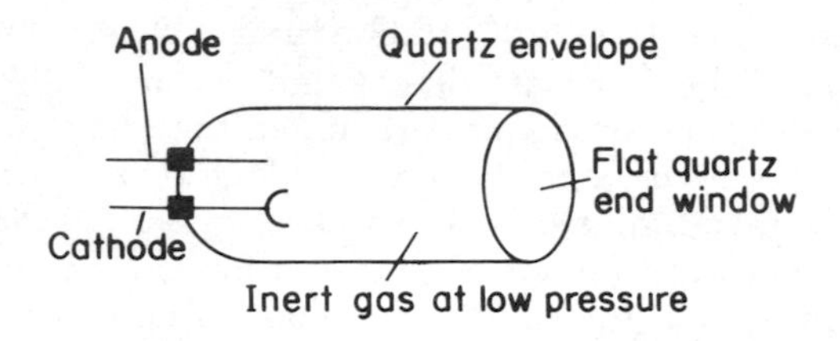

Figure 6.16a. The hollow cathode lamp

The cup shaped cathode is made of the element which is to be determined, the anode being of any suitable metal. When a direct current (100–200 V) is applied to the lamp it glows emitting the line spectrum of the cathode element. One line from this spectrum is used for analysis, the most appropriate line being isolated by the use of a narrow band pass monochromator of very high resolution. Interference from other lines can present a problem. Lines from the discharge of the inert gas atmosphere can be avoided by a suitable choice of gas although in some cases filters have to be used to eliminate interference. Sometimes the most sensitive line in the discharge spectrum is closely surrounded by other spectral lines and

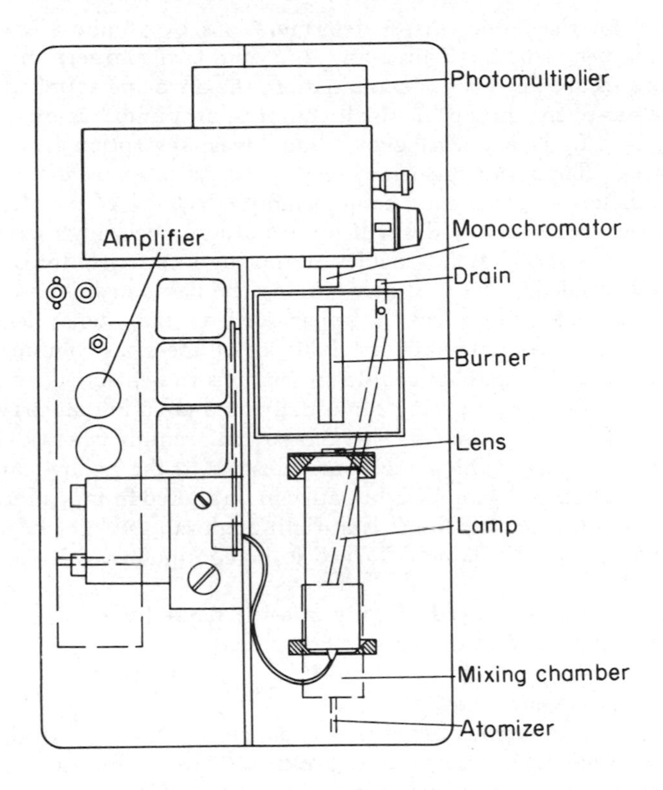

Figure 6.16b. Layout of the E.E.L. model 240 Atomic Absorption Spectrophotometer
[By courtesy of E.E.L.]

it is difficult to isolate the most useful one. In these circumstances a less sensitive line in a different section of the spectrum must be used in order to obtain accurate results. The lamp must be allowed to warm up for at least 15 minutes before analysis is attempted or stable readings will not be obtained. This can be a problem, especially when a number of elements have to be determined as there is a considerable delay on changing lamps. Several multi-element lamps are now available, notably calcium-magnesium for use in clinical chemistry and copper-nickel-cobalt-chromium-iron-magnesium for use in metallurgy.

The Burner

The flame-photometer usually employs coal gas/air or propane/air as fuel for the flame. Most determinations by atomic absorption use an acetylene/air flame and the standard burners on most apparatus are for this gas combination, the air being supplied from the atmosphere through a small compression pump. Some metals (Na, K, Pb, Bi and Zn) give rather lower absorption in the hot acetylene flame and dissociate readily in the propane flame. For these metals a propane/air flame is more effective. Many elements form very refractory oxides which must attain a very high temperature before dissociation is achieved and an acetylene/nitrous oxide flame is needed. The area surrounding the flame in this case must be water cooled to protect the apparatus from the intense heat.

Burners are commercially available for all these gas combinations and are readily interchangeable on most instruments.

The gas burner mounting can usually be rotated in the horizontal plane. The flame itself is flat and several centimetres wide. By altering the angle which the flame presents to the hollow cathode lamp, the effective sample light path can be altered from a minimum path length, when the flame is at right angles to the light beam, to a maximum as the flame is turned to be continuous with the light beam.

The sample is sprayed directly into the flame by a similar technique to that used in the flame photometer.

Monochromator and Detector

A sensitive narrow band pass monochromator is essential so that the monochromator pass wavelength can be adjusted to pass the element resonance line only. Most instruments are constructed so that the detector only responds to A.C. signals. The hollow cathode lamp is pulsed at mains frequency and can be detected but any extraneous D.C. signals from the flame are avoided.

Because of the narrowness of the bands passed by the monochromator, the amount of energy falling on the detector is small. A detector of high sensitivity is required and a photomultiplier with a suitable amplifier is used in most commercial instruments. Resonance detectors have recently been developed. Within these a cloud of atoms of the element to be determined is produced. These absorb the radiation of the resonance line from the hollow cathode lamp and are energized to give characteristic resonance radiation proportional to the energy absorbed. Thus a fixed wavelength monochromator is produced with a hand pass as low as 0·02 nm. The radiation from the resonator is picked up by a photomultiplier and amplifier system in the normal way.

In use the monochromator is adjusted to give the pass wavelength for the element resonance line noted on the lamp. A blank sample is sprayed through the flame, and the read-out meter is adjusted through the gain control to give an absorbance reading of zero (full scale deflection). Several different standard solutions are similarly burnt in the flame so that a series of absorbance readings can be used to plot a calibration curve. The 'unknown' solutions are then burnt in the flame and the concentration of each can be read off the calibration curve.

Advantages of Atomic Absorption Spectrophotometry

This technique must inevitably be compared with flame spectrophotometry to which it bears a superficial resemblance. The main advantage of atomic absorption is probably its high sensitivity. In emission spectroscopy only a very small proportion of the atoms which are sprayed into the flame actually emit light. These are the atoms whose electrons have been heated to a higher energy state and which emit light of a characteristic wavelength when the electrons release the extra energy and fall back to the ground state. In atomic absorption the emission of light by the flame has no significance. The absorbing atoms are those which are in the ground state in the flame. The concentration of absorbing atoms in the flame is many times the concentration of emitting atoms.

Flame emission spectrophotometry is limited to those atoms which are easily energized. A few, notably sodium, potassium and lithium, are very easily energized and can be estimated in a simple flame photometer. Although the flame emission spectrophotometer is a sensitive instrument, interference in the flame can lower the accuracy of results quite considerably. Atomic absorption readings are hardly affected by extraneous atoms or random atomic transitions from the ground state because of the high concentration of atoms in the ground state.

Readings of emission are on a linear scale and the range of readings obtainable is thus limited at any one instrument setting. Atomic absorption is a logarithmic function and the range of concentrations with which the instrument can cope after setting is considerably wider.

The versatility of the instrument is much greater as it is not necessary to raise refractory atoms above the ground state.

Because of the narrowness of the wavebands emitted by the hollow cathode lamp, the method is highly specific and interference from other elements in a sample presents no problem. Interelement effects which may lead to interference can be eliminated in the preparation of the sample before readings are made. Most commercial apparatus can be used for flame emission spectrophotometry.

The main disadvantage of the method is the need to use a separate expensive cold cathode lamp for each element to be estimated. Attempts have been made to produce continuous sources as used in ultraviolet spectrophotometry but none of the results has been commercially viable. The problem of time loss through the need to warm-up the lamp on changing from the assay of one element to another has been overcome in some instruments by providing multi-lamp holders so that one lamp is warming up while another is actually in use.

REFRACTROMETRY

A physical constant that is commonly employed when working with liquids is the refractive index. With modern instruments it is quickly measured and is an excellent criterion of purity.

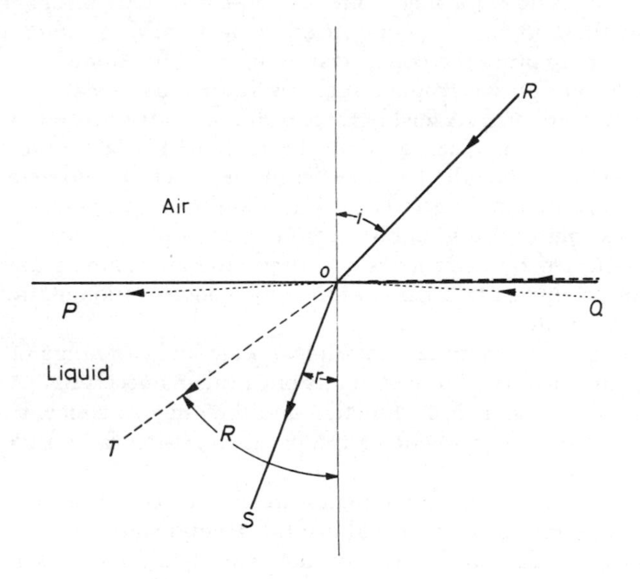

Figure 6.17. Refractive index for air to liquid

It was discovered by Snell in 1621 that the ratio of the sine of the angle of incidence to the sine of the angle of refraction when light passed from one medium to another was a constant. In *Figure 6.17* the ray *RO* is refracted *OS*, therefore $\frac{\text{sine } i}{\text{sine } r} = \text{air } \eta_{\text{liquid}}$ = refractive index for air to liquid. If the ray is made almost parallel with the interface, that is, 90 degrees, the angle of refraction is *R*. If the angle of the incident ray is increased further, it will be totally internally reflected as *QOP*. The angle *R* is the critical angle, and it will be seen that $\frac{\text{sine } 90°}{\text{sine } R} = \text{air } \eta_{\text{liquid}}$,

sine 90 degrees = 1, therefore $= \frac{1}{\text{sine } R}$ or cosec *R*.

Measuring the critical angle then is a method of estimating *n* If one observes the emergent ray when the incident ray is oscillated about 90 degrees, it will seem to disappear when greater than 90 degrees. This is because the reflected angle *T* is so much greater than the critical angle *R*.

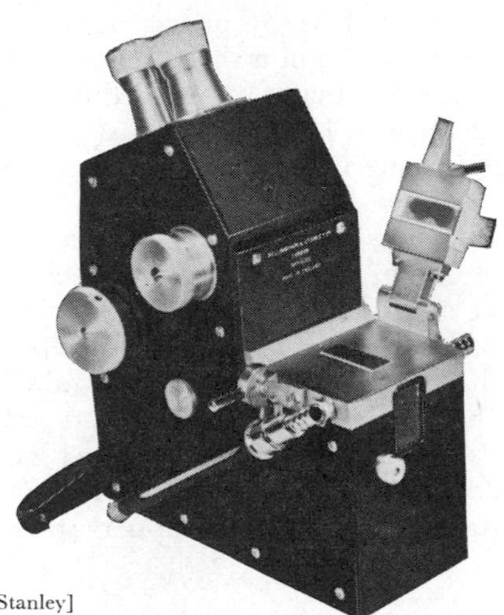

Figure 6.18. The Abbé refractometer

[By courtesy of Messrs. Bellingham and Stanley]

A telescope can be used to view this effect and a point will be reached when the field will be only half illuminated, the dividing line being called the border line. Unless monochromatic light is used the border line will be indistinct. The most accurate instruments for measuring refractive index use this critical angle principle. The refractometer to be described is the Abbé, perhaps the most popular type in use (*Figure 6.18*).

Abbé Refractometer

The liquid under examination is held in a narrow gap between two prisms. A beam of light passes through one prism then undergoes refraction through the liquid, and is again refracted through the second prism. The beam is finally reflected with a mirror into a telescope. The mirror is on a calibrated mechanism so that it may be moved until the border line can be seen. The illuminated scale is viewed through another eyepiece and is calibrated in refractive indices between 1·3 to 1·7.

Use of Refractometer

The refractive index varies with the wavelength of light and with temperature. Most instruments use daylight or a tungsten lamp for illumination but are calibrated for the sodium *D* line. This white light will be dispersed by its passage through the prisms giving a coloured borderline lacking in contrast. This dispersion is compensated for by prisms inserted in the beam. These prisms can be rotated until a sharp border line is seen.

The prisms that sandwich the sample are water-jacketed, enabling them to be used at constant temperature. The jackets are usually connected in series, enough tubing being used to connect the two boxes, allowing the upper one to hinge. The thermometer pocket is in the outflow. The temperature quoted in most textbooks is 20°C, but as modern laboratory temperatures tend to be higher due to air conditioning, 25°C may be preferred. Water at this temperature is pumped from a bath using apparatus described in Chapter 10. Temperature equilibration takes about 15 min. The hinged upper prism is raised and then 3 or 4 drops of the sample are applied to the polished face of the lower prism. Care should be taken to prevent this surface from being scratched by pipettes or spatulas. The upper prism is then lowered, the liquid

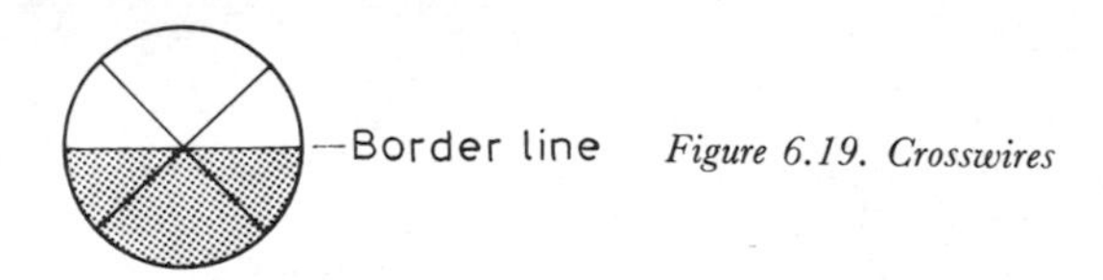

Figure 6.19. Crosswires

being trapped as a thin film. The shutter is partially opened on the upper prism. On looking down the telescope, a field with crosswires will be seen, *Figure 6.19*. The setting control is revolved until the border line appears, the upper part of the field should be light. Rotate the dispersion drum if the border line is diffuse and coloured until it becomes sharp, then adjust the border line so that it passes through the intersection of the crosswires. The scale is calibrated to 0·001 and readings are relative to the sodium *D* line. After noting the refractive index, raise the upper prism and remove the liquid sample with soft tissue or cotton wool, and wash the prism surfaces if necessary. Drain the water-jackets after use and place the instrument back in its box. A jolt or knock may throw the optical system out of adjustment. This can be rectified by using distilled water, $n = 1{\cdot}333$, or a glass test-piece supplied by the manufacturers. Adjust the field so that the border line passes through the intersection of the crosswires, then set the scale to the correct refractive index by turning the zeroing screw.

A temperature correction may have to be applied if the instrument is used at a temperature different from that at which it was originally calibrated. With the Bellingham & Stanley Abbé 60 Refractometer this is +0·0000078/°C rise.

POLARIMETRY

Many organic compounds have an effect on polarized light which is physically constant under defined conditions. This property may be used as an indication of the structure of a compound or its purity as melting points and boiling points would be used, or it may be utilized to estimate its concentration in solution.

Polarized Light

An ordinary ray of light is regarded as electromagnetic vibrations travelling in wave form. These waves vibrate at right angles to the direction of the ray but not in the same plane (*Figure 6.20a*). Light with all waves in the same plane (*Figure 6.20b*) is called 'plane polarized' and it is this that is used in polarimetry.

Certain pure crystals such as tourmaline are able to divide a ray of light so that it will vibrate in two planes only and the light in one of these planes is then removed by absorption or reflection. This was first noticed in 1669 by Bartholinus who found that the natural crystals of calcium carbonate, called calcite or iceland spa, showed double refraction and the two emergent beams were vibrating at 90 degrees to each other. In 1808, Malus discovered that light reflected at a particular angle was almost pure plane polarized. This was called the polarizing angle and Brewster showed that at this angle the reflected and refracted rays are at right angles to each other. A means of producing a usable beam of polarized light of high intensity was required. To do this Nicol in 1828 using the above knowledge designed a prism that is now universally used.

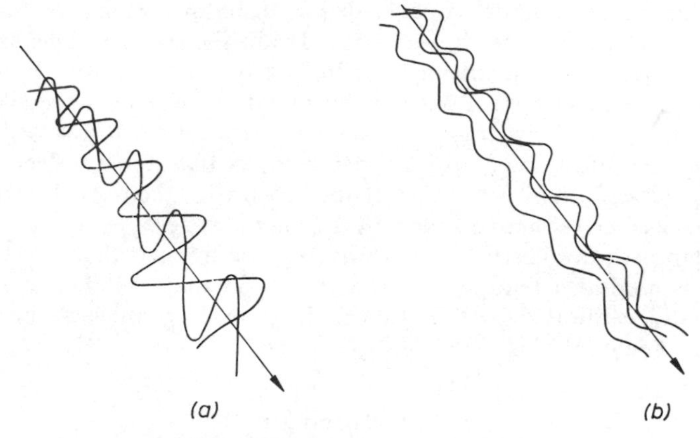

Figure 6.20. The polarization of light

The Nicol prism is a rhomb of calcite that is carefully sliced in two. The cut faces are then polished and rejoined using Canada balsam as an adhesive (*Figure 6.21*).

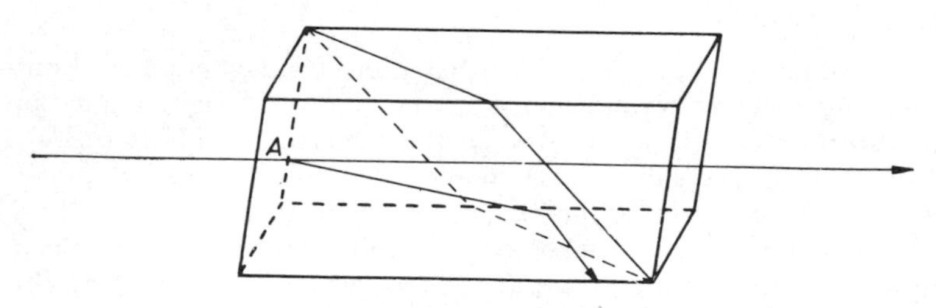

Figure 6.21 The Nicol prism

A beam of light incident at *A* is divided into two beams whose plane of polarization is at 90 degrees to each other. One beam passes through the prism and the other is internally reflected by the film of Canada balsam and is absorbed at the blackened side of the prism.

If a second Nicol prism is placed on the same axis but in a plane 90 degrees to the first, no light will emerge and the prisms are said to be crossed.

Optical Activity

Many substances such as sugars, quartz crystals and turpentine have the ability to twist or rotate a beam of polarized light. This occurs when the beam passes through the solid crystal and even when the substance is in solution. Substances that have this power are said to be optically active, the activity being due to the molecular structure and occurring when an asymmetric atom is present. Asymmetry is illustrated by supposing a compound to have a molecular structure whose atoms are linked together to form a tetrahedron (*Figure 6.22*). It will be seen that only two forms

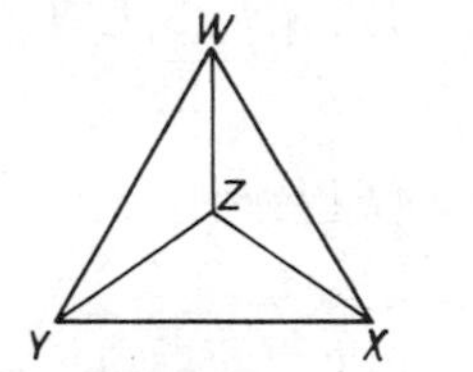

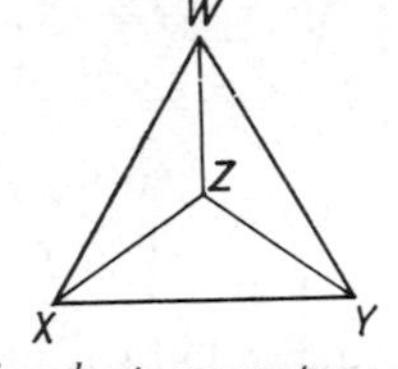

Figure 6.22. Geometrical isomerism due to asymmetry

can exist however the atoms are arranged, one the mirror image of the other. The two forms called optical isomers will rotate the polarized light in opposite directions. When the beam is rotated clockwise the substance is said to be dextro-rotary and if the beam is rotated anti-clockwise it is a laevo-rotary substance. Additional discussion will be found in Chapter 11, page 433.

Each optically active compound has a constant rotary power or specific rotation $[\alpha]$. This is derived from the following equation

$$\left[\alpha\right]_x^t = \frac{\phi}{lc}$$

(under defined conditions of temperature t°C and wavelength of light x).

Where ϕ is the observed rotation
l is the light path length through the solution in dm
c is the concentration of the substance in g/cm^3.

Polariscope or Polarimeter

The measurement of optical rotation is made in a polarimeter. This uses Nicol prisms as means of producing polarized light. A schematic lay-out is illustrated in *Figure 6.23*.

Light from source L passes through a filter and lens system producing parallel monochromatic light. This is polarized by prisms P_1, P_2, jointly called the polarizer. It then passes through the tube containing the material for analysis. The rotated beam then enters the prism A, called the analyser, and finally through the telescope and eyepiece.

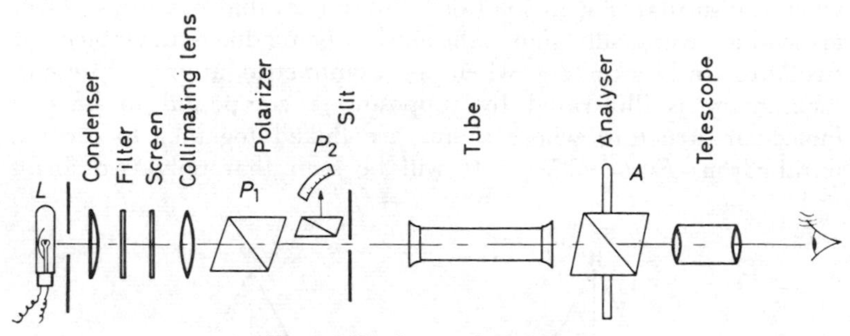

Figure 6.23. Layout of a polarimeter

Light Source

Monochromatic light is used as the ability of a solution to rotate polarized light diminishes, as the wavelength increases. The most common light source is a sodium lamp. This transmits two bands (set close together, and regarded as one) called the sodium D line of 589·3 nm wavelength. It is possible that a solution may absorb strongly at this wavelength. An alternative source is a mercury lamp whose spectrum contains a number of bands only two of which are sufficiently isolated for practical use. These are the blue 435·8 nm and green 546·1 nm the other bands being removed by filters.

For specialist studies a monochromator attachment can be used. The extinction when using ultra-violet light is obtained by the use of a photoelectric cell.

Polarizer

Light from the source passes through a collimating system of slits and lenses which contains a filter to absorb faint lines emitted by trace gases in the lamp. The resultant beam is parallel and even in illumination.

The polarizer consists of two Nicol prisms P_1P_2. P_1 is fixed, P_2 covers half the beam emerging from P_1 and is adjustable so that its plane of polarization may be up to an angle of 0–20 degrees to that of P_1, producing a field of view split in two when seen from the eyepiece (*Figure 6.24*). This device assists in the accurate setting at the balance or end-point, the two halves of the field being adjusted so that they are of equal intensity as in (*b*). In (*a*) and (*c*) the dark areas indicate that the respective prisms in the polarizer

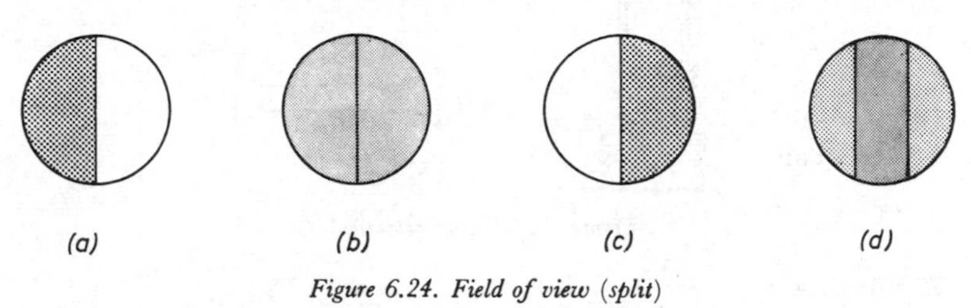

Figure 6.24. Field of view (split)

are 'crossed' or at 90 degrees with the analyser. This is called a 'half shadow' device. In some instruments the field is divided into three, the outer two portions receiving light from the same beam (*Figure 6.24d*).

The Tube

Tubes are from 25–400 mm long and up to 10 mm in diameter and are of several designs. The most common is a tube with ground end-faces enclosed by two optically parallel glass discs, held in place by a screw-cap on a threaded ring that is cemented to the glass (*Figure 6.25*). This is filled with solution by closing one end, setting it upright and filling with a pipette or by careful pouring to prevent bubbles adhering to the glass. The end cap is slid into position, excluding air, and the cap screwed on. The end caps must not be tightened unduly as strain may be set up in the glass, affecting the light beam. Care should be taken not to heat the tube by holding with the hand. The end faces must be clean and dry. The tube is then placed in the sample trough between the polarizer and the analyser. Other designs have a filler hole in the side of the tube, or a bubble to receive any dissolved gas that may come out of solution. A jacket may be incorporated to enable the contents to be temperature controlled.

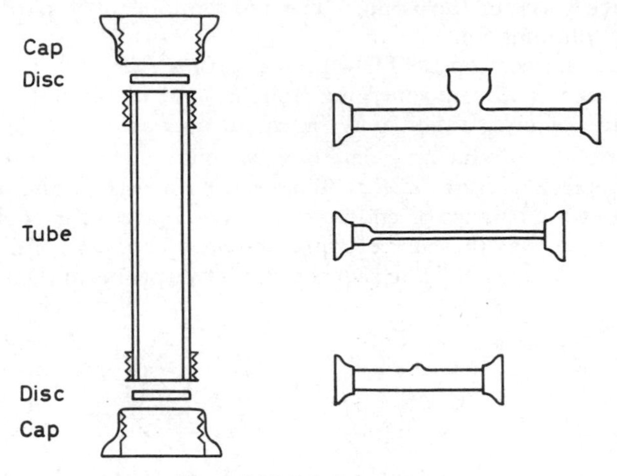

Figure 6.25. Polarimeter tubes

The Analyser

This is a Nicol prism mounted axially within a disc that is calibrated in degrees. Modern instruments have a fine adjustment. To improve accuracy a vernier scale is provided on either side 180 degrees apart reading to 0·01 of a degree. To facilitate reading they may be individually illuminated, each having a reading telescope.

Following the analyser is a telescope that must be focused on the exit slit of the polarizer.

The Estimation of the Specific Rotation

Choose a suitable light source, turn it on and allow time for warming up (about 15 min with a sodium lamp). Choose a tube of suitable length, this depends on the extent of light absorption by the substance and the rotation. Fill the tube with the solvent to be used and place it into the trough, and adjust the half-shadow angle to give suitable illumination by moving prism P_2. Focus the field by the adjustment at the eyepiece. Revolve the analyser till the field is uniformly dark. The final position is best found by oscillating the analyser about the end-point, gradually reducing the oscillations until an exact match is obtained. Note the reading on the disc. Make up a 0·5–2 per cent solution of the compound preferably in water although other suitable solvents are absolute alcohol and chloroform. Filter or centrifuge the solution to remove dust or particulate contaminants. Refill the tube with the sample and place it in the trough near the analyser. Close the lid of the trough and allow time for temperature equilibration. Revolve the analyser again to give a matched field, record the direction of rotation, and again note the reading on the disc. The difference in the readings is the observed rotation.

Example

Using the sodium D line and a 2 dm tube, a 2 per cent solution of a sugar gave a dextro-rotation of 2·1 degrees at a temperature of 20°C. Calculate the specific rotation.

$$\left[\alpha\right]_D^{20^\circ} = \frac{\phi}{lc}$$

$$= \frac{2{\cdot}10}{2 \times 0{\cdot}01} = +\ 52{\cdot}5^\circ$$

An unknown concentration of the same compound under the same conditions and tube length gave a rotation of +3·15 degrees. What is the concentration of the solution?

$$c = \frac{\phi}{[\alpha]_D^{20}\, l} = \frac{3{\cdot}15}{52{\cdot}5 \times 2} = 0.03\ \mathrm{g/cm^3}$$

A modification of the polarimeter used in industry is the saccharimeter. This has a 'sugar scale' on the disc and allows results to be expressed in terms of glucose concentration.

CHAPTER 7

CHROMATOGRAPHY AND ELECTROPHORESIS

CHROMATOGRAPHY

In 1906 the Polish botanist Tswett reported what has since been recognized as a classical experiment. In an attempt to separate the pigments contained in plant leaves, he filtered a petroleum ether extract of spinach through a long vertical column of adsorbent powder packed in a glass tube (*Figure 7.1*). The green pigment

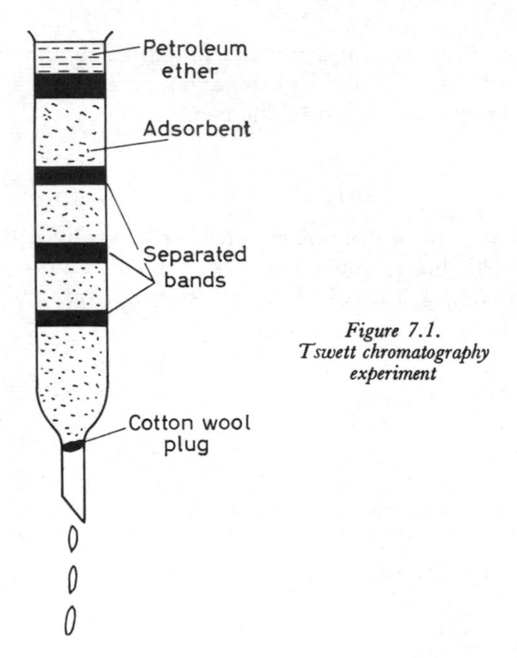

Figure 7.1. Tswett chromatography experiment

became adsorbed as a coloured band at the upper end of the column. As more petroleum ether was allowed to percolate through, the green band began to migrate slowly down the column, separating gradually into four distinct bands which moved down the column at different rates. These were the four pigments which form together the total pigment content of spinach leaves. Finally, Tswett pushed the wet column of adsorbent out of the tube, sliced out the coloured bands with a knife, and extracted them with petroleum ether. After the ether had been removed from the four solutions, he was left with four distinct pigments from the original green extract.

This simple method of separation was ideal for the labile and heat sensitive compounds with which Tswett was working, it treated them gently, and the separation was successful despite the fact that the four pigments were chemically very similar. Strangely enough it excited little attention, and virtually nothing more was heard of chromatography for a quarter of a century until in 1931 Kuhn and Lederer revived the technique in their important experimental work on the carotenoids, a group of pigments similar to the leaf pigments which Tswett had separated.

Because the early experiments were concerned with the separation of coloured compounds the technique was named *chromatography*, and this is still the term used even though the vast majority of compounds separated by the process are colourless. The separate bands of pigment on the column form a *chromatogram*. The separation of the components on the column is called *development* of the *chromatogram*, and the extraction of the pigments from the adsorbent material is termed *elution*.

Since the early nineteen-thirties the technique has become more and more significant and has been used in such diverse fields as the elucidation of the D.N.A. chain, and the purification of materials in the production of the first atomic bomb. Tswett's experiment relied upon the different partition coefficients of his four pigments between the stationary solid adsorbent phase and the mobile liquid phase. Chromatography has now developed so that in general it may be said to be a method for separating substances, depending on their partition coefficients between two immiscible phases. One of these phases is always stationary, and may be solid or liquid (a stationary liquid phase being bound to an inert solid support). The other is always mobile and may be either liquid or gas. Thus three broad classifications of chromatography may be stated:

1. Liquid-solid chromatography.
2. Liquid-liquid chromatography.
3. Gas-liquid chromatography.

The importance of this classification will become apparent as the experimental technique is described.

COLUMN CHROMATOGRAPHY

The experimental technique for column chromatography is similar for both solid and liquid stationary phases. Gas-liquid chromatography is a procedure which requires very special apparatus and will be discussed later in the chapter.

1. *Liquid-Solid Chromatography*

This technique may be considered under two main headings:

(*a*) Adsorbent stationary phase—solvent mobile phase.

(*b*) Ion exchange resin stationary phase—electrolyte solution mobile phase.

(*a*) *Adsorption chromatography*—Adsorption chromatography is the technique used by Tswett in his original work. A mixture of substances is adsorbed at the top of a column of solid powder and separated by the flow of a suitable solvent. Dozens of different adsorbents have been listed for various purposes, and give the choice of a suitable adsorbent for any particular separation. The main criteria for choosing an adsorbent are:

(i) The adsorbent must be completely insoluble in the solvent used;

(ii) It must be obtainable in such a pure state that different batches from different sources should have similar characteristics;

(iii) It must not react with the components of the mixture to be separated, nor may it catalyse their decomposition;

(iv) If coloured materials are to be separated and located visually it is convenient if the adsorbent is white or, at least, only very slightly coloured.

Another factor which must be considered in the choice of an adsorbent is particle size. Obviously very fine particles present a much greater surface for adsorption than coarse particles, and a column of fine particles would give greater separation and have a higher capacity. The flow rate through a column of very fine material may be so slow, however, as to be impracticable. A compromise must be reached empirically. On the other hand, the particles must not be so coarse as to create irregular zones or 'channelling' in the column which may lead to inefficient separation.

It is difficult to classify the more common adsorbents according to their adsorptive power as the solvent used has a profound effect on the capacity of a column. A rough guide is given in the table of adsorbents (Table 7.1). Special mention must be made of alumina, which is the most widely used of all adsorbents. This

Table 7.1. Adsorbents for Liquid-Solid Chromatography
(In order of increasing absorptive power)

1. Sucrose	7. Magnesia
2. Inulin	8. Calcium hydroxide
3. Talc	9. Silicic acid
4. Sodium carbonate	10. Magnesium silicate
5. Calcium carbonate	11. Alumina
6. Magnesium carbonate	12. Fullers earth

material may be readily purchased in suitable particle size (8–12 μm) and with a specified adsorptive capacity. The adsorptive capacity of alumina depends upon the degree of dehydration of the material. The most adsorptive grade is prepared by heating alumina to redness for 4–6 h and cooling in an evacuated desiccator. Four other grades are obtained by adding specific amounts of water to the dehydrated powder. Brockmann and Shodder have described this method together with a standardization technique using dyes for determining the absorptive power of alumina.

Numerous solvents have been described for adsorption chromatography, and although only one adsorbent is used, several solvents may be required successively to complete a chromatographic separation. The solvents perform three separate functions;

(i) The introduction of the mixture to the top of the column;

(ii) The separation of the mixture into its components on the column, i.e. development of the chromatogram;

(iii) The elution of the separated components from the column.

The choice of a solvent is governed by many factors. Some of these are similar to those by which the adsorbent is chosen, e.g. the solvent must not react with the components of the mixture to be separated. The two main considerations, however, are solubility and polarity. Again no hard-and-fast rule can be laid down but, in general, substances are adsorbed most strongly from non-polar solvents like petroleum ether and carbon tetrachloride, and least strongly from highly polar solvents like acids, bases and alcohols. A rough guide is given in Table 7.2 which is a list of solvents commonly used in increasing order of polarity. The lower numbers on this list may be suitable for introducing the mixture on to the column, solvents in the middle section may be suitable for development of the chromatogram, and the more polar solvents towards the end may be used for elution. Alternatively, if a non-polar solvent, say petroleum ether is used for introducing the

TABLE 7.2. Solvents in Solid-Liquid Chromatography
(Solvents are arranged in order of increasing eluting power)

1. Light petroleum	9. Acetone
2. Carbon disulphide	10. Propanol
3. Cyclohexane	11. Ethanol
4. Carbon tetrachloride	12. Methanol
5. Benzene	13. Water
6. Toluene	14. Pyridine
7. Chloroform	15. Organic acids
8. Diethyl ether	

mixture, a 1 per cent solution of pyridine in petroleum ether may be used for development and a 5 per cent solution of pyridine in petroleum ether for elution. A common technique is gradient elution. Two solvents A and B are used together, A being less polar than B. The solvent flow starts with pure A, the concentration of B in A being gradually increased until development and finally elution have been effected.

The polarity of the solvent is important, but even more important is the question of solubility. The concentration of mixture in solvent which is put on the column is rarely higher than 5 per cent, but as the chromatogram is developed separate components may reach a very much higher local concentration, and a large margin of solubility must be allowed in selecting the solvent.

A further characteristic which must be considered in the choice of solvent is volatility. It is obviously convenient to use a solvent of low boiling point as the recovery of materials from the eluate is made much easier, especially if they are at all heat sensitive. However, if the speed of solvent flow is being increased by reducing pressure at the end of the column, the solvent may boil and the column ruined.

Finally, it should be realized that the rate of flow of solvent will depend in some measure on its viscosity. Carbon tetrachloride, for instance, will flow through a column much more slowly than petroleum ether through a similar column.

There is already a vast literature of chromatography, and experimental details for the separation of a mixture can often be found. Even if an 'unknown' mixture is to be separated, a guide to the experimental procedure may be obtained from separations of mixtures from similar sources. In the final analysis, however, the technique for an 'unknown' mixture can be determined empirically by trial experiments on a small scale.

(*b*) *Ion-exchange chromatography*—Over a century ago it was recognized that if 'hard' water is filtered through certain naturally occurring minerals (zeolites, green sand) the water became 'soft'. This is because the minerals have the power of ion-exchange, i.e. they can exchange ions with their surroundings. The calcium and magnesium ions which are dissolved in the water are exchanged for the sodium ions which occur naturally in the mineral. (See 'Preparation of Conductivity Water', page 128.) Since the rise of the plastics industry, synthetic ion-exchangers have been developed some of which can be used for water softening. A large range of synthetic ion-exchangers have been developed which can be used in column chromatography.

A commercial ion exchange resin consists of tiny beads of an inert plastic (e.g. polystyrene, phenolic resin) to which a *functional group* is attached. The functional group is always an electrolyte, one ion of which is firmly attached to the resin, the other ion being mobile and exchangeable with ions of similar charge in the surrounding medium.

The beads of modern resins are porous so that liquid can percolate through them as well as around them. By increasing the effective surface area of the beads in this way the capacity of the material is considerably increased.

Amberlite IR-120 is a commercial resin widely used in biochemical separations. The chromatographic grade is sold as small beads of cross-linked polystyrene to which the electrolyte $NaHOSO_2$ is attached. The $HOSO_2^-$ ion is firmly attached to the resin while the Na^+ ion is mobile, and can exchange with cations in the surrounding medium. IR-120 is a cation exchange resin.

Amberlite IRA-400 is a commercial resin of cross-linked polystyrene where the functional group is $N(CH_3)_3Cl$, a quaternary ammonium compound. The $N(CH_3)_3^+$ ion is attached to the resin and the Cl^- ion is exchangeable with anions in the surrounding medium. IRA-400 is an anion exchange resin. A list of useful ion exchange resins together with their properties is given in Table 7.3.

TABLE 7.3. Some Useful Ion Exchange Resins

Some of the products of several manufacturers are shown. Resins on the same line are usually regarded as being equivalent in properties.

Permutit	*Amberlite*	*Duolite*	*Dowex*	*Type*	*Functional Group*	*Remarks*
Cation Exchange Resins						
Zeo-Karb 215	IR–1	C–10	30	Phenolic	—OH	Strongly acidic
	IR–100	C–3		Phenolic	$—SO_3H$	
Zeo Karb 225	IR–120	C–20	50	Cross-linked polystyrene	$—SO_3H$	Strongly acidic
Zeo Karb 226	IRC–50	—	—	Cross-linked methacrylic acid	—COOH	Weakly acidic
Anion Exchange Resins						
De-Acidite E	IR–4B	A–2	—	Phenolic	$—OH_1$ Nuclear amino groups	Weakly basic
De-Acidite FF	IRA–400	—	1	Cross-linked polystyrene	Quaternary NH_4 groups	Strongly basic
	IRA–410	—	2	Cross-linked polystyrene		
De-Acidite G	IR–45	—	—	Cross-linked polystyrene	Substituted amine	Weakly basic
De-Acidite H					$—N(CH_3)_2$,	Weakly basic

In considering the characteristics of ion exchange resins, several factors are important. The polystyrene of which they are mainly composed is always cross-linked, i.e. each polystyrene chain is linked to other chains to produce a three-dimensional network. The higher the degree of cross-linking the lower the solubility of the resin, and the lower the tendency to take up water and swell. High cross-linking also leads to a lower rate of exchange and a lower capacity of exchangeable ions. A compromise must be reached between these two factors. Commercially available cation exchangers have cross-linkage of between 4 and 20 per cent with water regain of 0·5 g/g to 2 g/g. Anion exchangers can have lower cross-linkage as water regain is not so high, but as cross-linkage is almost impossible to measure, water regain is the characteristic usually considered in choosing a resin for any particular purpose.

For most separations a 'strong' exchange resin of intermediate cross-linkage is usually most suitable. Where very large ions are to be chromatographed in biochemistry a lower degree of cross-linkage is desirable and where small metal ions are to be separated a high degree of cross-linkage would be suitable, always assuming that the cross-linkage is not so high as to slow the exchange of ions enough to prevent separation, nor so low as to cause excessive swelling of the resin.

'Weak' exchangers are unsuitable for the adsorption of the ions of weak acids or bases, but where the separation depends upon differences in dissociation constant weak exchangers may be suitable. Another advantage of weak acidic exchangers is that they may be buffered at a specific pH. This is not possible with strongly acidic exchangers.

Particle size is of some importance and, in general, the factors considered in adsorption chromatography apply equally to ion-exchange chromatography. In recent years the use of ion-exchange materials other than ion-exchange resins of the 'plastic' type have been increasingly employed in biochemistry. These materials are derivatives of cellulose, various acidic and basic groups being attached to the material. Modified celluloses have large capacities, up to 20 per cent of their own weight of protein being adsorbed upon them. The cellulose ion-exchangers so far marketed are:

(1) Diethylamino ethyl cellulose (DEAE).

(2) Triethylamino ethyl cellulose (TEAE).

(3) Carboxymethyl cellulose (CM).

(4) Sulphoethyl cellulose (SE).

(5) Sulphomethyl cellulose (SM).

(6) The result of the reaction of an epichlorcyanin bound cellulose with the basic groups of triethyanolamine (ECTEOLA).

These materials have been of enormous help in the separation of pure protein from mixtures, e.g. it is possible to obtain a pure sample of serum γ globulin on a column of DEAE with much more ease than the earlier method of preparation by fractional salt precipitation with sodium sulphate. Many manufacturers market these materials and provide specific instructions for their use.

Development of the Chromatogram

In both forms of solid-liquid chromatography there are three main methods of developing the chromatogram. These are useful for specified purposes.

The most common of these techniques is *elution analysis*. This is employed when complete separation of the components is required, and development is completed by zones of separated component interspersed with zones of solvent. The rate of movement of each component zone down the column with a specific solvent may give a clue to the identity of the component. The rate is expressed as the R value.

$$R = \frac{\text{Velocity of front edge of zone}}{\text{Velocity of solvent above the column of adsorbent}}$$

R is related to the retardation of a substance on the column and thus to the partition coefficient of the substance between the stationary and mobile phases. The more strongly a material partitions in favour of the stationary phase, the lower the R value will be, i.e. the slower the movement of the zone down the column. *Simple elution* with one solvent or solvent mixture gives the best results when it is applicable as the highest resolving power is achieved. Unfortunately, where large numbers of components are to be separated, simple elution is rarely satisfactory. Excessive 'tailing' may occur, or some components may leave the column at the solvent front very quickly, while others cling so tenaciously to the stationary phase as not to be eluted at all in a reasonable time.

Gradient elution has been mentioned above. This is the most satisfactory method for most purposes, especially in biochemical separations. One of the important advantages of gradient elution is that tailing is reduced as the rear part of each zone is in a more strongly eluting medium than the front, tending, therefore, to stay more compact.

Stepwise elution is a useful variation where 'unknown' mixtures are to be separated so that some idea of the conditions for separation can be obtained.

The main disadvantage of elution development is that only a small part of the total absorptive capacity of the stationary phase is taken up. About 1 per cent is normally used and 10 per cent is the maximum possible if reasonable results are to be obtained.

Frontal analysis (*Figure 7.2*) is the second main method of development. A much greater amount of the total capacity of the column is usable. A solution containing a mixture of components is passed through the column in a continuous flow. The pure solvent will emerge at the end of the column, followed by the least strongly

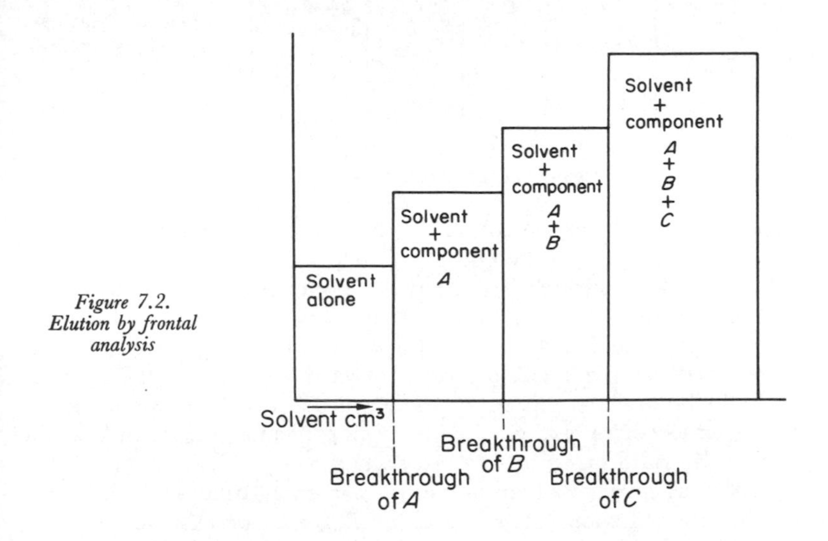

Figure 7.2. Elution by frontal analysis

adsorbed component as the column becomes saturated with this. As the column becomes saturated with a second component, this 'breaks through' and a mixture of the two components is eluted at the end of the column. As the column becomes saturated by each component, 'break through' occurs, and the component appears in the eluate at the end of the column. Frontal analysis is used in qualitative analysis as the 'break through' fronts are sharp, and changes in the eluate may be easily detected. It may also be used for the deionization of materials (including the preparation of deionized water) or for the separation of a strongly adsorbed component from a weakly adsorbed one.

Displacement analysis (*Figure 7.3*), the third general method of development, allows up to 50 per cent of the total capacity of column to be used. After the mixture has been introduced on the column it is followed not by pure solvent or solution as in simple elution or frontal analysis, but by a new solution containing a solute more strongly absorbed than any of the components in the mixture. As the new

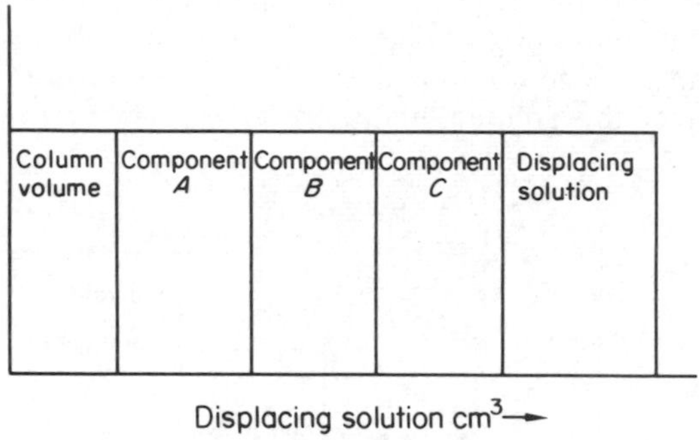

Figure 7.3. Elution by displacement analysis

solute moves down the column it displaces the components already on the column and sweeps them forward one behind the other. Thus the least strongly adsorbed component is eluted first followed closely by the next component and so on. Tailing is eliminated because if a tail is formed it is quickly displaced by a more strongly adsorbed component and swept forward again. In theory there should be a sharp separation between successive components, but in practice there is always some mixing of one zone with the next as equilibrium takes time to be reached.

The practical disadvantage of displacement analysis is that there is no 'clear' region between zones. This can be eliminated by the addition to the mixture of substances which have intermediate adsorption and which will separate the zones of the original mixture. These substances should be easily removable from the eluate. For instance, amino acids may be mixed with a mixture of homologous alcohols which interpose zones between the acids on a charcoal column. The alcohols are volatile and are easily separated from the amino acids. This variation is called carrier displacement. The practical details are often very difficult to work out, but it is such a useful method for the preparation of large quantities that the effort of evolving a practical procedure is often worth while.

(2) *Liquid-Liquid Chromatography*

Liquid-liquid chromatography was first reported by Martin and his co-workers in the early nineteen-forties. They found that silica gel contained up to 40 per cent of bound water, even when a seemingly dry, free-flowing powder. Martin *et al.* separated acetylated amino acids on columns of silica gel using 1 per cent butanol in chloroform as solvent. The separation depended upon the partition coefficients of the various acetylated amino-acids between the bound water and the butanol/chloroform. The silica itself took no direct part in the separation, merely acting as an inert carrier for the water which forms the stationary phase in the separation.

Silica gel was used in the original experiments, other common support media are cellulose, celite and porous glass. Starch and inulin have also been used. Until recently column liquid-liquid chromatography was confined to preparative and 'clean-up' work. With the advent of more sensitive detection devices for monitoring eluates more use of the technique is now made in quantitative analysis, and in fact rivals G.L.C. in some spheres.

The majority of liquid-liquid chromatography is carried out on sheets of paper (paper partition chromatography) and layers of support media spread on a rigid sheet such as glass (thin layer chromatography). These are discussed later in the chapter.

The support material is coated with the stationary phase and ideally this should be of uniform thickness. Porous glass has distinct advantages in that it has no absorption characteristics and the stationary phase may be chemically bound to the glass so that it does not leach out.

The coating of the support media with the stationary phase is carried out either by (*a*) shaking measured amounts together in a flask or (*b*) shaking a solution of the phase in a volatile solvent with the support media, and then evaporating the solvent. The other practical techniques are similar to those for solid-liquid chromatography.

In liquid-liquid chromatography the stationary phase is often water and the mobile phase some solvent only slightly miscible with water. Often the mobile phase consists of a mixture of two or more solvents which together give the correct conditions for satisfactory separation, e.g. the separation of sugars has been effected using water saturated phenol, water saturated collidine, or the top layer of a mixture of *n*-butanol, acetic acid, and water in the proportion 4:1:5.

(3) *Gas-Liquid Chromatography*

Since the invention of gas liquid chromatography (G.L.C.) about twenty years ago, it has become the most widely used of all analytical instrumental methods and has assumed profound importance in every branch of industry and research. G.L.C. is a form of column chromatography in which the stationary phase is a liquid, usually adsorbed on to a solid support, and the mobile phase is a gas. The sample is vapourized in a stream of carrier gas and passes through a column packed with stationary phase. The components of the sample separate on the column and are eventually eluted in the carrier gas which is monitored by a detector. The detector is connected to an amplifier which transmits the signal from the detector to a chart recorder or print-out reader. The chart recorder gives a series of peaks, each peak relating to one of the separated components of the mixture. The print-out reader records the area of each peak over the chart base line, this being proportional to the concentration of each component in the sample.

The main components of a simple gas chromatograph are shown in *Figure 7.4*.

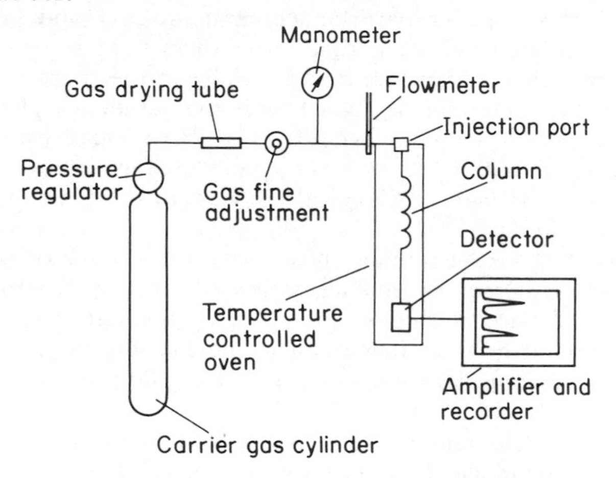

Figure 7.4. A simple gas-liquid chromatograph

Carrier Gas—The two main carrier gases used are helium and nitrogen (which must be oxygen free). Both are inert and are capable of producing excellent chromatograms. Helium is probably the more commonly used, especially in the United States. Its advantages are:

(*i*) Because of its low density, high flow rates are possible at comparatively low pressures. This reduces the time required for a separation.

(*ii*) The detectors used often depend upon gas density and thermal conductivity. Helium has a low density and variations through dissolved sample components are easily detected. It has a high thermal conductivity which also increases the sensitivity of detectors. Helium is expensive, however, and for many purposes, especially where other types of detector are used, oxygen-free nitrogen is adequate for the successful operation of a gas chromatograph. In some special cases argon or hydrogen are used as carrier gases.

The gases are obtained in cylinders from the commercial suppliers. Gas flow through the system must be very carefully regulated and stabilized. Most chromatographs have built-in flowmeters for this purpose.

Sample Application—One of the main advantages of G.L.C. is its ability to use very small samples, usually 0·1–50 mm^3 of liquid. The most common way of applying this is with a micro-syringe. The sample is injected directly on to the column through an injection port which is sealed with a replaceable septum made of neoprene or silicone rubber. The microsyringe is fitted with a long fine needle which pierces the septum when the sample is applied. The elastic nature of the septum allows self-sealing on the withdrawal of the needle.

A more accurate but less convenient method involves the sealing of a weighed amount of sample into a fragile glass ampoule. The ampoule is inserted into a heated chamber at the head of the column. After the chamber is sealed off, carrier gas is allowed to flow through it and the ampoule is crushed mechanically releasing the vapourized sample into the gas stream.

Samples of gas can be applied with gas micro-syringes, or can be taken from a gas stream through specially designed valves which fill a chamber of known volume with sample gas and then open the chamber to the column.

The Column—Columns for G.L.C. are usually made of glass or copper tubing of up to 6 mm i.d. The column is heated in use so that the tubing is coiled or bent in order to fit conveniently inside a temperature controlled oven. In most machines columns of 1·3 or 2 m length are standard. The packing consists of a solid granular support with the stationary phase, a non-volatile liquid, held on its surface.

The more common supports are porous inert materials based on diatomaceous earths and crushed firebrick. For even and consistent results, the support powder is sieved to give a fairly uniform powder size with no 'fines' which tend to slow the passage of carrier gas. Common sizes are 60–80 mesh (0·25–0·18 mm), 80–100 mesh (0·18–0·15 mm), and 100–120 mesh (0·15–0·13 mm). Before coating with substrate, the support is acid and alkali washed and thoroughly dried. Some supports have a slightly adsorbent effect which leads to slowing and tailing of components. This can be overcome by pre-coating the powder with dimethylchlorosilane. This hydrophobic material renders the support virtually inert. The process is called *silanizing*.

Several hundred liquid phases or *substrates* have been described in the literature for specific purposes. In fact most G.L.C. is carried out on about a dozen well documented substrates. The two main criteria which govern the choice of a substrate, apart from those common to all forms of chromatography, are polarity and the upper useful limit of temperature (Table 7.4).

The polarity of a substrate liquid is not defined in absolute terms as its dielectric constant but is decided empirically by its ability to separate the components of a mixture type. Non-polar components will usually separate better on a non-polar substrate; on a polar substrate the peaks will tend to crowd closer together. A mixture of polar compounds will be dealt with on a polar substrate.

There is always a tendency for the substrate to vaporize slightly especially at elevated temperatures. The upper useful limit represents the highest temperature at which the column can be used without this vaporization interfering with the operation of the detector. It is not a definite figure but depends to some extent on the type of detector in use. Most substrates can work efficiently for short periods at temperatures which exceed the quoted upper useful limit.

The substrate concentration on the column is specified in terms of percentage by weight and is referred to as the *loading* of the column. The required weight of substrate is dissolved in a volatile solvent, the solution being mixed with the prepared support material. The volatile solvent is removed under vacuum leaving an evenly coated dry granular column packing. This is poured into the glass or copper tube being tapped or vibrated with an electrical device to aid even packing. Glass tubes are packed after coiling but copper tubes are usually packed straight and coiled afterwards.

A recent development is the capillary column which eliminates the support medium. This consists of a very long (50 metres) very narrow (0·25 mm i.d.) tube in which the stationary phase is coated directly on the walls. These columns are highly efficient and require very small samples.

Before use all columns must be conditioned by flushing with carrier gas for several hours at the upper permissible temperature limit.

Detectors—The number of useful detectors which have been described for G.L.C. runs well into double figures. They have been grouped into two major types or 'families':

(*i*) First family detectors respond to the concentration of the component as a solute in the carrier gas.
(*ii*) Second family detectors respond to the flow rate of the solute in the carrier gas.

The detectors of the second family tend to be both more accurate and more sensitive than those of the first family for quantitative measurements. Their main disadvantage is that they destroy the components of the sample when detecting them. Modern G.L.C. is often used as a preparative as well as an analytical device, and even the analytical method is extended by linking the apparatus directly to a mass spectrometer so that individual components may be further characterized. This is not possible with second family detectors unless special stream-splitting devices are used.

In the space available here discussion of the different detectors must necessarily be limited and only a few examples can be given.

The *thermal conductivity detector* or *katharometer* is an example of the first family and is the device used on the earliest G.L.C. apparatus. The carrier gas leaving the column, passes through a small chamber containing either a heated resistor or a thermistor. The temperature of the resistor will depend upon the rate at which it can dissipate heat. This, in turn, will depend upon the thermal conductivity

TABLE 7.4. Stationary Phases for G.L.C. (Supplied by B.D.H.)

Stationary phase	*Approx. density* g per cm³ *at* 20°C	*Recommended working temperature* °C	*Solvent*	*Typical Applications*
Apiezon grease L	Solid	up to 280	2, 4, 6, 7	Many, including alcohols, aldehydes, essential oils, esters, fatty acids and derivatives, phenols, etc.
Benzylbiphenyl	Solid	80–150	1, 3, 4	Alkyl halides, aromatic amines, fatty acids, oxygenated compounds
Di(2-cyanoethyl) ether	1·05	30–150	5, 6	Hydrocarbons, C_1—C_6 olefines, oxygenated compounds
Diethyldigol	0·91	room temp.	1	Aliphatic aldehydes, alcohols, amines
Diglycerol	1·3	20–100	5, 6	Hydrocarbons, alcohols, aldehydes, ketones, carboxylic acids
Di(2-methoxyethyl) adipate	1·08	up to 100	5	Light hydrocarbons
Dimethylformamide	0·95	—20 to 20	5	Hydrocarbons including olefines (up to about C_5)
Dimethylsulpholane	1·14	room temp.	1, 3, 5, 6	Alkyl halides, hydrocarbons, xylenes
Dinonyl phthalate	0·97	20–100	1, 3, 6	Many, including halogenated compounds, organic sulphur compounds, hydrocarbons, etc.
Dodecylbenzene sulphonic acid sodium salt	Solid (not normally coated)	up to 225		Phenols
n-Hexadecane	0·77	20–50	2, 4, 7	Fluoro compounds, hydrocarbons up to C_8
2, 6, 10, 15, 19, 23-Hexamethyltetracosane (squalane)	0·81	20–150	2, 4, 6	Alcohols
Poly(1, 4-butane-diol succinate)	Solid	up to 200	3	Fatty acids, essential oils, etc.
Poly(diethyleneglycol succinate)	Solid	up to 180	1, 3	C_5^+ fatty acid esters, amino acid esters
Poly(2,2-dimethylpropane-1,3-diol succinate)	Solid	up to 200		Fatty acids, essential oils, etc.
Polyethylene glycol 400	1·12	up to 100	3, 5, 6	Ethanol in blood, C_1—C_5 alcohols, esters

Polyethylene glycol 1540	Solid (m.p. 43–46)	up to 160	3, 5, 6	Alcohols, esters, aldehydes, ketones, phenols, aromatic compounds
Polyethylene glycol 20 M	Solid (Softens 50–55)	up to 250	3, 5, 6	Alcohols, esters, aldehydes, ketones, phenols, aromatic compounds
Polyethylene glycol adipate	Solid	up to 200	1	Fatty acid methyl esters
Polyethylene glycol succinate	Solid	up to 200	1	Fatty acids, esters, essential oils
Propylene carbonate	1·2	room temp.	1, 3, 6	IP Method for buta-1,3-diene
Silicone fluid MS 550	1·07	up to 225	1, 4, 6	Boranes, esters, hydrocarbons C_5 and above
Silicone gum XE-60	Solid	up to 250	1	Separation of TFA + amino acids, steroids, bile acids
Silicone oil MS 200/200	0·97	up to 250	4, 7	General
Silicone OV-1	Gum	up to 350	3, 4, 7	Very stable DMS gum, useful for preparative work
Silicone OV-17	Viscous liquid	up to 350	1, 3, 4, 7	A stable DMS polar fluid for high temp. work; for polar derivatives of low volatility
Silicone OV-25	Viscous liquid	up to 300	3	A DMS fluid, highly polar, for high temp. work with flame detectors
Silicone OV-101	Viscous liquid	up to 350	3, 7	A dimethylsilicone useful for capillary work
Silicone OV-210	Viscous liquid	up to 300	1, 6	A highly stable fluorosilicone, useful for separation of ketones and alcohols
Silicone OV-225	Viscous liquid	up to 300	1, 3	Highly polar with excellent temperature stability
Silicone QF-1	1·26	up to 250	1, 3	TFA-amino acids, steroids, halogen compounds
Silicone QF-1, 1·5% solution in acetone	0·82	up to 250		
Silicone SE-30 5% w/v solution in chloroform	1·45	up to 250		Oxygenated or halogenated compounds
Tritolyl phosphate	1·16	20 to 150	1, 3	Alkyl halides, hydrocarbons, solvents, organic sulphur compounds, octenes
Versamid 930 (polyamide resin)	Solid	up to 250	8	High boiling polar materials
N-Trifluoroacetyl-*L*-*β*-phenylalanine cyclohexyl ester	Solid	up to 120		Resolving optical isomers

1 = Acetone
2 = Benzene
3 = Chloroform
4 = Hexane
5 = Methanol
6 = Dichloromethane (methylene chloride)
7 = Toluene
8 = Hot chloroform: *n*-butanol = 1:1 or chloroform: methanol = 85:15

of its surroundings. If the resistor is in the gas flow and the rate of flow remains constant the temperature of the resistor will not change while surrounded by pure flowing gas. If the flowing gas contains some organic sample component as a solute, the thermal conductivity of the surroundings will drop, and the temperature of the resistor will rise. This will increase the resistance of the resistor. A second, similar reference resistor in a second chamber is surrounded only by pure flowing carrier gas. The two resistors form opposite arms of a Wheatstone bridge circuit. As long as both are surrounded by pure carrier gas the bridge is balanced. If a component leaves the column to affect the temperature of only one of the resistors the bridge will become unbalanced. This is normally detected by a galvanometer or can be displayed on a chart recorder. *Figure 7.5* shows a simple arrangement for using a thermal conductivity detector. If thermistors are used the resistance will fall as the detector thermistor is heated relative to the reference thermistor.

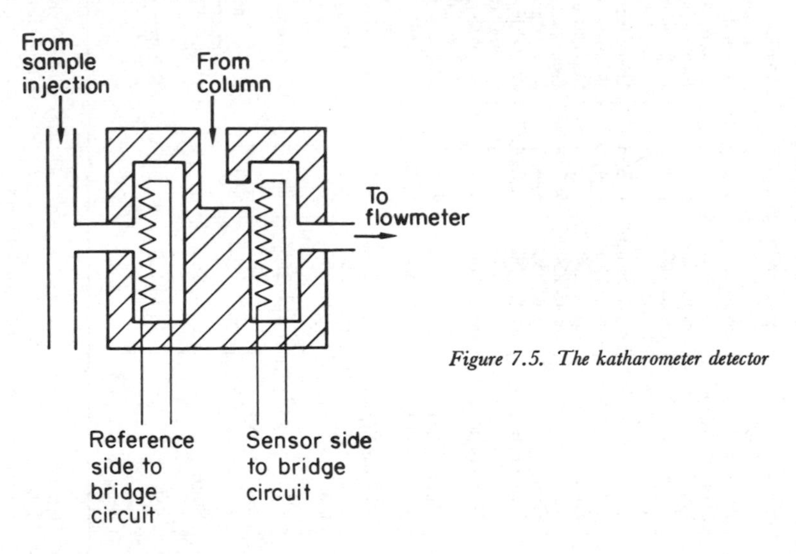

Figure 7.5. The katharometer detector

The electron capture detector (*Figure 7.6*) is also a first family device. The gas leaves the column and passes through an ionizing chamber which contains a small permanent radioactive source. Strontium 90, Yttrium 90 or titanium foil saturated with tritium are often used. These are pure β emitters and the radiation hazard is negligible.

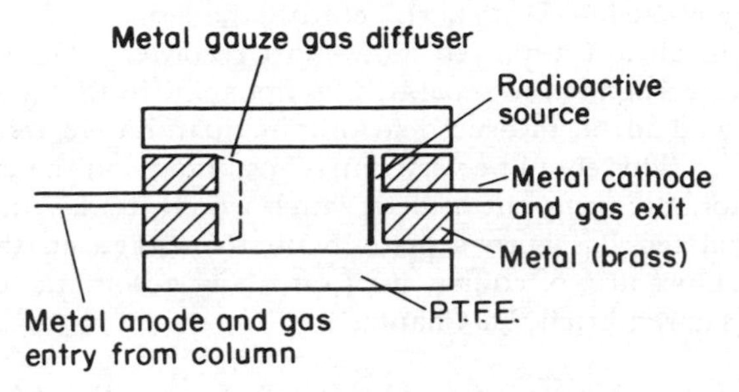

Figure 7.6. The electron capture detector

As the carrier gas passes through the chamber, it is subjected to a constant bombardment of β particles. Electrons are attracted to a postively charged collector grid thus sending a signal current through an amplifier to a recorder. As long as only pure carrier gas is passing the signal will be constant. As an organic sample component, which contains electron absorbing molecules, passes through the chamber the signal current falls and a peak appears on the recorder. Hydrogen and nitrogen are used as carrier gases.

The Flame Ionization Detector (*Figure 7.7*)—This is an example of the second family detector. The sample component dissolved in the carrier gas is pyrolysed in a small oxy-hydrogen flame. The carbon in the sample is ionized, and the ions are collected on a small positively charged metal ring supported just above the flame. The collection of ions on the charged ring sends a signal through an amplifier and produces peaks on a chart recorder.

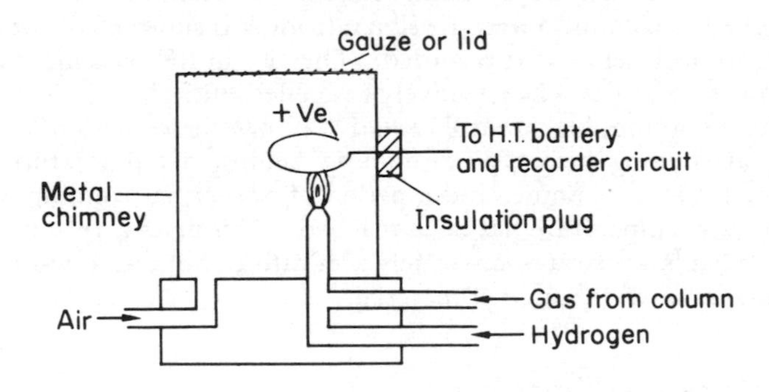

Figure 7.7. The flame ionization detector

The Display of Results—Until fairly recently the signal from the G.L.C. detector was always displayed on a chart recorder. The peaks on the chart corresponded to separated components in the sample and the peaks had to be measured to obtain quantitative results (see Chapter 8). This could be very time-consuming and the introduction of automatic print-out devices which would display the results ready calculated has been a great boon to workers in this field. These machines are, of course, used on many automatic analysers and are discussed briefly in Chapter 8.

Programmed Temperature G.L.C.—In the foregoing discussion it has been assumed that the column temperature is kept constant throughout the chromatographic run. If the sample contains a number of components of similar type but widely differing volatility, the components of high volatility would be eluted quickly and would crowd together, while high boiling point compounds would be eluted very slowly and would produce shallow spread peaks on the chart.

In programmed temperature G.L.C. the oven temperature is increased at a fixed rate throughout the run. The peaks are then more evenly distributed and the run is finished more quickly. Unfortunately the rise in temperature during the run increases the rate at which substrate vapourizes from the support material. This increased *bleeding* of substrate gives rise to a steadily rising baseline on the chart record. Sensitivity and accuracy are both considerably reduced under these circumstances.

The answer to this problem is the use of a pair of matched columns usually marked *A* and *B* with associated matched detectors. The bleeding signal will be the same from both columns. The sample is applied to column *A* and the signal from *B* is subtracted from the signal from *A* before it is recorded. The rise in the baseline due to column bleeding is thus effectively cancelled out.

The programming control usually allows for a period of isothermal running before programming begins, a temperature rise rate of 1–64° per minute, and a period of isothermal running when the higher temperature has been reached. This gives a very flexible and versatile apparatus on which accurate quantitative work can be done on a wide range of materials.

The Apparatus of Column Chromatography

For the majority of uses of column chromatography, the basic apparatus is simple and inexpensive. For many experiments all that is required is a suitable glass tube in which the column may be

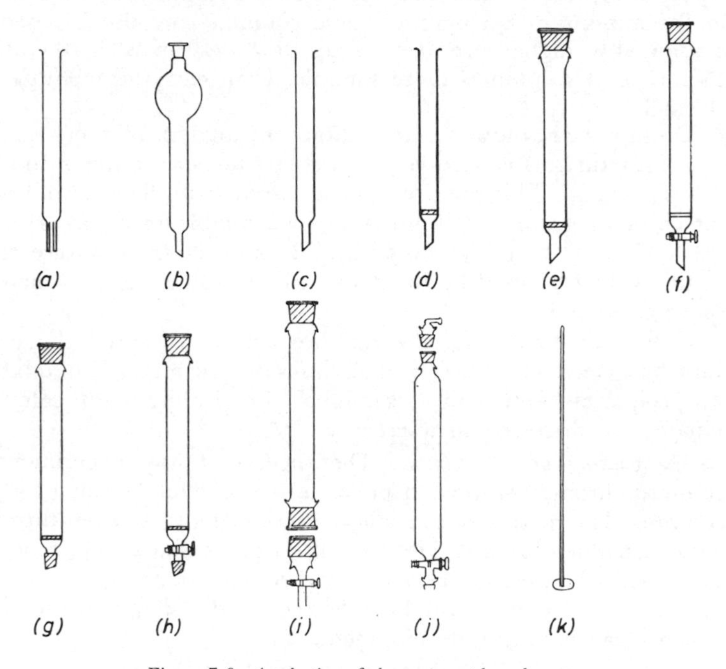

Figure 7.8. A selection of chromatography columns

packed. A selection of tubes designed for various purposes is shown in *Figure 7.8.* Extensions to columns having ground glass sockets at the upper end are useful during the packing of the column. To ensure even packing a tamping rod (*k*) is often used. This consists of a disc of perforated stainless steel very slightly smaller in diameter than the inside diameter of the tube, attached to a long stainless steel rod, so that the disc can be pushed the length of the tube.

It is usually convenient for the developing and eluting solvents to be held in reservoirs above the column. Globular or Squibb's type separating funnels make useful reservoirs. The reservoir may be attached directly to the tube by a ground-glass joint, or may be held some way above the column and connected by flexible tubing, thus creating a fairly constant slightly positive flow pressure.

Several commercial companies now market more sophisticated

glass columns. These have piston seals at each end enclosing the column between porous discs. Dead space, where re-mixing of eluate may occur, is eliminated, and the sample may be applied in the minimum of volume. These columns can also be used at considerably higher pressures of up to $7 \times 10^5\ Nm^{-2}$ (100 p.s.i.). This type of column is more suitable when carrying out upward elution.

To improve resolution and facilitate the analysis of micro quantities of mixtures, it is necessary to use narrow bore columns and fine grain packing. This requires higher pressures to develop and elute the chromatograms. Columns are now available for use at pressures up to $7 \times 10^6\ Nm^{-2}$ (1000 p.s.i.). Because of their nature these columns may be used horizontally. Higher pressures also increase speed of analysis.

If the components of the mixture are all coloured the eluate may be collected in tubes or small flasks but the majority of mixtures are colourless and some techniques for dealing with colourless eluates are discussed in a later section.

The packing of the column—The success of any experiment in column chromatography depends upon careful packing of the column. The density of the solid material should be even throughout. Channels or air bubbles will prevent sharp separation of the zones. The upper surface of the column must be smooth and flat; any unevenness will be reflected in the shape of the zones which separate during development.

There are three main methods of packing the column:

(1) *Dry packing*. The dry method of packing is used for dry adsorbent materials, i.e. alumina. The column is packed in sections of up to 2 cm length, the wider the column the shorter the section. The adsorbent powder is poured into the tube and gently tapped down with a tamping rod. The tamping rod can be formed using a cork whose wider end is slightly less in diameter than the bore of the tube, with a glass rod stuck into the narrower end. After each section is evenly packed a further amount of powder is poured into the tube and tamped down gently until a column of the required height has been built. The tamping of each section must be even so that some layers are not packed more tightly than others.

For very small narrow columns it may be possible to dispense with the tamping rod and to settle each section by gently tapping the end of the tube on the bench.

The upper surface of the freshly packed column may be protected with a disc of filter paper or by the addition of a further section of some inert practically non-adsorbent powder such as one of the common filter-aids. This protection will prevent the upper surface of the adsorbent column being made uneven when liquids are poured on to it.

(2) *Wet packing.* The simplest, and most common, method of packing columns is simply to make a mixture of adsorbent and solvent to the consistency of a thin cream; this 'slurry' is poured into the tube, and the solid allowed to settle. If the total volume of the slurry is greater than the volume of the tube, an extension tube is attached.

Before pouring the slurry it may be convenient to let the mixture settle in the flask in which it is prepared so that the 'fines' which remain in the supernatent may be sucked off. The solvent removed with the 'fines' is replaced by fresh solvent and the homogeneous slurry reformed. In general the more dilute the slurry, the more even the column packing, but, of course, the limit of dilution is controlled by the volume which can be poured conveniently.

(3) In another method of wet packing the vertical tube, closed at the lower end, is filled with solvent. The solid is gently sprinkled on the solvent surface and allowed to settle evenly until a column of the required height has been formed.

A well-built column can usually be used several times for similar separations. If this is necessary the column must never be allowed to become dry or it will shrink away from the walls of the tube and develop cracks and channels. The glass wool or sintered material at the lower end must not be allowed to dry out or uneven flow may result.

Application of the mixture—The solvent on the newly packed column is allowed to run out at the lower end until the meniscus is just in the upper surface of the column.

The mixture to be separated, dissolved in a very small quantity of solvent is run very carefully on to the adsorbent so that the surface is not disturbed. For this purpose a pipette with a bent tip is convenient so that the solution can be run down the inner wall of the tube. The solution is allowed to soak into the surface of the adsorbent. The mixture may be 'washed in' with two or three further small quantities of solvent. When the meniscus has just soaked into the surface of the adsorbent, the tube above the column is filled with the developing solvent and a reservoir attached if this is to be used.

Collection of fractions—If the mixture is composed entirely of coloured materials, it may be possible to push the column out of the tube when the zones have separated and slice out the coloured bands with a knife. The coloured material may be extracted from each band with a suitable solvent. This is by no means a simple process as a column may disintegrate on extrusion through being either too wet or too dry, and the ideal degree of moistness which will hold the cylinder together is not easy to gauge.

It is much more usual to wash the coloured material out of the column and collect each band separately in a suitable container.

If the mixture contains colourless substances, however, and this is more likely to be the case, special measures must be taken to collect each substance in a separate container. Various techniques have been developed for doing this and some of the more common are briefly discussed.

(*a*) Before chromatography is attempted the substances in the mixture are converted to coloured derivatives which can be easily reconverted to the original material. For example, colourless carbonyl compounds may be converted to 2:4-dinitrophenyl hydrazones which are orange. The carbonyl compound may be recovered by the action of diacetyl.

A disadvantage of this technique is that yields may be low both on conversion to and conversion from the coloured compound, and it may be difficult to prevent considerable loss of material.

(*b*) The column itself may be treated with a reagent which is strongly adsorbed and which locates individual bands during development. For example, some acids may be located by pre-treatment of the column with methyl orange indicator.

(*c*) The extruded column may be painted along its length with a thin streak of a locating reagent. This will indicate the position of the bands. The small portion of coloured material can be discarded after the zone has been cut out of the column and before extraction. For example, Vitamin A can be located by painting a thin streak of a chloroform solution of antimony trichloride along the column. The reagent gives a blue colour with Vitamin A.

(*d*) Many colourless substances fluoresce in ultra-violet light, and the zones on a column can be located in the illumination of the discharge from a mercury vapour lamp in a darkened room. This technique is especially useful when dark adsorbents such as charcoal or ion exchange resins are used.

(*e*) Radioactive tracers may be followed down the column. If one particular compound is to be isolated from a mixture it may be possible to synthesize a separate sample of the compound in which one of the elements is radioactive. This is added to the mixture. The labelled and non-radioactive compounds behave identically on the column and the zone containing the tracer may be followed from application to elution with a Geiger counter.

[By courtesy of Aimer Products Ltd]

Figure 7.9a. The 'Central Ignition' fraction collector

The use of tracers is especially useful in the chromatography of inorganic compounds as simple labelled ions are readily available.

(*f*) Dye markers are often used. Methods have been developed in which a dye is added to the mixture, the progress of the dye down the column marking the progress of some other component. For example, Sudan III is added to mixtures containing Vitamin D to mark the position of the Vitamin D zone on the column.

(*g*) The eluate is collected in small quantities in a large number of test-tubes. If a small sample is taken from each tube and tested for the components of the mixture the course of the separation may be followed. As many chromatographic separations take hours, even days, to complete, it would be plainly impossible to separate the eluate manually in this fashion. Automatic fraction

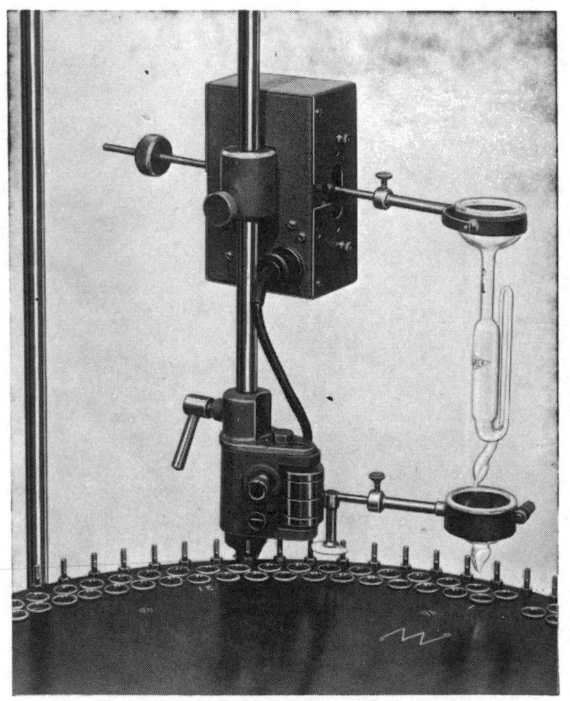

[By courtesy of Aimer Products Ltd]

Figure 7.9b. Siphon arm, Central Ignition fraction collector

collectors are used for this purpose. The fraction collector shown in *Figure 7.9* is representative of this apparatus. A circular turntable has test-tubes in holes around the edge. The first tube is placed so as to collect the eluate from the column. At a predetermined point the turn-table jerks so that the first tube moves from under the column and the second takes its place. The process continues until the turn-table has moved completely round and the tubes contain successive small portions of the eluate. The turn-table may be actuated in a number of ways. The most common ways are:

(*i*) By a time switch, so that each tube is collecting eluate for a predetermined time. This has the disadvantage that the rate of flow from the column tends to slow down as the solid phase packs down when liquid filters through it. The rate of flow at the end of the experiment may be only half that at the beginning and the tubes will not contain comparable amounts of the liquid.

(*ii*) By drop counter. The drops leaving the end of the column are counted by a photoelectric device and the turn-table is actuated when a pre-set number of drops have fallen from the column. This is useful when very small fractions are to be collected. A popular drop-counter is marketed which can be set to actuate the fraction collector for any number of drops between 1 and 999. One disadvantage of the drop-counter is that it is expensive. Another is that the size of the drops may vary during the separation if the eluant alters.

(*iii*) The siphon balance (*Figure 7.9b*) is a metal balance arm with a mercury switch at the centre. At one end is a siphon pipette, at the other end weights are fixed so that the arm is horizontal when the siphon is about two-thirds full. The funnel of the siphon is placed under the column so that the eluate drips into it. When the siphon is full it overflows into a tube on the collector turn-table and as it empties the balance arm is tilted, tilting the mercury switch in turn and actuating the motor to bring the next tube under the siphon. Siphons may be obtained in sizes from 1 cm^3 upwards although it may be difficult to balance a siphon of less than 3 cm^3 capacity. The siphon balance is perhaps the most popular way of using a fraction collector. The two main disadvantages are the delicate structure of the balance mechanism and the small amount of mixing that may occur from residues which do not drain completely from the siphon.

(*h*) Continuous monitoring of the eluate is often employed and is essential for quantitative analysis. Several systems are used, the most common are:

(*i*) The measurement of absorbance. The column eluate is

passed through a flow cuvette in an absorbance monitor. This may in fact be a spectrophotometer but usually is an instrument with a fixed wavelength from an interference filter or from an ultra-violet source. The optical path length may be 10 mm with a cell volume of less than 10 mm^3.

(*ii*) The differential refractometer. This monitors the refractive index changes between the eluate and a stream of pure solvent. Again the cell volume is a few mm^3. The sensitivity may be 1×10^{-7} R.I. units.

(*iii*) The moving wire system. A coil of wire is fed successively through a cleaning oven, the column eluate, a solvent vapourizing oven then a pyrolysing oven where the remaining solids on the wire are vapourized and swept into an F.I. detector by an inert gas.

(*iv*) Continuous chemical analysis. In this system the eluate is continuously analysed in an automatic analyser.

Results from these systems are recorded on a chart recorder. The results may be simultaneously fed into an integrator and the concentration calculated. The eluate may be fed into a fraction collector. Each fraction may be automatically noted with an event marker on the recorder. Some systems are able to detect and collect the peaks only, by-passing eluate that is without chromatographed compounds into a 'dump' reservoir.

The Uses of Chromatography

Although the use of chromatography for straightforward separation of mixtures for preparative or analytical purposes is obvious, there are many other problems which may be solved by the use of this technique. It may be helpful, therefore, to conclude this section with a list of some of the more common uses of chromatography.

(*i*) The simple separation of mixtures into constituents for qualitative and quantitative analysis.

(*ii*) The preparation of pure substances and the purification of the products of reaction.

(*iii*) The proof of the homogeneity of a substance. If only one spot is obtained after paper chromatography of a substance with several solvents it is likely that the material is homogeneous. Impurities will be indicated by other spots, which may lead to identification of the impurities.

(*iv*) The identification of substances. The position of spots on paper and the reaction with location agents is a common method of identification of substances. The monitoring of column eluates performs a similar function.

(*v*) Monitoring of other separatory processes. The fractions from other procedures, e.g. zone electrophoresis, countercurrent extraction, may be examined by chromatography.

(*vi*) Monitoring of reactions. The course of a reaction may be checked by taking a sample of the reaction mixture and running a chromatogram. Thin layer chromatography is especially useful for this because of the great speed with which results are obtained. This is a common procedure in manufacturing where fermentation reactions are carried out.

PAPER CHROMATOGRAPHY

Liquid-liquid partition chromatography on columns has already been mentioned above. Chromatography on sheets of filter paper or on thin layers of support material is probably even more common. Filter paper, which is composed almost entirely of cellulose fibres, acts as the inert carrier for the stationary liquid phase, usually water, while the solvent mobile phase is drawn through the paper by capillary action.

The procedure may be summarized as follows: a drop (a few mm^3) of the mixture to be analysed is carefully placed about 25 mm from one end of the strip of filter paper and dried. The end of the filter paper nearest the spot is dipped into the developing solvent which soaks into the paper and flows over the 'spot' by capillary action. As the solvent front wetting the paper progresses it is followed by 'spots' of the individual substances in the mixture which move forward at different rates, the rate for each substance depending upon its partition coefficient between the solvent and aqueous phases. This dependence may be related to the relationship between the distance of the spot from the origin or starting position and the distance of the solvent front from the origin. The factor is termed the R_F value of the substance for the particular solvent, i.e.

$$R_F = \frac{\text{Distance of spot from origin}}{\text{Distance of solvent front from origin}}$$

The R_F value is, of course, specific for a combination of solvent and substance. Tables of R_F values for a very large number of substances in various solvent combinations have been worked out and are available in the specialist literature.

Techniques of Paper Chromatography

Several methods of producing a paper chromatogram have been described, many of them only applicable to special cases. Most separations depend on one of the two main procedures.

In *descending* paper chromatography, the reservoir of solvent is supported and the paper inserted so that it hangs free from the reservoir and the solvent front moves down the paper. In *ascending* paper chromatography the solvent reservoir is at the bottom of the apparatus with the paper suspended above it so that the solvent front moves upwards. Most commercial apparatus for paper chromatography may be adapted for either technique.

Where two or more of the substances in a mixture have similar R_F values for a particular solvent, the technique of two dimensional chromatography is often used.

The spot is placed towards one corner of a square sheet of paper about 25 mm from either edge, and a chromatogram developed along one side with a suitable solvent. The paper is dried and turned through 90 degrees so that the chromatogram is horizontal and a second solvent applied which may fully resolve spots which have been incompletely separated (*Figure 7.10*).

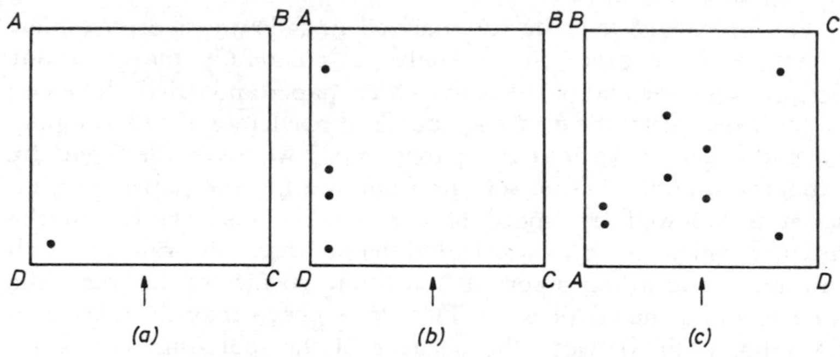

Figure 7.10. Two-dimensional paper chromatography
(*The direction of solvent flow is indicated by the arrows*)

Paper—Several grades of Whatman paper are available for chromatography and the careful selection of the suitable grade of paper for any particular separation is very important. Table 7.5 gives the characteristics of some of the most commonly used papers and it will be seen that there are variations in rate and uniformity of flow, strength, and carrying capacity. Some papers do not exhibit uniform flow in all directions and individual packets of paper are marked with an arrow showing the direction of machining in manufacture, which is also the recommended direction of solvent flow.

TABLE 7.5. Whatman Filter Papers for Chromatography

No.	*Weight* g/m^2	*Upward flow rate index* (*water*)	*Characteristics*
1	87	55	Standard qualitative paper
3	185	63	Thick rough paper
3MM	180	52	Thick smooth paper. High sample capacity
4	92	105	Same thickness as 1 but much faster
15	140	102	Thick open texture, fast flow rate
20	93	25	Very slow paper producing very compact spots

Apparatus—The apparatus of paper chromatography is, if anything, even simpler than that commonly used in column chromatography. The necessary characteristics of the apparatus are:

(*a*) The solvent reservoir must be big enough to hold sufficient solvent to complete the chromatogram.

(*b*) The paper must be freely suspended.

(*c*) Large temperature changes during the 'run' must be avoided. Although R_F values do not change appreciably for a variation in temperature of a degree or two, gross changes in temperature can produce quite large variations in R_F.

(*d*) The solvent and vapour phases must remain in equilibrium. This means that chromatography is usually carried out in closed tanks containing an excess of solvent at the bottom.

A typical chromatography tank is shown in *Figure 7.11*. For class work a gas collecting cylinder with a greased lid is often convenient. The samples are spotted along one edge of a sheet of paper which is then rolled to form a cylinder with a paper clip to hold it in shape. The rolled paper is placed in the gas cylinder which has solvent at the bottom with the samples arranged around the base.

Where large numbers of papers are to be run together they may be clamped side by side in a metal frame which is then inserted in the chromatography tank. Up to 50 samples may be run simultaneously in this manner.

Preparation of sample. Desalting—Mixtures of amino acids and sugars are often prepared by the hydrolysis of large molecules.

This process leads to the mixture of organic compounds being contaminated by large amounts of the acid or alkali used in hydrolysis. These contaminants would interfere drastically with chromatography on paper both by waterlogging the paper, forming spots of high water content, and by interfering with locating agents, e.g. $AgNO_3$ is used as a locating agent for sugars and would react with Cl^- instead of sugar if HCl were present.

[By courtesy of Shandon Scientific Coy Ld]

Figure 7.11. Shandon tank for paper chromatography

Several methods of *desalting* have been described.

(*a*) Passing the hydrolysate through a column of suitable ion exchange resin.

(*b*) Extracting the aqueous hydrolysate with organic solvents which will take up the organic material and separate it from the salts.

(*c*) Electrolytic desalting.

(*d*) Precipitation by chemical means.
(*e*) Gel filtration.
(*f*) Dialysis.

(*a*) To desalt a sample two resins may be required, one to remove inorganic anions, the other cations. The simplest method is to simply shake the sample with a small quantity of a mixture of resins which will leave the organic material in the filtrate after the mixture has been filtered.

A more efficient procedure is to run the sample through a *mixed bed* resin, i.e. a mixture of anion and cation exchange resins. Alternatively two columns may be used consecutively. Another common method is to pass the sample through a column which retains the organic materials but not the inorganic salts.

In clinical chemistry the paper chromatography of the amino acids in urine is important as a diagnostic technique. Urine contains large quantities of salts which must be removed before chromatography can be attempted. 2 cm^3 of urine can be desalted on a column 15 cm $\times$ 1 cm, of Zeokarb 225. After the 2 cm^3 of urine has been run on to the resin surface and soaked in, the column is washed with water. All the anions and sugars are in the first 30 cm^3 of eluate. The amino acids are eluted with 2 M ammonia which are eluted in the first 30 cm^3. The inorganic cations remain on the column. The ammonia is evaporated off in a rotary evaporator to provide a sample free of interfering inorganic ions.

The disadvantage of this method is that the final sample is in a rather large volume of liquid which must usually be evaporated down to give a more easily handled amount. Occasionally it is found that decomposition of the sample is catalysed by the resin.

(*b*) Another method of desalting urine is to extract the amino acids with three volumes of absolute alcohol. The salts and proteins are precipitated and removed by centrifugation. In other cases, the organic materials are extracted from aqueous solutions which also contain salts with non-miscible organic solvents which leave the salts in the aqueous layer. The organic solution is then evaporated down and the residue taken up in a solvent suitable for application to the paper.

(*c*) Electrolytic desalting is perhaps the most commonly used method. A typical apparatus is shown in *Figure 7.12*. By using a circulating mercury cathode and a platinum anode the inorganic ions can be removed electrolytically and washed away. A modification of this method is the electrodialysis cell. This is shown in *Figure 7.13*. The membranes are only permeable to ions of similar charge and reject ions of opposite charge. The electrodes cause continuous regeneration of the ion exchange material.

(*d*) If sulphuric acid has been used for hydrolysis it may be precipitated by the addition of a quantitative amount of barium

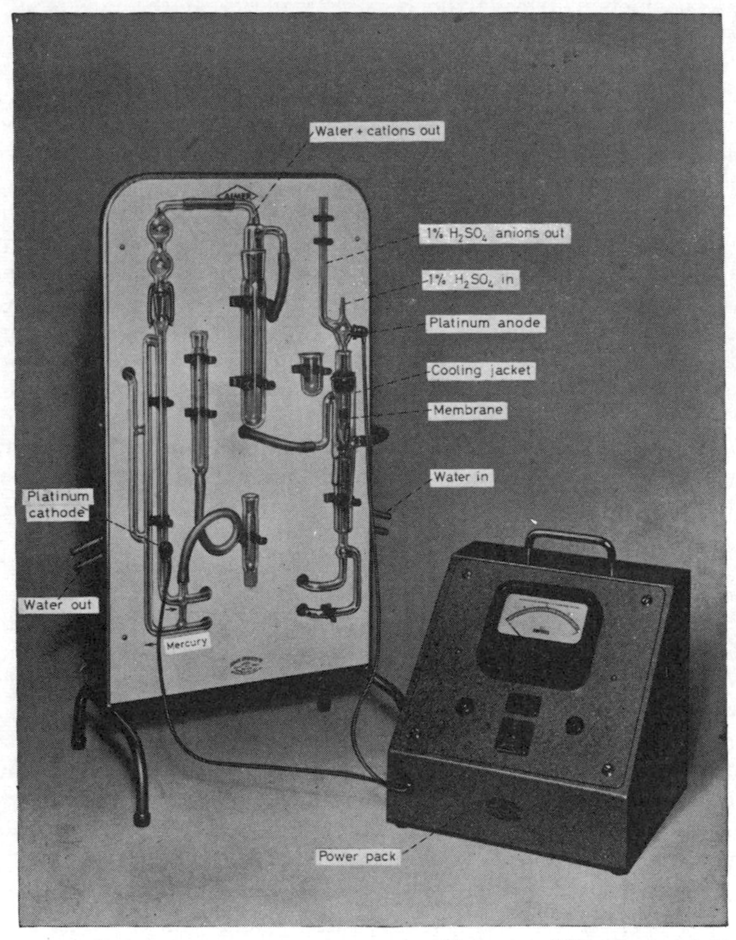

[By courtesy of Aimer Products Ltd]

Figure 7.12. The Aimer electrolytic desalter

carbonate. This type of chemical precipitation is often used when large amounts of material are to be desalted.

(*e*) This method is described in the literature issued by Messrs. Pharmacia (see page 306).

(*f*) This technique is discussed in Chapter 10, page 423.

Application of the sample—Successful paper chromatography depends upon the area of the 'spot' of mixture applied to the paper being as small as possible. Micropipettes ranging in size from 1 to 10 mm^3 are obtainable.

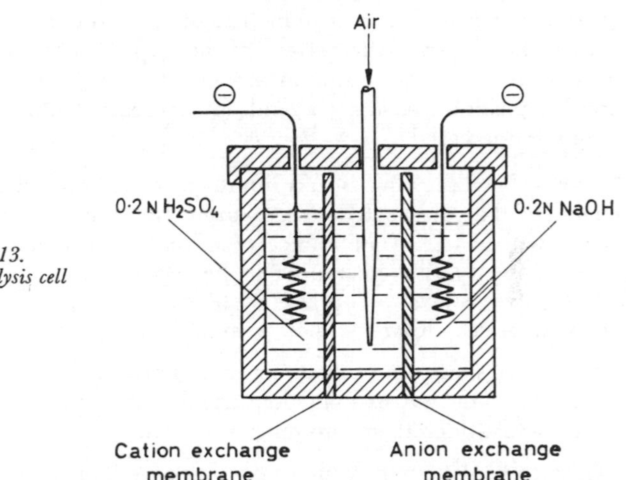

Figure 7.13. The electrodialysis cell

The pipette is used to apply a sample to the paper and the sample is dried. A commercial hair-dryer clamped in a stand will be found useful for this purpose. If it is desired to apply more than is contained in the pipette, further amounts can be applied to the same spot, each aliquot being dried before the next is applied. Some workers prefer to 'spot' the paper with a platinum wire loop, although it is difficult to work with quantitative accuracy using this device. A 'zig-zag' of heavy glass rod is useful as a 'spotting' platform to keep the paper off the bench surface when the wet spot is being applied.

In order to measure or compare R_F values it is usual to draw a faint pencil line at the origin before the paper is 'spotted'.

It is usually recommended that not more than 5 mm^3 of sample be applied to the paper in one aliquot. The larger the area of the 'spot' the more diffuse the pattern of the final chromatogram.

Before the paper is fitted into the solvent trough it is important to check that the trough itself is absolutely horizontal using a spirit level. Unless the paper is vertical the chromatogram may not run straight and if more than one sample is on the paper the chromatograms may run into one another.

Developing the chromatogram—The paper is placed into the tank or cabinet so that the end is dipping into the solvent trough. If the descending method is to be used, the paper is weighted down in the trough with a glass rod. The paper passes over another glass rod which acts as a siphon breaker and prevents the solvent merely siphoning over the surface of the paper. The spot is placed as near as possible to the siphon breaker, i.e. as close to the solvent source as possible, as the R_F tends to decrease if the solvent-spot distance is increased.

Solvents—In discussing liquid-liquid partition on columns it has been stated that most solvents are substances which are only slightly miscible with water. The choice of solvent is, of course, governed by the nature of the substances to be separated.

Most solvents are two-phase systems or mixtures of miscible organic solvents. One or two general principles can be stated.

(*a*) The more slowly the solvent moves the rounder and less diffuse the spots. This means that viscosity, density, and surface tension must be considered in choosing a solvent.

(*b*) In a solvent system which specifically contains water the R_F is increased by increasing the water content.

The characteristics desirable in a solvent are, in the main, similar to those required in column chromatography (page 269). Because a paper chromatogram is run in air, however, a few other considerations enter.

(*i*) It should not be difficult to remove the solvent from the paper after the chromatogram has been run. The paper must always be dried before the spots are sprayed and dipped.

(*ii*) The mixture must remain homogeneous for small changes of temperature.

(*iii*) The solvent should be stable in air.

(*iv*) Ideally the separated mixture should spread the whole length of the paper from origin to solvent front, i.e. the R_F values should be between 0·05 and 0·95. The individual substances should vary in R_F value by at least 0·05 from each other to separate completely.

Several solvent systems have been found to be useful for many biochemical separations. Water saturated phenol, *n*-butanol, and collidine are commonly used binary mixtures and *n*-butanol/acetic acid/water or *n*-butanol/ethanol/water are useful ternary mixtures.

Detection of spots—Many of the techniques of column chromatography are obviously applicable in detecting the separated zones on paper chromatograms. There are however some methods which are specific for separations on paper.

It is much easier to display the separated spots by colour reactions than it is on columns. The paper is removed from the tank or cabinet and dried, preferably by hanging it in a forced air oven. The paper is suspended by stainless steel clips from a glass rod and hung in the oven until it is completely dry. After removal from the oven it is hung in a hood or fume cupboard and sprayed with a suitable reagent from an atomizer of the scent-spray type. Several of these are marketed and the all-glass models with rubber blow ball and reservoir are the most suitable. One firm sells an aerosol atomizer with a glass reservoir for the reagent. The construction of a glass spray should not be beyond the scope of a moderately skilled glass-blower from one of the many published designs. After spraying, the paper is once again dried and the spots should show clearly against the white background of the paper.

Alternatively the dried paper may be dipped into a vessel containing the colour reagent. This method is especially suitable for small strips or sheets. Several advantages are claimed for this technique and where poisonous or offensive solvents are used it is certainly much safer.

If the sample has been radioactively labelled (Chapter 9) the spots can be recorded on an x-ray film by autoradiography.

The strip may be soaked in oil to render it transparent, and passed through a photoelectric densitometer which measures or records peaks corresponding to the absorbance of each spot. Other scanners measure the radioactivity of spots when the original mixture has been labelled. Details of these methods are given in the reference textbooks in the bibliography (Appendix II).

Some Modifications of Paper Chromatography

Although the term, paper chromatography, usually refers to liquid-liquid partition on paper sheets where the stationary phase is water, there have been several interesting modifications of the standard procedure which may be briefly mentioned.

(*a*) *Reversed-phase chromatography* has been described in which the stationary phase on the paper is not water but some organic solvent. For instance, esters of fatty acids have been separated on paper impregnated with rubber, and D.D.T. derivatives on Vaseline impregnated paper. This is a fruitful field of investigation for the separation of mixtures where no suitable solvent/water system can be found.

(*b*) Ion-exchange papers have been developed in which the cellulose of the paper has been modified to an ion exchange form (see DEAE, CM, etc.). This offers the opportunity to chromatograph samples too small to be separated on a column.

(*c*) Papers impregnated with aluminia have been used for liquid/solid chromatography in the manner of paper chromatography.

THIN LAYER CHROMATOGRAPHY

A technique which has rapidly become accepted is *thin layer chromatography*. Samples are chromatographed on plates of glass or plastic which have been coated with a very thin even surface of some suitable material. Alumina and silica gel are two common coatings.

Until recently the plates were invariably prepared in the laboratory, a typical apparatus for spreading glass plates being shown in *Figure 7.14*.

Many manufacturers now market ready prepared plates for chromatography and it must be admitted that these are generally more satisfactory, the coating being tougher and more even than can usually be achieved in the laboratory. Most workers now prepare their own plates only when a coating of a special nature or of particular instability is required.

A large number of designs for spreaders were marketed in the early days of the technique and some had only doubtful efficiency. They fell into two main groups:

(*a*) Apparatus in which a slurry spreader is moved over a series of fixed glass plates.
(*b*) Apparatus in which glass plates are pushed under a fixed slurry spreader.

In the authors' experience type (*a*) is much easier to use and gives more evenly coated plates than type (*b*), although one expensive motorized version of the latter has proved suitable.

One version of type (*a*) which has been much quoted is that designed by Stahl who put the method on a working basis in 1958. This is the apparatus, manufactured by Messrs Desaga (British agents Camlab (Glass) Ltd) which is shown in *Figure 7.14*.

Several of the larger manufacturers market chemicals specially prepared for thin-layer chromatography and a number of these sell a range of ready prepared plates. Some of these are on glass although plastic and aluminium foil are also used as supports for the thin-layer. The latter two versions are particularly useful as the plates can readily be cut to odd sizes or shapes for particular purposes. Care must be taken, however, that the support does not react with either the sample, the mobile phase, or any locating reagent which may be used.

Thin-layer chromatography has many advantages over paper chromatography and has become the method of choice in many fields. Among its advantages are:

(*i*) Speed—most chromatography separations take less than 1 h.

(*ii*) The separated zones tend to be far more discrete than those achieved on paper. There is a noticeable absence of spreading and tailing.

(*iii*) A much wider variety of reagents can be used, e.g. strong acids, which would attack paper, can be used in thin-layer chromatography.

(*iv*) Much smaller quantities can be detected than in paper chromatography.

In the last few years the method has been extended to include preparative-layer chromatography. A much heavier layer of stationary phase is used and components can be separated, located, and scrapped off the support to prepare appreciable quantities of useful materials. The technique is reminiscent of the earliest work of Tswett.

Other variations include the preparation of gradient layers where the composition of the stationary phase varies across the support.

Several quantitative methods of thin layer analysis have been developed. If the spots can be colour developed these may be removed into tubes by a suction device where the coloured component is extracted and measured in a photometer. One commercial device scans the plate with a flying spot of light, detecting areas of absorption photometrically.

Several systems produce thin layers on rods or narrow plates. After development these are pyrolysed and the product fed through an ionization detector similar to the moving wire system in liquid column chromatography.

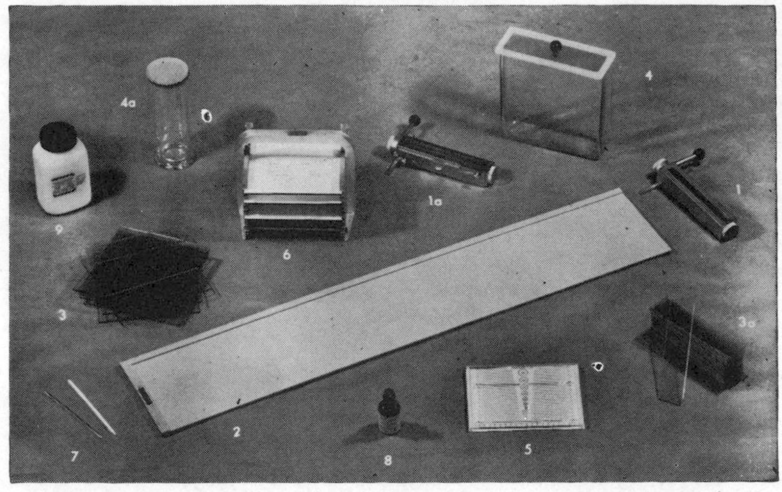

[By courtesy of Camlab (Glass) Ltd]

Figure 7.14. Desga apparatus for thin layer chromatography.
1. 1a. Spreaders. 2. Template to hold glass plates while spreading. 3. 3a. Glass plates. 4-4a. Tanks for different sizes of plate. 5. Spotting template. 6. Rack for drying plates in oven. 7. Micro pipettes. 8. Test mixture for comparing activation of plates. 9. Adsorbent powder

GEL PERMEATION CHROMATOGRAPHY

An extension of the molecular sieve concept which is of great importance, especially in biochemistry, is *gel permeation chromatography*. Although this is not a chromatographic method according to the most rigid definition, the methods and apparatus used are essentially the same and for all practical purposes it is considered to be a further chromatographic technique.

Separation of compounds is based here on the size or shape of the molecule rather than partition coefficients, and the older term *gel filtration* is perhaps more descriptive of the method.

Although some early work had indicated the feasability of separation on this basis, it was Porath and Flodin in the late nineteen-fifties who first produced materials which could be used practically and reproducibly as stationary phases. Artificially cross-linked dextrans were designed which, in gel form, could specifically exclude molecules of larger size than a particular molecular weight, while allowing smaller molecules to permeate the particles of the gel, and thus be retarded in the flow of mobile phase. The cross-linking can be carefully controlled so that the molecular weight of substances excluded from the particles of gel can be defined: the higher the cross-linking, the lower the molecular weight exclusion limit.

Each particle of gel may be considered as a bead with a three-dimensional network of pores within it. The aqueous gel is packed in a column and the sample applied. As the mixture proceeds down the column in an aqueous solvent, the larger molecules percolate around the beads of dextran gel because they cannot enter the porous network of the particles. The smaller molecules can actually enter the beads and are retained within them to a greater or lesser degree depending upon their size. Thus molecules larger than the molecular exclusion limit appear in the eluate virtually with the solvent front. Smaller molecules will separate on the gel column, their position as zones in the eluate depending upon their size and shape.

The dextran gels of Porath and Flodin are marketed by Messrs Pharmacia under the trade name 'Sephadex'. Several grades of Sephadex are available differing in molecular weight exclusion (Table 7.6). These are all for use with aqueous media. Recently grades of Sephadex have been produced which may be used with organic solvents. The manufacturers have published a very large range of leaflets describing their products and giving detailed instructions for their use. Technical experts are available to help with individual problems and may readily be consulted. Pharmacia have companies in U.S.A., Britain and other European countries for distribution of their products and for consultation.

A series of ion-exchange materials, based on Sephadex and similar to cellulose ion exchangers, has been developed. These are rather easier to use than the cellulose products and have a much higher capacity. They have no gel permeation properties, however, and must not be confused with the materials described above.

Another material of different composition but similar properties to Sephadex is Bio-Gel P, a polyacrylamide prepared for gel filtration by Messrs Bio-Rad Laboratories. These materials cover a different range of molecular exclusion limits and are especially useful in the higher range.

Both companies also market agarose gels for very high molecular exclusion limits under the trade names Sepharose and Bio-Gel A. Agarose is the non-ionic component of Agar and can be used for the separation of components with molecular weights up to several million. These are especially useful for separating nucleic acids, viruses, and other very large molecules.

TABLE 7.6. Grades of Sephadex for use in Aqueous Media

Sephadex type	*Particle diameter* μm	*Water regain* H_2O/g *dry Sephadex*	*Bed volume* ml/g *dry Sephadex*	*Fractionation range for*	
				peptides and globular proteins M_w	*dextrans* M_w
Sephadex G-10	40–120	1·0 ± 0·1	2 – 3	– 700	– 700
Sephadex G-15	40–120	1·5 ± 0·2	2·5– 3·5	– 1 500	– 1 500
Sephadex G-25 Coarse	100–300	2·5 ± 0·2	4 – 6	1 000– 5 000	100– 5 000
Sephadex G-25 Medium	50–150	2·5 ± 0·2	4 – 6	1 000– 5 000	100– 5 000
Sephadex G-25 Fine	20– 80	2·5 ± 0·2	4 – 6	1 000– 5 000	100– 5 000
Sephadex G-25 Superfine	10– 40	2·5 ± 0·2	4 – 6	1 000– 5 000	100– 5 000
Sephadex G-50 Coarse	100–300	5·0 ± 0·3	9 –11	1 500– 30 000	500– 10 000
Sephadex G-50 Medium	50–150	5·0 ± 0·3	9 –11	1 500– 30 000	500– 10 000
Sephadex G-50 Fine	20– 80	5·0 ± 0·3	9 –11	1 500– 30 000	500– 10 000
Sephadex G-50 Superfine	10– 40	5·0 ± 0·3	9 –11	1 500– 30 000	500– 10 000
Sephadex G-75	40–120	7·5 ± 0·5	12 –15	3 000– 70 000	1 000– 50 000
Sephadex G-75 Superfine	10– 40	7·5 ± 0·5	12 –15	3 000– 70 000	1 000– 50 000
Sephadex G-100	40–120	10·0 ± 1·0	15 –20	4 000–150 000	1 000–100 000
Sephadex G-100 Superfine	10– 40	10·0 ± 1·0	15 –20	4 000–150 000	1 000–100 000
Sephadex G-150	40–120	15·0 ± 1·5	20 –30	5 000–400 000	1 000–150 000
Sephadex G-150 Superfine	10– 40	15·0 ± 1·5	20 –30	5 000–400 000	1 000–150 000
Sephadex G-200	40–120	20·0 ± 2·0	30 –40	5 000–800 000	1 000–200 000
Sephadex G-200 Superfine	10– 40	20·0 ± 2·0	30 –40	5 000–800 000	1 000–200 000

ELECTROPHORESIS

In Chapter 3 the phenomenon of electrolysis has been discussed. The migration of a charged ion or group towards the electrode of opposite charge is called *electrophoresis*. The rate of migration varies for different ions and this difference can be used to separate the components of mixtures. Thus chromatography and electrophoresis are both mainly concerned with the separation of very small amounts of substances which may not be easily separable by classical methods. The two techniques have developed side by side and are now considered to be complementary to one another, when chromatographic separation is not practicable electrophoretic separation may be used and vice versa.

The simplest form of electrophoresis apparatus is shown in *Figure 7.15* The two beakers contain a buffer solution and are connected by a strip of filter paper soaked in the buffer and supported on a glass plate. The carbon electrodes are connected via a switch to a high-voltage battery. If a spot or streak of a mixture of proteins is applied to the paper and the switch closed, a direct current will flow along the paper and the individual proteins in the spot will move towards the electrode of opposite charge at different rates. After a suitable period the current is switched off, the paper removed, dried, and dipped into a locating reagent when the individual spots of protein will show up in a similar manner to the display on a paper chromatogram.

Theory of Electrophoresis

There are three main factors controlling the movement of charged particles in an electric field:

1. The nature of the charged particle itself; the sign and magnitude of the charge and the size, mass and shape of the particle.
2. The nature of the buffer solution which forms the environment of the charged particle; its pH, electrolyte concentration, ionic strength, and viscosity are significant. The presence of large non-polar particles which may interfere with the movement of the ions also has an effect.
3. The nature of the electric field itself; its purity (absence of a.c. components), intensity, and distribution along the migration path must be considered.

The Nature of the Charged Particle

The three factors which govern the behaviour of a particle

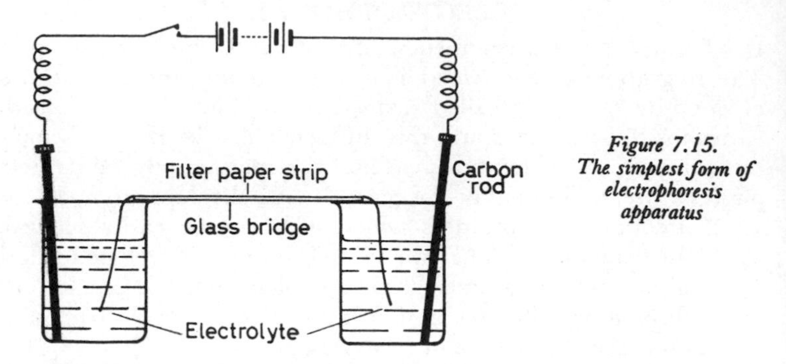

Figure 7.15. The simplest form of electrophoresis apparatus

in an electric field to the greatest extent are:

1. The physical size of the ion (its molecular weight).
2. The size of the charge it carries.
3. The sign of the charge.

The first two features act in opposition, the larger the particle the slower its travel, the larger the charge the faster the migration. The sign of the charge will decide the direction of migration from the point of origin.

Other factors may also have some effect on the course of the separation. The shape of the molecule is of some importance, i.e. whether it is globular or elongated. Crystalloids (small molecules) being small are unaffected by the inert paper support and move very quickly. Colloids (macromolecules) like proteins move more slowly and tend to be adsorbed to some extent. This causes a dragging effect on the migrating band and tailing may result.

The Environment of the Particle

The pH of the buffer solution may have a profound effect upon the migration of *ampholytes*, substances like amino acids or proteins which carry both positive and negative charge. The choice of buffer pH can determine the total charge on the particle. An amino acid would be a cation at acid pH and an anion at alkaline pH. At the isoelectric point the total charge would be nil and no migration would take place toward either anode or cathode except for interfering effects discussed below. The mobility of the ion is also affected by the concentration of salts in the buffer. In general, the higher the concentration of the buffer the greater the ionic mobility. Higher buffer concentration increases the conductivity and the current which passes through the buffer. The undesirable effects of too high a current are discussed below.

A further complication is the *endosmotic* (*electro-osmotic*) flow, which causes movement of all particles both polar and non-polar. When a liquid flows through a capillary it is charged relative to the capillary. In this particular system the water in the buffer becomes positively charged with respect to the paper capillary system. The water is free to flow along the fixed paper strip and when a current flows through the buffer the water moves towards the cathode carrying with it all the dissolved substances. Thus a non-polar solute, or a protein with an iso-electric point exactly similar to the buffer pH, will not remain fixed at the origin during electrophoresis, but will migrate slightly towards the cathode because of the endosmotic flow.

Diffusion is another interfering factor in electrophoresis experiments. The spreading of the sample by diffusion goes on throughout the experiment. In zone electrophoresis diffusion rarely presents a problem as long as reasonable precautions are taken. The sample should not be allowed to stand on the paper for a long time before electrophoresis is begun, and the paper should be dried as soon as the current has been switched off. If a high voltage is used the time required for adequate separation of the components will be short, and the diffusion effects will be limited.

In paper chromatography the R_F expression is used to help both in the identification of components and in the design of experiments, e.g. the choice of solvent. Electrophoresis has no solvent front on which R_F values can be calculated. *Migration velocities* are sometimes quoted for electrophoretic systems. This factor is defined as the migration of a substance in centimetres per second where the potential drop is 1 V/cm. Allowances must be made for the effect of endosmotic flow. Most workers are not concerned with migration velocities, however, and results are generally given in terms of the experimental conditions as a whole. The optimum conditions of separation are determined experimentally and the experiment is reported together with the voltage, the movement of ions in centimetres, and the time during which current flows.

In general, endosmotic flow is ignored in most electrophoresis experiments. It may be demonstrated or measured directly by 'spotting' a suitable non-polar compound on the origin line adjacent to the sample. For mixtures of small molecules urea or glucose may be used as non-polar markers while dextrans are used as markers when larger molecules are being separated.

The Nature of the Electric Field

The intensity of the direct current which causes the movement of ions in an electrophoretic separation is limited by several factors. It has already been stated that, the higher the voltage between the ends of the paper support, the faster the migration of ions, and the faster the separation the less interference there will be from diffusion, etc. Unfortunately, the passage of an electric current through a conductor converts the electrical energy to heat energy. Where a strip of resistance R ohms has a current of I amperes passing through it for t seconds, the heat generated in the strip will be I^2Rt joules. A doubling of the current will increase the heat produced four times. If the strip gets hot the buffer water will evaporate. At low voltage this evaporation is fairly small but at high voltages the heat generated may be such that the strip becomes completely dry. As the ends of the strip are dipped into the buffer, more buffer will soak into the paper by capillary action to replace the evaporated water. The salts in the buffer do not normally evaporate and the salt concentration in the strip increases. This in turn lowers the resistance of the strip which further increases the current, increasing the heating effect still further. It is clear that to keep the experimental conditions as constant as possible and to achieve separation in the shortest acceptable time, a compromise must be reached. Usually a voltage of 2–10 V/cm of paper is considered to give a tolerable evaporation and a reasonably fast separation for most systems.

The power supply itself can help to keep conditions stable. Although it is possible to use high voltage batteries for electrophoresis, electronically controlled power-packs are usually employed as a source of high-voltage direct current. A detailed account of these is outside the scope of this volume. For most purposes a power-pack which converts the a.c. from the mains supply to a d.c. of up to 500 V at up to 100 mA is adequate. These may be constructed to automatically control the current or the voltage to a constant value throughout the electrophoresis 'run' despite any changes in the resistance of the circuit due to evaporation and the concentration of salts in the strip. Suitable power-packs may be obtained from the usual laboratory suppliers. They are rather expensive pieces of equipment and may cost from £50–£100. The technician with some knowledge of basic electronics should not find it too difficult to construct a suitable power-pack for £15–£20. *Figure 7.16*

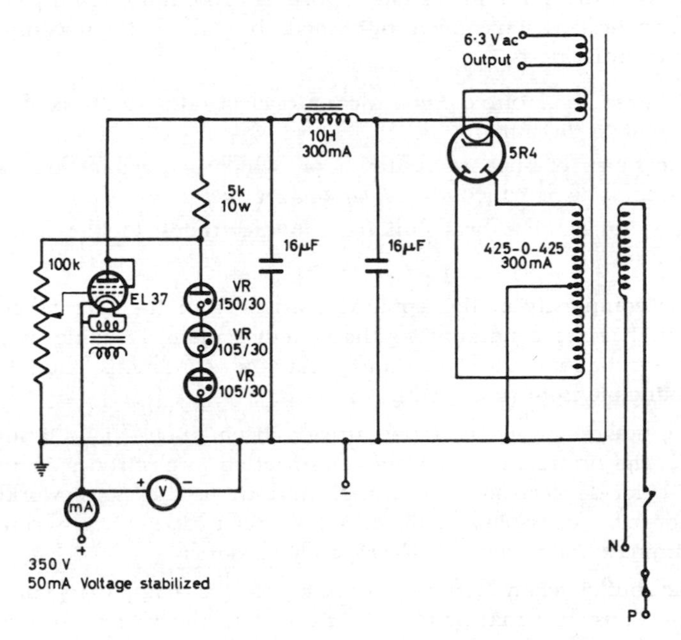

Figure 7.16. Circuit diagram of an electrophoresis power supply

shows the circuit diagram of a power-pack constructed in the author's laboratory which has been in use for several years without breakdown and which has been found suitable for a variety of electrophoretic systems.

In *high-voltage electrophoresis* where 20–100 V/cm may be used, special measures are taken to minimize the effects of evaporation.

(*a*) The whole paper strip is immersed in an inert organic solvent except, of course, for the ends immersed in the buffer; or,

(*b*) The paper, wetted with buffer, is clamped between non-conducting flat cooling plates.

High-voltage electrophoresis is discussed briefly at the end of this section.

Practical Electrophoresis

The simple electrophoresis apparatus shown in *Figure 7.15* may be suitable for a class demonstration of the phenomenon, but would be inadequate for use in a working laboratory. Many quite complex designs have been published, but all of them comprise three essential parts:

1. A stable d.c. source and adequate electrodes to transmit the current to the buffer.
2. The block or strip of material on which the separation takes place. This is called the *stabilizing* or *support medium.*
3. The buffer chambers linking the electrodes to the support medium.

The complexity of the apparatus arises from the care taken to ensure that the conditions in the system remain as stable as possible throughout the run, and that the conditions should be reproducible from run to run.

The power supply itself has already been discussed. Platinum wire is the best material for the construction of electrodes. Silver/silver chloride electrodes are suitable and are used by some workers. Carbon electrodes tend to disintegrate over a period and seriously contaminate the buffer in the electrode chamber.

The buffer chambers usually add to the complexity of the design. It is clear that besides the electrolyte effect which promotes electrophoretic migration of the constituents of the sample, the buffer will be decomposed by electrolysis. This can radically alter the pH of the buffer solution and effect the course of the separation. One way of minimizing this affect is to use a large volume of buffer solution in each chamber so that the proportion of decomposed buffer is small. It has been suggested that a minimum volume of 300 cm^3 in each buffer chamber is necessary for a normal 24 h analytical run. Another common procedure is the reversal of the electrodes, i.e. the direction of current flow, for each run so that the buffer decomposition is even. Most modern electrophoresis tanks are designed so that each buffer compartment is in two sections, one for the electrode and one for the paper strip, the two sections being connected by a wick of cotton wool or white lint soaked in buffer. The wick gives good electrical connection between the two compartments. Any electrolytic decomposition takes place in the electrode compartment, the paper dipping into buffer which remains fairly constant in composition. The level of buffer on both sides of the tank must be identical or buffer will siphon along the paper from one compartment to the other.

The Apparatus of Paper Electrophoresis

Electrophoresis tanks are usually constructed of plastic. Commercially constructed apparatus is often of plastic pressings, but very satisfactory apparatus can be made of sheet perspex.

There are two main types of tank for paper electrophoresis.

1. The horizontal or flat-bed electrophoresis tank (*Figure 7.17*).

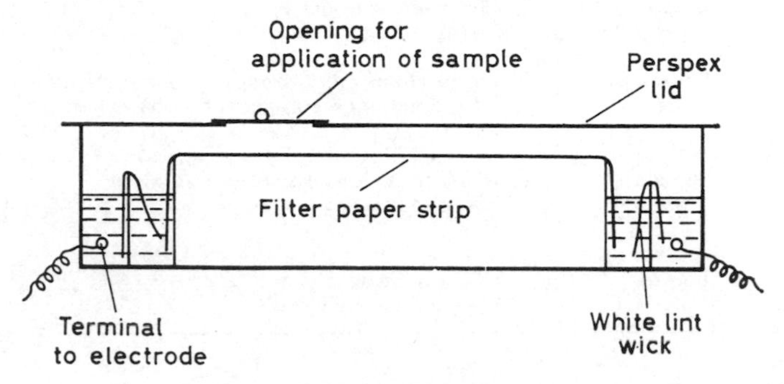

Figure 7.17. Apparatus for horizontal paper electrophoresis

2. The vertical or inverted V electrophoresis tank (*Figure 7.18*).

Both types give almost identical patterns with similar mixtures.

Most workers use Whatman No. 1 or 3 MM paper in strips. Both these papers have been found to give good separation. 3 MM is the heavier of the two, has greater wet strength, and permits greater loading.

The Buffer

In general the buffer has two functions:

1. To wet the strip with electrolyte so that current will flow.
2. To maintain the correct pH throughout the experiment so that the conditions of migration remain as constant as possible.

The correct buffer is found empirically, the best pH and ionic strength being determined experimentally. The other necessary characteristics are comparable with those for a chromatographic solvent, e.g. the sample must be soluble in the buffer and the buffer must not react with the components of the sample. Some examples of buffers commonly used in paper electrophoresis are shown in Table 7.7.

TABLE 7.7. Some Buffer Solutions for Paper Electrophoresis (To make 1·0 dm^3 in water)

1. *For serum proteins*	
Barbiturate pH 8·6	1·84 g diethylbarbituric acid 10·3 g sodium barbiturate
Borate pH 8·6	8·8 g sodium borate 4·65 g boric acid
Borate pH 9·0	7·63 g sodium borate 0·62 g boric acid
Phosphate pH 7·4	0·6 g sodium dihydrogen phosphate (H_2O) 2·2 g disodium hydrogen phosphate (Anhyd.)
2. *For amino acids*	
Phthalate pH 5·9	5·10 g potassium hydrogen phthalate 0·86 g sodium hydroxide
3. *For carbohydrates*	
Borate pH 10·0	7·44 g boric acid 4·0 g sodium hydroxide

Application of the Sample

Before the sample is applied the paper is moistened with buffer and blotted between filter paper. The damp paper is fixed to the apparatus. As in paper chromatography it is very easy to overload the paper with sample. Most workers use sample solutions of 2–10 per cent concentration and about 5 mm^3 of sample is applied as a streak for each centimetre width of paper. Application is carried out with a smooth tipped micro-pipette. It is essential that the surface of the paper is not torn or abraded by the pipette tip. The applied sample should be clear of the edges of the paper.

The voltage across the paper should be 5–10 V for every centimetre of strip length.

It is only possible to determine the time for which the current should flow empirically. Sometimes it is possible to add a marker dye which will bind with the components of the sample through which migration can be followed in a trial experiment, e.g. a little bromophenol blue is added to serum albumin in preliminary studies of its migration behaviour.

Detection and Estimation of Sample Components

The methods of detecting the zones in a completed electrophoretogram are similar to those used in paper chromatography.

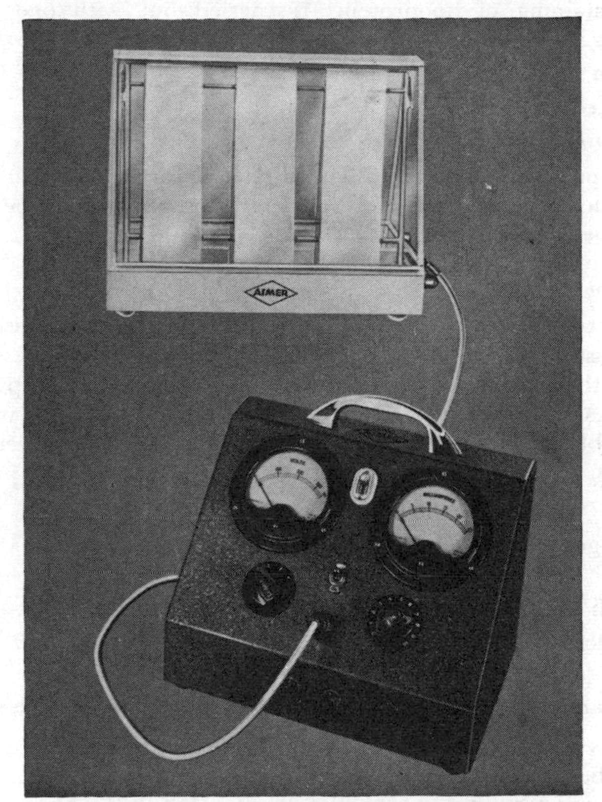

[By courtesy of Aimer Products Ltd]

Figure 7.18. Apparatus for vertical paper electrophoresis

The most usual methods are by dye staining. Several dye solutions have been described for staining protein zones. Among the most common are:

1. Amido Schwartz 10B (Naphthalene Black 12B.200).
2. Bromophenol Blue.
3. Lissamine Green (Light Green. I.C.I. SF 150).

Quantitative estimation of the zones may be carried out with moderate accuracy by dipping the stained strip in oil to render the paper transparent and passing it through a photometric scanner. Alternatively the coloured zones may be cut out, the coloured material extracted and the colour measured in solution.

The staining of lipoproteins is carried out with one of the following:

1. Sudan Black.
2. Oil Red O.
3. Oil Blue N.

Glycoproteins are detected with diphenylamine.

The details of methods for staining pherograms may be found in the specialist literature.

Other Stabilizing Media

So far only electrophoresis on filter paper strips has been discussed. In the last 20 years many other support media have been described, some of them more useful than paper for the satisfactory separation of particular mixtures. These have ranged from glass wool to foam rubber sheets. Among the more widely used at present are:

1. Starch grain block (Kunkel and Slater. *J. clin. Invest.* 31 (1952) 677).
2. Agar gel (Graber and Williams. *Biochim. biophysica acta* 17 (1955) 67).
3. Starch gel (Smithies. *Biochem. J.* 61 (1955) 629).
4. Cellulose acetate strip (Kohn. *Clinica chim. Acta* 2, 297 (1957) 120).
5. Acrylamide gel (Leaflet from Messrs. Kodak Organic Chemicals).

Details of these methods will be found in the literature. The gel methods have the great advantage that besides their useful characteristics purely as support media, the gels appear to act as molecular sieves so that the movement of components is controlled by the size and shape of molecules in addition to the charge. In the electrophoresis of serum, for example, many components may be seen on a starch gel which would not be separated on a simple paper strip.

Cellulose acetate strips (Oxo Ltd.) may be used instead of paper strips for electrophoresis. These have the advantage of being thinner, more homogenous, and chemically purer than paper. In consequence smaller samples may be used and 'tailing' is minimized. Cellulose acetate may be cleared more easily than paper to produce transparent strips for automatic scanning.

Electrophoretic separations are carried through with much greater speed than on paper.

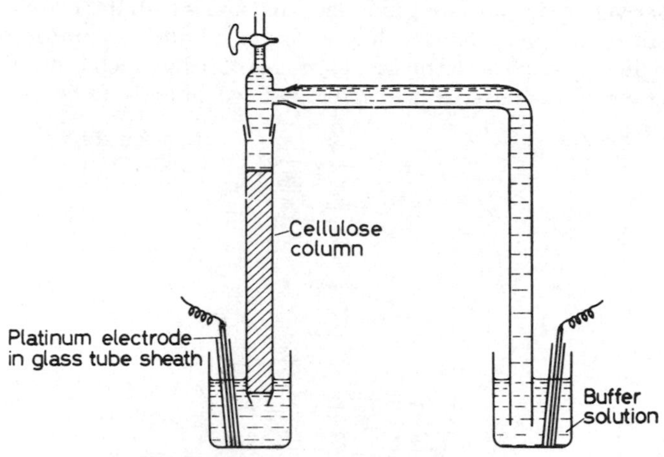

Figure 7.19. Column electrophoresis

Column Electrophoresis

Electrophoresis on columns has the obvious advantage that the zones may be recovered by the normal elution techniques of column chromatography using a fraction collector. Porath (*Ark. Kemi* 11 (1957) 161, 259) has described the most widely used technique for column electrophoresis. A simplified version of the apparatus is shown in *Figure 7.19.* The stabilizing medium may be either pure cellulose powder or one of the 'Sephadex' cross-linked dextran molecular sieve materials marketed by Messrs. Pharmacia. The commercial version of the Porath column produced by Messrs. L.K.B. can separate as much as 20 g of dry substance. Fairly high voltages are required and it is necessary to cool the separation column with flowing water while electrophoresis is in progress.

High-voltage Electrophoresis

The techniques so far described come under the heading of low-voltage electrophoresis. This may be considered something of a misnomer as potentials as high as 500 V are often encountered and careless handling of equipment may give lethal electric shocks. For the electrophoretic separation of low molecular weight materials (e.g. amino acids and small peptides) a potential gradient of 5–10 V/cm is unsatisfactory as the high diffusion rate of the components prevents the formation of sharp zones.

It has been shown that potential gradients of 50–200 V/cm

produce very sharp and well differentiated zones with these mixtures. Because of the very high voltages employed the apparatus must be modified to cope with the great amount of heat which is produced during electrophoresis, and to make the technique safe for normal handling.

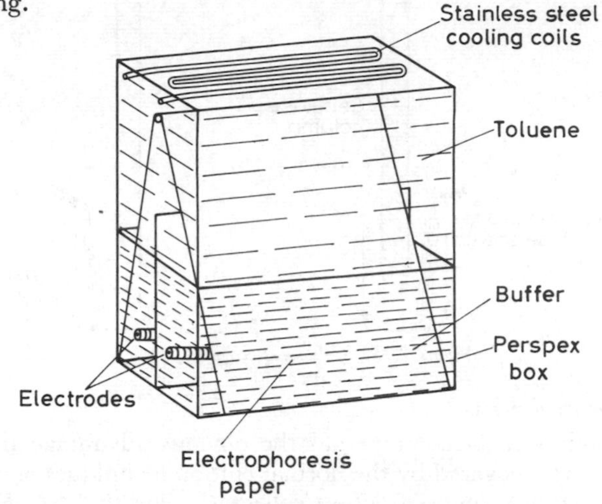

Figure 7.20. High-voltage electrophoresis

Paper strips are used in high-voltage electrophoresis. The strip may be placed horizontally between two cooled plates of insulated metal or glass, or held vertically in a large volume of organic solvent (*Figure 7.20*). The power-pack should be capable of supplying up to 10 kV at 500 mA although currents as high as this may give rise to difficulties in heat dissipation. Needless to say, apparatus of this type should not be constructed by anybody but an electronics expert. The dangers of faulty apparatus at kilovolt levels cannot be over emphasized. If possible the apparatus should be set in a separate room with entry restricted to those using it.*

Details of the technique would be inappropriate in a volume of this type but may be found in the textbooks listed in the appendix.

Moving Boundary Electrophoresis

The separation of materials on solid supporting media is called *zone electrophoresis*. The individual components of a mixture can be separated and removed from the supporting medium for analysis. An earlier technique which is still used is *moving boundary*

* Suitable equipment is available from: Messrs. Iver Scientific Instruments Ltd., Iver, Bucks.

electrophoresis. In this method the components of the mixture move, under the influence of an electric field, through a completely liquid medium contained in a cell. Analysis is carried out by the observation of the boundaries of the component zones as they move through the liquid medium. As the components are usually colourless, special optical apparatus has to be employed as the boundaries may only be differentiated by the effects due to differences in refractive index. Until comparatively recently the Tiselius electrophoresis apparatus was widely used to obtain data in the investigation of proteins.

CHAPTER 8

AUTOMATION IN THE LABORATORY

Between the two world wars laboratory instruments became increasingly important in the day-to-day work of the chemist. By 1945 photoelectric colorimeters, pH and conductivity meters etc. were common in the working laboratory. Because of the introduction of these devices, much human error was removed from analytical work and there was an improvement in the reliability of experimental results. As these became more precise, the significance of small differences between samples increased and more and more analytical work was demanded of the laboratory chemist. The increasing work load, together with a general shortage of skilled personnel, gave an impetus to research into automatic instruments which would perform part or all of an analytical process mechanically. The main aim was to relieve the chemist of a large proportion of his routine analytical burden. By relying on the machine for this he could deploy his skills on non-routine work and more interesting and exacting tasks. As a bonus the machine could work much more quickly and for much longer hours than the human, and the time of processing a sample was reduced very considerably.

The ingenuity which has been applied to the development of automatic laboratory equipment over the last twenty or thirty years has led to a situation where every known analytical technique could be automated using our present knowledge of electronics, mechanical engineering and computers. The obvious demonstration of this is our ability to send automatic analytical equipment to the moon which can sample the moon's surface, analyse it, and send the results back to earth by radio signals.

Speed and time saving are not the only advantages of automatic analysis. There is no doubt that modern automatic instrumentation is more discriminating and more precise than the older manual apparatus and that random variation in results can be considerably reduced.

It would be impossible in this chapter to give a comprehensive survey of automatic instruments but an attempt will be made to discuss some of the general principles with examples drawn from specific pieces of laboratory equipment. Initially it will be necessary to decide the meaning of the term 'automatic laboratory instrument'. Although automation has reached every branch of chemistry in the laboratory and in industry, there is one common feature in all the wide diversity of equipment. In every case there is some mechanical movement within the apparatus *which is controlled by the apparatus itself*. The movement may be:

1. Of the sample to be tested (e.g. in a flow cell colorimeter).
2. Of a reagent to be added to the sample (e.g. by an automatic burette).
3. Of a component of the apparatus which applies some physical stress or stimulus to the sample (e.g. in the potentiometer of an automatic polarograph).

The gathering of information by which the instrument can control its own activities is called the *feed back* of information, and the whole process of feed back and subsequent control is called a *servo-loop*. The requirement for automatically controlled mechanisms has initiated most of the research into automatic instrumentation.

CLASSIFICATION OF AUTOMATIC INSTRUMENTS

Having established what automatic equipment is, the different classes of apparatus must be indicated. There are several headings to be considered and any individual set-up must fall into one or more of the categories discussed below. The different classifications can be considered in pairs, several complimentary pairs being distinguishable.

1. Some instruments automate classical manual methods replacing the human operator with machinery which performs the manipulations required mechanically. An example of this is the automatic burette which 'turns itself off' when a pre-set end point has been reached in a titration. Another example is the automatic fraction collector (Chapter 7, page 292) which collects the eluate from a chromatography column in a large number of separate tubes. Both of these are comparatively simple devices, but quite complex manual procedures can be automated by using trains of separate pieces of machinery (*modules*) each of which performs an individual operation in carrying out the analysis.

There are, however, several instruments which are automatic in their original conception. The classic example of this is Heyrovsky's polarograph which was designed as a machine to automatically record polarographic waves on a photographic plate. It was largely through the automatic character of Heyrovsky's instrument that polarography was neglected for so long. Analysts in the 1920s found it difficult to accept that an instrument could perform both qualitative and quantitative analysis and present the results automatically in the form of a graph. The manual polarograph was designed after the automatic machine had been described. Gas-liquid chromatography (see Chapter 7, page 284) has never been a manual technique, indeed it would be impossible to design a non-automatic gas-liquid chromatograph. This is another example of a commonly used method which was automated *ab initio*.

2. Automatic equipment is very expensive and as more operations are built into a particular analytical set-up the price rises steeply. Careful examination, on a work-study basis, of the individual manipulations of an analytical method can show that *partial automation* can give nearly all the advantages of *full automation* at a small fraction of the cost. S. M. Hardy has described 'Limited Function Automation' (LFA) which consists of a series of automatic dispensers and automatic diluters operated through motor driven syringes by switched handsets. Essentially the normal manual methods are used but the reproducibility and accuracy compare well with fully automatic systems and the volume of work which may be handled is much higher than can be expected with manual techniques. Patient in Annals of the N.Y. Academy of Sciences (Vol. 87, Art. 2, July 1960), describes the investigation of an analytical technique in an industrial pharmaceutical laboratory when Messrs Baird and Tatlock were asked to develop a partially automated system.

Plainly in some areas full automation is essential, but often fully automatic machinery is installed in laboratories where the workload does not justify the enormous expense involved and where partial automation would be an adequate solution.

3. The fully automated apparatus available can be divided into two main classes:

(*a*) Continuous flow equipment.
(*b*) Discrete sampling equipment.

The most common example of the former is the Technicon Auto Analyser, the most successful of the automatic machines developed to date. Liquid samples are pumped through a series of tubes and manifolds where they are automatically mixed with appropriate reagents and heated etc. before being estimated in an automatic colorimeter or some other suitable physical measuring device. All the samples follow one another in sequence through the system and precautions have to be taken to avoid mixing or contamination of one sample with traces of its predecessor. There is a continuous flow of samples through the same system of tubes.

In discrete sampling each individual analysis is performed separately in separate vessels. The addition of reagents by automatic pipettes, heating of samples etc. is all carried out on the sample contained in its own individual test tube or sample cup and no problems of mixing or contamination of samples arise.

The importance of the differences between these two classes will become apparent as individual types are described in detail below.

PRIME MOVERS, PUMPS, DISPENSERS AND DILUTERS

In all fully automatic systems there must be some means of moving liquids to give an accurate and controlled flow. In discrete sampling, samples are aspirated and reagents and diluents added in aliquots by very carefully constructed dispensers. These are usually electrically driven piston pumps or syringes. The Association of Clinical Biochemists has published a very valuable critique of commercially available dispensers and diluters (Automatic Dispensing Pipettes: an assessment of 35 commercial instruments, P. M. G. Broughton, *et al.*, A. C. B. Scientific Report No. 3, 1967) which evaluates these instruments. Most of the dispensers can pipette with far greater accuracy and reproducibility than can be obtained by using a normal bulb pipette. In some cases the dispensing stroke of the piston is electrically operated but the filling stroke is spring operated. This leads to one of the main problems associated with these instruments; the sticking of the piston. If this should happen errors arise through incomplete filling or emptying of the syringe. This is often not immediately apparent as the syringe mechanism is hidden from view within a metal casing. On the better instruments, the piston is motor-driven in both directions. The danger here is that a sticking piston might actually be driven through the wall of the syringe barrel by the strong impetus of the motor. One way of avoiding this problem is illustrated by the automatic burette manufactured by Messrs Radiometer (Auto Burette ABU). The piston of this syringe burette is very carefully machined so that its volume is known. The motor drives the piston into a loose barrel (i.e. the piston is not in contact with the wall of the barrel) and displaces liquid from the barrel in proportion to the length of piston inserted. The accuracy of this instrument depends upon careful construction of the piston which leads to the Auto Burette being a comparatively expensive device.

Diluters are two syringe devices: one for sample and one for diluent, and the chance of a piston sticking is thus doubled. A further problem with diluters is the chance of contamination of a sample by the remains of the previous sample in the syringe. In the better instruments, the sample is completely washed out by diluent and a small amount of the next sample is taken up to rinse the sample syringe before the full aliquot is taken.

In continuous flow apparatus, piston pumps are sometimes used especially in machines used for chromatography, e.g. the Moore and Stein automatic amino acid analyser. These work through a constant speed motor of very high stability. In order to keep the speed of the motor constant, special solid state triac type circuits are used which include a thyristor controller and which usually have a tacho feedback device which notes any tendency for the motor speed to vary and makes the appropriate correction. The pumps themselves are usually double piston systems with, in effect, two pumps operated in a reciprocating manner, i.e. one barrel fills while the other empties. This tends to overcome pulsation in the pump flow. The output from these pumps is controlled by altering the length of the piston stroke in the barrel and there is usually a micrometer control which may be adjusted to give a very accurate and stable flow. The rate of flow of liquid does not alter within very wide ranges of pressure within the system and reproducibility from run to run is assured. Needless to say these devices are very expensive, and more than one pump is required in the system, e.g. the Moore and Stein analyser requires a minimum of three.

This type of pump is usually very stable and can work for months on end with little attention. The main problem is the loss of small particles of material from the substance of both piston and barrel by the constant friction of one against the other. Any variation in flow caused by this can easily be corrected but these microscopic shards of metal or P.T.F.E. circulate in the pumped fluid and may eventually penetrate the chromatography column through which it flows. Efficient filtering of the input and output of each pump with some inert material prevents this. Obviously no liquid can be pumped which will react with the materials from which the piston or barrel are constructed.

The more common type of pump in continuous flow analysis is represented by the Watson Marlow 'Delta' pump, or by the proportioning pump of the Auto Analyser. In its simplest form the *peristaltic pump* consists of a flexible tube connected to a reservoir of liquid, clips to hold a short loop of tubing in a fixed position and an eccentric roller which squeezes the tube and thus forces its contents along it in one direction. The movement of liquid may be

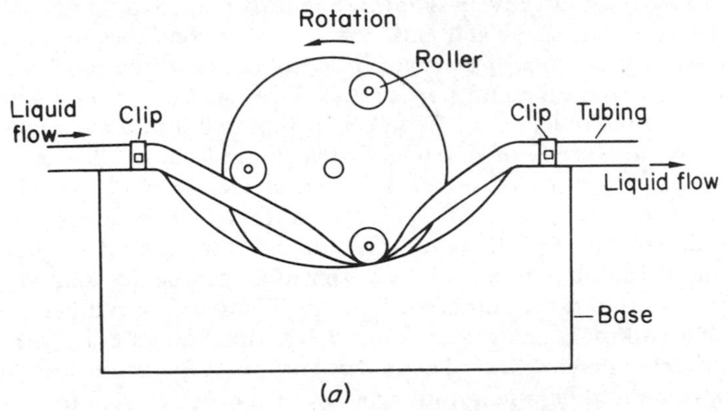

Figure 8.1. The peristaltic pump
(*a*) *Principle*
(*b*) *Watson-Marlow 12 channel peristaltic pump*

likened to the peristalsis in the animal alimentary canal. The action can be understood by examining *Figure 8.1.* The roller comes into contact with the tube and squeezes a short portion of it against a hard baseplate, effectively isolating the proximal (reservoir) portion of the tube from the distal end. As the roller continues to move it squeezes liquid along the distal portion of the tube and more liquid flows along the proximal portion to fill the tube up to the point of contact of the roller. As this process is repeated liquid moves along the tube in a series of spurts giving a pulsatile flow. The rate of flow can be controlled either by altering the speed of the motor, which is closely controlled as described above, or by using tubing of different internal diameter. In the more sophisticated versions of the peristaltic pump, up to a dozen tubes can be accommodated to pump up to twelve different liquids. The motor speed being the same for all the tubes, the only way of varying flow rates is by using tubing of different i.d.s. In the proportioning pump of the Auto Analyser the different liquids which are flowing are controlled in this way so that the appropriate solutions come together in the correct proportion for the analytical method. The main advantage of this type of pump is the fact that the liquid only comes into contact with the inert flexible tube. This means that materials can be transmitted which would corrode the piston or barrel of a metal piston pump. In general, however, the liquid flow in the peristaltic pump cannot be controlled with the very high precision of the piston type. One problem, which has largely been overcome, is the distortion and stretching of the flexible tubing. Plainly if the internal diameter of the tubing is altered by the regularly applied stress from the roller, the flow rate of liquid will change. This has been prevented by very careful design of the roller so that stretching effects are reduced to a minimum and by formulating plastic tubing which is particularly suitable for this purpose. A second problem, which has proved more intractable, is the turbulence and surging effects within the tubing in the vicinity of the roller. There is always a tendency for liquid to flow in *both* directions along the tube as it is squeezed and this is largely uncontrollable and causes surging effects which in turn alter flow rates. Turbulence within the tube at the point at which the impulse is applied can also affect the flow. These effects become more noticeable as the tube diameter increases. Attempts have been made to control surge effects by special devices but, on the whole, these have not been very effective. It must be emphasized, however, that for practical purposes these distortions have only a minor effect on analytical results.

In any automatic analytical system accurate pumping, whether to dispense or to dilute liquids, is essential and in all the commercial systems a large part of the cost is due to the pumping systems used.

Continuous Flow Systems

Apart from very special machines, such as the Moore and Stein Amino Acid Analyser the only commercial apparatus on the market for general analysis is the Technicon Auto Analyser, undoubtedly the most successful apparatus to date. Basically the machine consists of separate modules, each of which operates to perform a separate analytical function and the sample flows in a common system of tubes from one module to the next until eventually it enters a flow-cell absorptiometer where the signal for the absorbance of the treated sample is marked on an associated chart recorder.

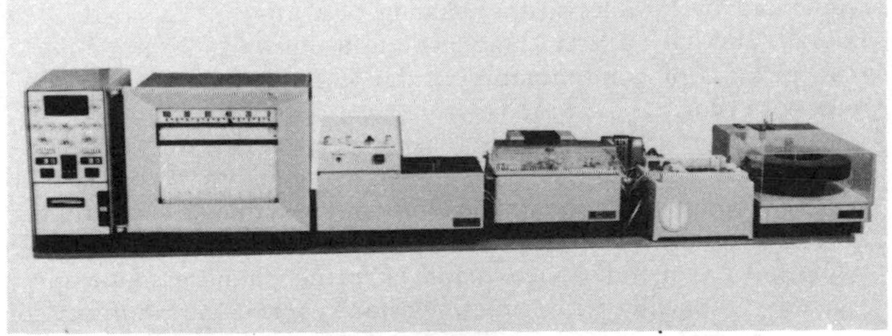

Figure 8.2. The Auto Analyser Mark II. The modules are shown in progression from the sampling plate on the extreme right to the recorder and print-out on the left
[By courtesy of Technicon Ltd]

CONTINUOUS FLOW ANALYSIS

The Auto Analyser is made up of separate modules through which each sample flows in sequence, treatment and reading of results following sequentially. The modules are connected using inert plastic tubing, the actual connections between individual tubes being designed to allow the confluence of streams of sample and other reagents at appropriate points during the flow of a sample through the system.

The basic modules are as follows:

1. The Sampler Module—This is a circular plate with (usually) 40 peripheral holes. Each hole contains a 2 cm^3 cup made of polystyrene. Each cup is roughly filled with one of the samples to be analysed so up to 40 individual samples can be treated at a time.

The plate is mounted on a motor which turns it at regular intervals so that each cup in turn is presented to a sampling crook, a bent tube of metal with a polythene tip. The tip of the crook dips into each cup in turn and aspirates an aliquot of the sample which is thus introduced into the system.

2. *The Proportioning Pump*—A manifold of plastic tubing is fitted on to the platen of a peristaltic pump so that chain driven rollers can move over the manifold to compress the tubes and draw fluids through them. Each tube of the manifold is connected to a reservoir of reagent except the one tube which is connected to the sampling crook and a second which deliberately introduces air bubbles into the line. The purpose of these is twofold. Individual samples are separated by the bubbles and, in fact, each sample is internally segmented by bubbles so that mixing of samples is avoided. The bubbles also have a 'scrubbing' action on the walls of the tubing so that pick-up of contaminants as the samples follow each other does not occur.

The proportioning pump is a fixed-speed device and the motor speed cannot be controlled. As the relative volumes of all the fluids flowing through the system depend upon the diameter of the tubing provided for each, the proportion of liquid A which will mix with liquid B is a function of these diameters rather than the motor speed and this proportion will remain constant even if the pump motor speed increases or decreases slightly. A large range of tubing sizes is available (see Table 8.1).

At appropriate points in the flow system tubes will join at Y or T pieces to allow mixing of two liquids. At these points mixing is facilitated by allowing the liquid to pass through vertical glass coils which have the effect of shaking the mixed materials and presenting a homogeneous mixture to the next part of the system. Coils are also used for delaying the flow of liquid in the system.

3. *Dialyser Module*—In biochemistry and clinical chemistry it is often necessary to deproteinize a sample before analysis is performed. Manual methods often demand the precipitation of protein by suitable reagents (e.g. trichloracetic acid, sodium tungstate) followed by centrifugation. The supernatant is then taken for analysis while the precipitate is discarded. This is not possible in a continuous flow system and the dialyser unit performs this function. It consists of two channelled perspex plates bolted together and separated by a sheet

TABLE 8.1. Tubing used in the Auto Analyser

Colour	*Abbreviation*	*Bore Diameter* (inches)	*Approximate delivery* (ml per min)
Orange and black	O/Blk	0·005	0·015
Orange and red	O/R	0·0075	0·030
Orange and blue	O/B	0·010	0·048
Orange and green	O/G	0·015	0·096
Orange and yellow	O/Y	0·020	0·159
Orange and white	O/W	0·025	0·235
Black	Blk	0·030	0·32
Orange	O	0·035	0·42
Clear	C	0·040	0·60
Red	R	0·045	0·80
Yellow	Y	0·056	1·20
Blue	B	0·065	1·60
Green	G	0·073	2·00
Purple	P	0·081	2·50
Purple and black	P/Blk	0·090	2·90
Purple and orange	P/O	0·100	3·40
Purple and clear	P/C	0·110	3·90

of dialysing membrane. The sample containing protein enters one channel and the non-protein constituents cross through the membrane to the other channel where they can be analysed. The protein is discarded to waste. Complete dialysis is not achieved, but as all samples and standards are treated in exactly the same way and are subject to exactly the same conditions correlation between samples and standards can be achieved.

It is usual practice to enhance the rate of dialysis by adjusting the salt concentration and hence the osmotic pressure on either side of the membrane. The dialysis unit is contained in a 37° water bath.

4. *Heating Bath*—In most analyses, especially where absorptiometry is involved, the final stages of the method require heating of the treated sample. For most purposes the Auto Analyser uses a glass coil immersed in a thermostatically controlled oil bath. The flow of solutions through the bath may take 5–8 minutes depending upon the length of the coil.

For enzymology a 37° heating bath is used in the same way.

5. *Measuring Module*—In most cases the treated sample flows finally through a flow-cell absorptiometer. The air bubbles are removed

before the stream of liquid enters the cell. Several different absorptiometers are available for different purposes ranging from simple instruments with selenium barrier layer cells to fairly complex dual beam instruments. Other measuring modules are available and on-line flame photometers and fluorometers are often used.

6. *Recording systems*—Various recording systems are available for the Auto Analyser ranging from simple chart strip recorders to on line computers. Recording systems and their use will be discussed later in this chapter.

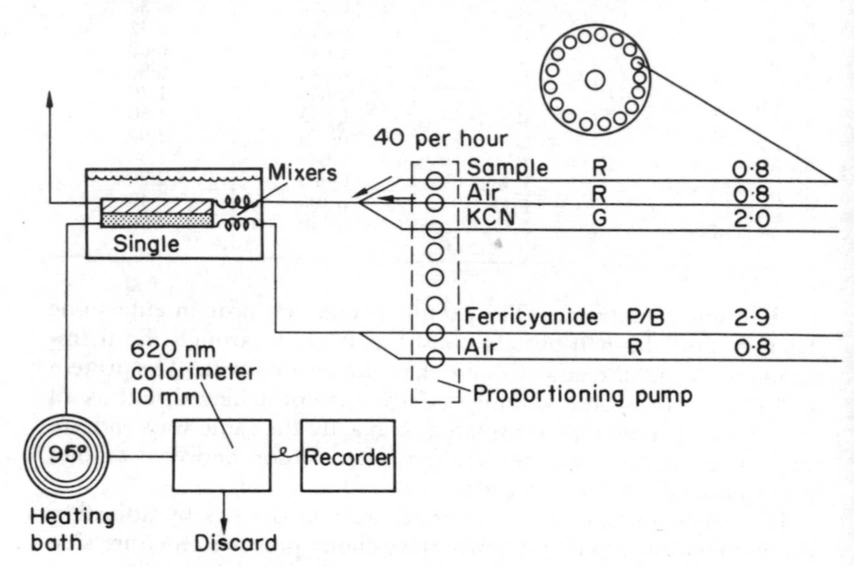

Figure 8.3. Flow chart for the determination of urea by Auto Analyser
[By courtesy of Technicon Ltd]

Figure 8.3 gives a typical example of the way the Auto Analyser is used. A manifold is kept in the laboratory ready to use for each particular analysis and it is the work of a moment to change the manifold from one method to another.

Since the introduction of the Auto Analyser it has become a very common feature in clinical laboratories where the workload is constantly rising. To cope with this, multichannel analysers have been developed which will deal with up to 12 estimations simultaneously on one sample. In earlier days complex multichannel analysers were fitted up using the standard modules of the single channel (one estimation at a time) system. The maximum number of channels which can be obtained in this way is seven or eight. Any increase leads to a cumbersome and virtually unmanageable system.

In 1967 Technicon introduced their SMA 12/60 multichannel analyser. Each estimation of the twelve has an individual cartridge associated with it which may be plugged into the controller when the estimation is required. Changing cartridges takes very little time and each one carries appropriate blanks for the determination. The estimations require 1·8 cm^3 of serum and 60 samples may be estimated per hour. In multichannel analysis accurate standardization is vital and each tenth sample is, in fact, a control serum of known concentration in each of the estimations being performed. The machine is initially standardized on this serum, the signals on the pen recorder being adjusted to relate to the actual concentrations in the control. In this way it is very easy to see if a particular sample presents an abnormal biochemical picture or if the machine itself is working efficiently.

One of the great advantages of the Auto Analyser is its versatility. Virtually any standard absorptiometer technique can be adapted to give satisfactory results using the machine and other detector modules are easily fitted where appropriate, e.g. flame photometer or fluorimeter. Even the standard Kjeldahl technique can be performed automatically using an ingenious flow through digestion module.

DISCRETE SAMPLING SYSTEMS

These can vary from comparatively simple devices to expensive bulky and complex machines. The Autochemist for example costs nearly £200 000 and will perform over quarter of a million analyses in 24 hours in its 12 channel version. It includes a PDP.8 on line computer.

The simpler, and more common, machines are all similar in basic design. A number of sample cups are suspended in a water bath from a turntable. As the turntable moves round each cup is placed in turn under a series of diluters and dispensers which add the appropriate reagents for the particular estimation being undertaken. The timing of the turntable is appropriate for the reaction involved. When the reaction is complete the contents of each tube in turn is aspirated into a flow through absorptiometer. Usually there is an automatic print-out device which gives the result for each sample directly in terms of concentration so that no further calculation is required.

Although it is possible to build multichannel discrete sampling systems, the technical problems are much greater than are found with the simple stream-splitting devices of the Auto Analyser. Several manufacturers have produced multichannel machines, notably the Autochemist mentioned above, the Vickers Multi-channel 300, and the Beckmann DSA-560.

In the smaller laboratory these are usually considered unnecessary and far too expensive as the discrete sampling system can usually work at a much faster rate than the Auto Analys er, processing about 300 samples per hour compared with the 60 achieved on the continuous flow machine.

The greatest disadvantage of the discrete flow system is its lack of versatility. This is especially noticed in the inability of most machines to separate precipitated material, e.g. protein. There is no way of incorporating a dialyser system, and special techniques have been evolved where the precipitation or dialysis step can be omitted. Some machines have special devices incorporated to overcome this disadvantage notably the Quickfit analyser which has a flow-through centrifuge for this purpose. Unfortunately the machine will only cope with 100 samples per hour which is a rather low output for a discrete sampling apparatus. In the Mecolab system the turntable for the sample tubes may be fitted directly on to a conventional centrifuge and spun. After the precipitate has been centrifuged down, the head is placed on a different module which automatically aspirates a measured aliquot of each supernatant for further automatic processing. Other systems work in a similar manner.

It must be accepted that these methods of dealing with a precipitation step in an analytical technique do tend to increase the time spent on each sample and to involve the operator in more manual work. It may be more satisfactory to devise methods which are especially suited for discrete sampling methods in which precipitation is eliminated. Some progress has already been made in this direction.

A Comparison of Discrete Sampling and Continuous Flow Systems

Both systems have their own advantages and disadvantages summarized in Table 8.2. In order to decide which type of apparatus is more suitable for a particular laboratory, these have to be weighed in the light of the laboratory's needs. It would be uneconomic, for instance, to purchase an expensive discrete sampling machine with a 300 samples per hour output if the maximum number of analyses required were only 300 per day. On the other hand, a busy laboratory with a very large workload may prefer the speed and high output of an Autochemist rather than find the space and staff to run a battery of Auto Analysers. Needless to say, the capital cost will be an important factor in making such decisions.

TABLE 8.2. Comparison of Discrete Sampling and Continuous Flow Methods of Automatic Analysis

Discrete sampling	*Continuous flow*
Sample retains identity throughout analysis	Samples follow one another in sequence in a flowing stream
Detector almost exclusively limited to absorptiometer	Versatility of detectors (absorptiometer, flame photometer, polarograph, fluorimeter modules in common use)
Can deal with 100–300 samples per hour	Maximum 60–70 samples per hour
Difficult to build in protein separation	Protein separation by simple dialysis
Many complex moving parts (Lower reliability)	Few moving parts
Difficult to detect random errors	Virtually self-monitoring
Low standard deviation on within-batch results	Higher standard deviation on within-batch results (sample pick up? back flow?)
Direct print out in terms of concentration	Chart record which must be evaluated (except where on-line computer is used)
Reaction usually goes to completion	Reaction not complete but similar treatment of all samples gives comparability
Probably most satisfactory where a large number of repeated simple analyses required	Probably most satisfactory where a large variety of analyses performed on comparatively small batches

THE EVALUATION OF RESULTS

Results from automatic analytical machinery may be presented in a number of ways. The most common is still the chart recorder which presents each result as a peak on a graph. When most detectors are absorptiometers and in the case of the gas chromatograph, the area under the peak is directly related to the concentration of the substance being monitored. In absorptiometry, of course, this will only be true if the system obeys the Beer-Lambert Law within the range of concentration being used.

The chart recorder is usually part of a bridge circuit which is in balance when the pen of the recorder is at zero. One arm of the bridge is connected into the apparatus so that a signal from the detector of the apparatus unbalances the bridge. A mechanical potentiometer on the recorder turns automatically to rebalance the circuit and this causes a pen to travel across the moving chart paper and make a recording. In modern chart recorders the response to a signal is almost immediate and the time taken for the pen to traverse the chart paper completely is less than 0·25 sec.

The response of the pen to a signal can be altered by altering the amplification which the signal receives. The recorder is usually specified as having a maximum voltage/FSR (Full Scale Recording) ratio this being the minimum voltage signal which will produce a complete traverse of the pen with maximum amplification.

The signal is compared against some standard voltage. In simple recorders this may derive from a zener diode circuit or even a dry battery, but the more modern devices employ a complex miniaturized integrated circuit for this purpose.

It is usual for the recorder circuit to be electrically isolated from the device which produces the signal. This may be simply achieved by passing the raw signal through a chopper device or oscillator and then through the primary coil of a transformer. Other, more sophisticated methods have been devised, especially for very sensitive recorders including photoelectric systems which achieve complete electrical isolation.

For the analytical chemist, the main problem with the chart recorder is the actual computation of results from the graph traced on the paper. Let us consider a sample moving through a continuous flow system, e.g. an amino-acid analyser or a gas chromatograph. Each component is in the form where it will initiate some response from the detector. We can consider each component as a 'slug' of material travelling at constant speed towards the detector in a neutral gas or liquid flow. The extremities of the 'slug' will be more dilute than the centre as there will tend to be slight diffusion and mixing with the carrier. The detector will respond to each part of the 'slug' as it moves through and, under ideal conditions, will produce a Gaussian curve on the chart. The more compact the slug and the less mixing has occurred, the sharper, taller and narrower the peak will be. Among other factors this will depend upon the time that the component has spent actually travelling in the medium, leading to the common phenomenon that the early peaks tend to be narrower and sharper than the later ones. The total area under the peak will be directly related to the quantity of the component which has flowed through the detector and by running accurately known standards through exactly the same analytical procedure, factors can be produced which will be used to quantitate 'unknown' samples.

Several methods have been described for doing this. Most of them depend upon treating the Gaussian curve as a triangle and measuring the peak height and the width at half the peak height. Multiplied together these two factors give the area of the triangle approximating to the Gaussian curve. If the peaks are very sharp and narrow with no noticeable 'width', peak height alone is often taken as analagous to concentration. Where peaks overlap slightly the notional continuation of overlapping peaks is drawn in with a pencil and the area calculated in the normal way. Better results can sometimes be obtained by increasing the speed of the chart paper and thus widening the peaks to make measurement of the width at half peak height more accurate.

A second common method is to cut out the peaks on the chart carefully with scissors and weigh the small pieces of paper on an analytical balance. Surprisingly consistent results can thus be obtained as long as the chart paper is consistent. Even greater precision is possible if a Xerox photocopier is available as the paper used is of very consistent density and cutting and weighing the peaks from a Xerox photocopy gives better results than calculating triangles.

TABLE 8.3. Comparison of Precision of Integrating Methods

Method	*Precision* (σ REL)
Planimeter	4%
Triangular (H × B)	4%
Triangulation (H × W at ½H)	2·5%
Cut and Weigh	1·8%
Disc Integrator	1·3%
Electronic Digital Integrator	0·5%

The planimeter is a mathematical instrument used for measuring the area of irregular plane surfaces and many workers find it suitable for calculating peak areas. It requires some skill in use and takes some time for each measurement.

Where peaks appear slowly a special type of recorder is used which does not trace a continuous line but produces a dotted record, the pen marking a dot at a fixed time interval. The peak height is measured, the half-height marked off and the dots forming the upper half of the peak are counted. As the width of the peak is dependant on time and the number of dots marked is similarly dependant, the two figures are analagous.

The disc integrator is a useful device which is fixed to the recorder and responds to movements of the potentiometer shaft. A secondary zig-zag trace is recorded below the baseline for each peak and by counting the points on the secondary line the area of the peak can be ascertained. Later versions of this device have an associated digital print out.

The trace reader has a flat metal sheet to which the chart is fixed. There is a potential difference between the lower and upper edges of the metal sheet. A stylus is provided and by touching the tip of each peak with its point a voltage proportional to peak height is tapped off and displayed on a digital voltmeter. Associated equipment will calculate concentrations etc. from this figure on the basis of standards.

All these instruments and devices are comparatively cheap, and in the field of data acquisition the term can be applied to apparatus costing less than £500. For full automation, automatic print-out devices, automatic integraters, and computers lead to the spending of very large sums. Let us consider the steps which have to be taken for a full data acquisition system to be effected using as an example an automatic colorimeter.

(*i*) The absorbance or per cent transmission must be read.

(*ii*) The signal from (*i*) must be converted to figures (this is called analogue to digital conversion).

(*iii*) Often the figures in (*ii*) are not linear, the signal from (*i*) not being directly proportioned to concentration. The logarithm of the figures must be used and this conversion must be performed.

(*iv*) Converting the figure from (*iii*) into concentration units which would involve a multiplication step.

(*v*) Correcting for any instability in the machine, e.g. base-line drift.

(*vi*) Printing out the results on to paper in terms of concentration units with an automatic typewriter.

It is clear that a reliable and stable device to perform all these steps needs to be carefully designed and constructed and will necessarily be of some complexity. It is not surprising that such apparatus is very expensive.

Even this level of automation does not exhaust the possibilities. We have dealt with the aquisition of data from automatic systems

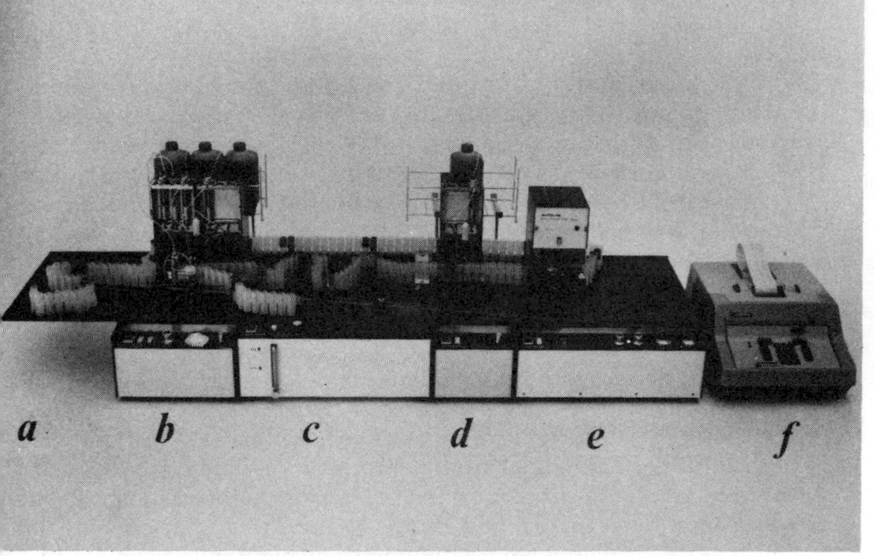

Figure 8.4. The Autolab. This is a modular discrete sampling analytical system
[By courtesy of Grant Instruments (Cambridge) Ltd]

(*a*) Sample table. Batches of ten samples at a time are inserted in individual test tubes or sample cups carried in plastic links forming the 'sample chain'.
(*b*) Sampler unit. An aliquot of each sample is taken from the tube and transferred quantitatively to a second tube in the 'analysis chain' together with a metered quantity of reagent. Up to four other reagents can be added at this stage with stirring.
(*c*) Incubator. Samples can be incubated for between 3 and 30 minutes from room temperature to 100°C.
(*d*) Reagent unit. Up to 3 more reagents can be added with stirring.
(*e*) Analysis unit. A narrow band automatic absorptiometer which can be set between 340 and 620 nm or 600 to 1000 nm depending on the photocell.
(*f*) Printer. Each sample is identified by a number which is printed on the paper tape together with its result in the required concentration units (e.g. mg/100 cm^3, mol/kg, absorbance units).

The machine can handle 240 samples per hour.

and its conversion to an intelligible form. In many laboratories data processing and data transmission systems are required, so that information regarding a sample is automatically handled from the time it enters the department to the time a report is sent out. The use of computers, punched card systems, etc. is thought to be beyond the scope of a book on laboratory techniques although some of the titles in the appendix deal with these matters.

Conclusion

There is no doubt at all that automation has entered the laboratory in a very big way, and in many departments where very large batches of routine analyses have to be performed, the work load would be impossible to cope with if automatic machinery were not available. In many disciplines traditional skills, painstakingly acquired, are becoming out-of-date at an alarming rate and one can foresee laboratories staffed by semi-skilled machine-minders with a trained technician occupying the posts of foreman, office manager and trouble-shooter. It must be acknowledged that automation on this scale is valid only when very large numbers of similar tests have to be performed. There will always be the special samples for analysis where the numbers involved do not justify full automation and where the ability to perform a bench analysis will still be needed. Indeed as multichannel instruments become more common the need for accurate standards including large numbers of materials which must be evaluated will increase, otherwise the machines will churn out their daily quotas of hundreds or thousands of results, all very precise—and all wrong. To realise this is to realise that the introduction of machinery is not a challenge to the skills of a profession but merely involves the mastery by the professional of more analytical techniques.

CHAPTER 9

SOME SPECIAL TECHNIQUES

HIGH VACUUM TECHNIQUE

Many laboratory processes need to be carried out under conditions of reduced pressure. These low pressures are usually produced by three classes of pump:

(1) the water pump,

(2) the rotary vacuum pump, and

(3) the vapour diffusion pump.

Other types of pump designed for specialist purposes and ultra-low pressures will not be covered here, but suggested sources of information on these will be found in the bibliography.

The types of pump mentioned are for use over different pressure ranges. The water pump (see Chapter 2) is limited to the vapour pressure of water. This is rarely below 1300 Nm^{-2} (10 mm Hg) although it is possible to improve performance a little by including a good desiccant such as phosphorus pentoxide, in the system between the pump and the apparatus being evacuated. The rotary pumps are used for pressures between 130 and 0·1 Nm^{-2} (1 and 10^{-3} mm Hg). The vapour diffusion pump, which is used in conjunction with a rotary pump, will reach 10^{-5} Nm^{-2} (10^{-7} mm Hg).

Theory

Some basic information will be found useful in understanding modern low-pressure techniques. The units of pressure are described on page 29.

The law relating pressure and volume of a gas was propounded by Boyle in 1660. This states that at constant temperature, pressure $P \times$ volume $V =$ a constant. For example, a pressure of 1000 Nm^{-2} on 40 cm^3 of gas when increased to 2000 Nm^{-2} will compress the volume of gas to 20 cm^3. This law only holds good over a limited range at a particular temperature for each gas, but for practical purposes in vacuum work differences at other temperatures and pressures can be ignored.

Charles' law states that at a given pressure the volume is proportional to the absolute temperature.

Dalton's law of partial pressures states that the pressure exerted by a given volume of a gas mixture is equal to the sum of the pressures exerted by the individual gases if they occupied the volume alone. The partial pressure, therefore, is the pressure exerted by a component in a gas mixture. The total pressure is the sum of the partial pressures.

The '*mean free path*' is the average distance one molecule may move before it collides with another. This varies with pressure. Table 9.1 gives the mean free path of air at various pressures. It will be remembered that in molecular distillation the condenser must be at a shorter distance than the mean free path to the distilling liquid.

TABLE 9.1. The Mean Free Path of Air at Various Pressures

Pressure (Nm^{-2})	*Mean free path at* 15°C
1300	0·005 mm
13	5·0 mm
13×10^{-2}	50 cm
13×10^{-4}	50 m
13×10^{-6}	5 km

The manner in which a gas flows in a vacuum system varies with pressure. Under the majority of chemical laboratory vacuum conditions (i.e. from atmospheric to a pressure of 0·1 Nm^{-2}), gas flow is turbulent and the flow *rate* is considerably influenced by the design of the piping and connections. Greatest efficiency will be obtained by the use of short connections, gentle bends, maximum tube diameter and the minimum number of obstructions to gas flow. At lower pressures the flow is modified as the mean free path becomes of the same order as the diameter of the pipe, and the flow is viscous. At very low pressures the mean free path is considerably greater than the diameter of piping and the flow becomes molecular.

Pumps

The mechanical pumps make use of Boyle's law. The volume of the apparatus to be evacuated is increased by opening up a

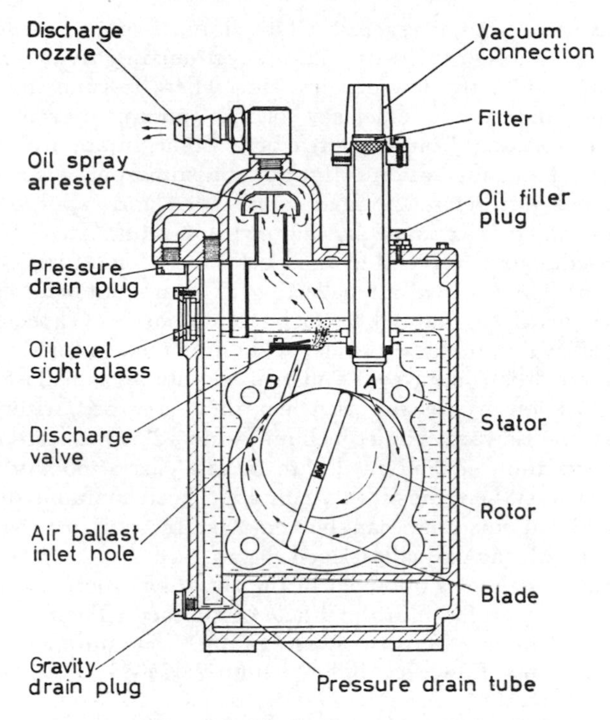

Figure 9.1. The Speedivac rotary pump

chamber in the pump. The gas expands into this chamber which is then closed off, the gas is expelled from it and then the chamber is opened into the apparatus again. This process is continuous and many pumps make use of this principle. The Edwards High Vacuum Ltd. Speedivac Rotary Pumps (*Figure 9.1*) have a rotor that revolves eccentrically in a cylindrical chamber or stator such that its circumference is always in contract with the stator wall. Two vanes stressed by compression springs slide in a slot cut across the diameter of the rotor. As the rotor revolves the vane sweeps round and gas from *A* fills the space created behind the vane. The second vane then traps this volume of gas and pushes it out through the non-return valve *B*. The pump is immersed in a low-vapour pressure oil and the gas is exhausted through it.

Some vapours, for example, water, condense just as they are forced through valve *B*, due to the fact that under compression the saturation point is reached. This liquid is then expelled into the oil and emulsifies with it, and on recirculating in the pump is again exposed to the low-pressure side where it vaporizes. This obviously impairs the efficiency of the pump, preventing the desired low pressure from being reached. Some pumps are modified to prevent this condensation by introducing atmospheric air into the exhaust side just before the final compression and expulsion. The extra air which is retained by another non-return valve *C*, when compressed, causes valve *B* to open before vapour saturation point is reached. This is called air ballasting. The amount of air entering may be varied so that when such conditions are expected, the air ballast valve may be opened wide and then gradually closed down as conditions improve. Pumps used with a gas ballast cannot produce as low a pressure as when they are used without, for example, the Edwards High Vacuum pump 1SC50 can produce an ultimate vacuum of 0·6 Nm^{-2} with ballast closed, 60 Nm^{-2} with ballast open. Where the oil in a pump has been contaminated with condensed vapours they can be removed by running the pump overnight with the gas ballast open.

A better vacuum is obtained in the situation where the exhaust from one pump is fed into the intake of another. These conditions have been arranged in a 'two-stage' pump. The ultimate vacuum of such a pump (Edwards High Vacuum 2SC50) is 0·006 Nm^{-2}.

Vapour Diffusion Pumps

These pumps work on an entirely different principle, the gas molecules from a vessel being exhausted are caught and trapped in a stream of oil or mercury vapour through which they cannot return. The pump employs a reflux system, the oil or mercury is boiled in a cylinder, the vapours rise up a tube and are ejected as a cone of vapour against the cooled walls of the cylinder. The vapours condense and fall to the bottom of the chamber to be vaporized again.

The original pumps were made of glass or silica and are used for small-scale high vacuum work. One such pump is made by Fisons and is illustrated in *Figure 9.2.* Mercury is placed in the boiler *A* which is shaped to take a bowl-fire type of heating element. The mercury is boiled and the vapours pass up tube *B* and out of the jets at *C*. The vapour from the vessel, which is attached at *D*, is trapped and compressed towards the exit at *E*. Mercury vapour is condensed by water in the jackets *F*. This type of pump has no great pumping capacity so the metal cylinder pump of the basic design illustrated in *Figure 9.3* is more commonly used. These vary in diameter from 5–50 cm or more. Oil or mercury in *A* is heated by a flat electric element *B*. The vapour rises in the chimney and is ejected at the jets *C*. Gas outflow is through the baffle *D*.

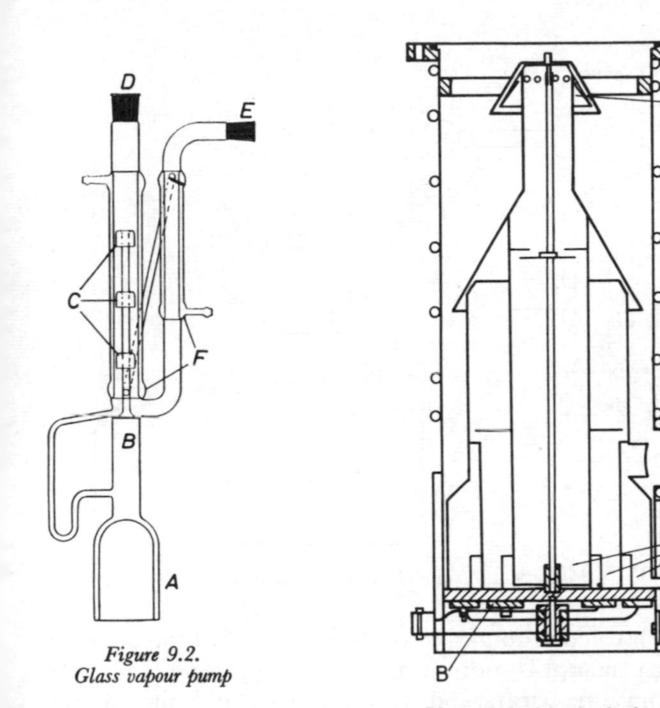

Figure 9.2. Glass vapour pump

Figure 9.3. Metal cylinder vapour pump

The oil vapour is ejected at great velocity and the trapped gas is highly compressed at a ratio of about 100 000 to 1. This rapid build-up of gas has to be taken away at speed or the pump stops working. In fact all vapour pumps must have the internal pressure reduced to below 60 Nm^{-2} before they function. This is done by attaching a mechanical pump having the requisite pumping speed to the baffle outlet *D*. This is called a backing pump and it follows that the apparatus being evacuated must also have its pressure reduced by the backing pump before the diffusion pump can be brought into use.

A schematic diagram of a complete high vacuum pumping unit is shown in *Figure 9.4*. The chamber for evacuation is connected to the pump via a large baffle valve, which is designed to give the

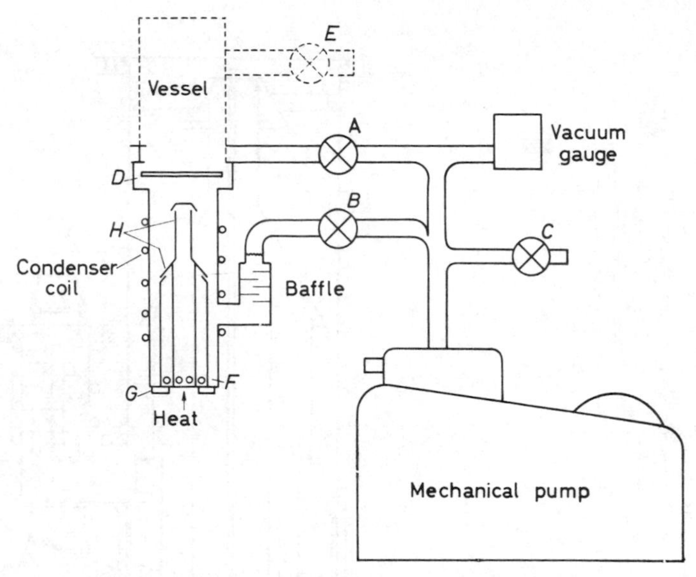

Figure 9.4. High vacuum pumping unit

minimum of obstruction to gas flow. Poor results will be obtained if the backing pump is not in peak working condition. All connections must be clean and vacuum tight and the oil uncontaminated. The unit is operated in the following order:

(*a*) Turn on the cooling water for the vapour pump.

(*b*) With all valves closed, turn on the mechanical pump. The pressure registered in the vacuum gauge should fall quickly to below 60 Nm^{-2} (0·5 torr).

(*c*) Open valve *B* and wait until the pressure registered is again below 60 Nm^{-2} (0·5 torr).

(*d*) Switch on diffusion pump heater and allow time for it to warm up.

(*e*) Close valve *B*.

(*f*) Open valve *A* and pump down the vessel to below 60 Nm^{-2} (0·5 torr.)

(*g*) Shut valve *A* and open valve *B*.

(*h*) Open baffle valve *D* slowly.

If it is necessary to open the vessel and then re-evacuate it, valve *D* is closed and air is admitted at *E*. Close valve *E* and proceed as above from sub-para (*e*).

The unit is turned off in the following order:

(*a*) Close valve *D*.
(*b*) Switch off the diffusion pump and allow it to cool for 15 min.
(*c*) Close valve *B*.
(*d*) Open valve *C* to admit air.
(*e*) Turn off mechanical pump and cooling water.

Note that the vessel can be left evacuated.

There are a variety of attachments and modifications available. The fractionating pump is a diffusion pump with a modified chimney and jet assembly, which is able to partly separate the oil components so that the more volatile vapour does not come near the vessel being evacuated. Other means of preventing pump fluid vapours entering the evacuated vessel are the use of specially designed baffles and guard rings. Mercury vapours must be retained by a liquid air or nitrogen trap situated above the pump. A phosphorus pentoxide trap attached to the backing pump is essential.

Measurement of Vacuum

No single device is capable of measuring the complete low-pressure range. The several gauges needed work on totally different principles.

Bourdon Gauge

This type is useful for measuring pressures of 1300 Nm^{-2} (10 mm Hg) upwards and is only accurate to a few Nm^{-2}. It consists of a bent hollow tube or capsule sealed at one end and held firmly at the other which is also connected to the vacuum system. When the pressure is lowered, the radius of the bend in the tube is reduced, thus causing the closed end to move. This movement is magnified and transmitted by a rack and pinion mechanism to a pointer on a scale. This gauge is useful as a permanent accessory on a water pump.

Manometers

These are discussed in Chapter 2. For more accurate measurements in the range 2500 Nm^{-2} (0–20 mm Hg) fluids other than mercury may be used, such as Apiezon oil. For pressure differentials at higher pressures, where water vapour does not interfere, Brodie's fluid (Chapter 12) may be used.

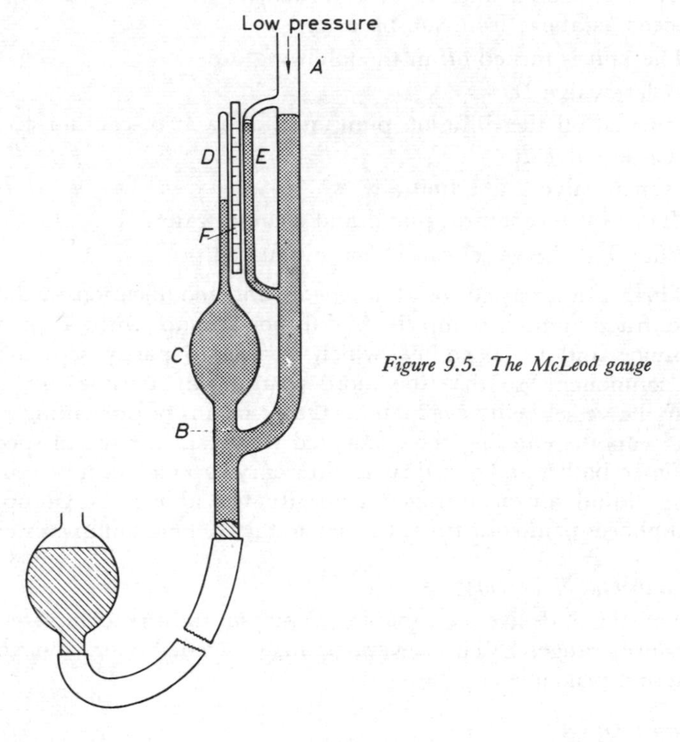

Figure 9.5. The McLeod gauge

McLeod Gauge

This gauge is very accurate provided that the pressure of the gas being measured does not contain condensable vapours. It is a partial pressure gauge. A known volume of low-pressure gas is compressed into a closed limb of a manometer with mercury and the limb calibrated mathematically by applying Boyle's law. This type of gauge may be used for pressures down to 10^{-3} Nm^{-2} (10^{-5} torr) but several are required to cover the complete range.

The gauge is attached to the evacuated system at *A* (*Figure 9.5*). The reservoir of mercury is adjusted so that the level in the gauge is not above point *B*. When a reading of pressure is taken, the reservoir is raised and mercury traps the portion of low-pressure gas in bulb *C* and compresses it into capillary *D*. Adjustment is made until the mercury in E is level with the top of capillary D. This being calibrated, the pressure can be read at F.

The Vacustat Gauge manufactured by Edwards High Vacuum Ltd. is a modified form of McLeod gauge and is much more convenient to handle but suffers from the same disadvantage of not giving a continuous reading. It is suitable for measuring pressures down to 0·1 Nm^{-2} (10^{-3} torr).

Pirani Gauge

This gauge works on the principle that if the loss of heat from a wire varies, so will its resistance. The loss of heat will depend on the thermal conductivity of the surrounding gas. Over a certain range this is proportional to pressure.

A tungsten element is sealed into the closed end of a vertically held glass tube, the open end being attached to the vacuum system. The element is connected into a Wheatstone bridge circuit across which is applied 3 or 4 V. When the gas pressure surrounding the element drops, less heat is carried away by molecular conduction and the temperature of the element rises. This alters its resistance and the calibrated galvanometer will be deflected, and the pressure read. These gauges are continuous reading and are available in several modified forms, the major one being an instrument in which filament temperature is actually measured by an attached thermocouple. The filament is heated by a constant current supply. The Pirani gauge measures pressures between 1300 and 0·0013 Nm^{-2} (10 and 10^{-5} torr).

Thermionic Ionization Gauge

This is a device similar to a triode valve which has an opening for connection into the vacuum system. The grid is held at a high voltage with respect to the anode and cathode. The cathode to grid electron emission is kept constant and any ionization caused by collision with gas molecules causes a current to flow through the anode. This effect depends on the number of gas molecules present and, therefore, the pressure of the surrounding gas. The ionization current is amplified and is linear with gas pressure over the range 1 to 10^{-6} Nm^{-2} (10^{-2} to 10^{-8} torr).

These gauges cover the vacuum range likely to be met with in normal laboratory practice.

Freeze Drying

Water can be transferred from one vessel to another by joining them together with tubing and surrounding the empty one with a freezing mixture. The water vapour freezes on the cold surface and is replaced by evaporation from the surface of the water. This continues until all water is transferred. Similarly, ice can be sublimed under reduced pressure on to a cooler surface or into an efficient desiccant. Sufficient vacuum is necessary to maintain an evaporation rate that will keep the sample frozen until all the water is removed. This is a useful way of drying a solution of a substance from the frozen state with the preservation of tissue cells and large molecular compounds such as antisera intact. Freeze drying overcomes the problem encountered with normal evaporative drying where the salt strength of a solution increases to saturation before all the water is removed. This may change the structure of large molecules and rupture cells.

The two methods used have already been indicated. The first is rapid evaporation of the samples thereby freezing them and the absorption of the moisture on to a desiccant, usually phosphorus pentoxide. This method is used for drying small samples of up to 2·5 cm^3 in volume, the samples being placed into ampoules. Because of the presence of dissolved gases the tubes have to be centrifuged throughout the initial freezing to prevent frothing and subsequent loss of material from the tube. After the tubes have been subjected to a low pressure for about 5 min, the samples will have frozen and the centrifuge can be stopped. The vacuum is maintained until the samples are dry, the time taken varying from 2 to 10 h. To ensure complete dryness secondary drying can be achieved by placing the tubes on to a manifold connected into the vacuum system. Fresh desiccant is then placed in the trays of the trap and a vacuum maintained for several hours. The use of the manifold makes possible the storage of the samples under vacuum as the tubes can be sealed with a hand torch while the pump is still maintaining the vacuum.

The second method is used for larger volumes, the liquid being pre-frozen before connection to the freeze drying apparatus. The solution can be frozen to the walls of a round bottomed flask by rotating the flask in an alcohol or acetone/CO_2 freezing mixture. For efficiency it is important that the solution is spread over as large an area as possible within the flask (*shell frozen*). For freeze drying larger volumes (up to several litres) the liquid is poured into large trays which are then placed on refrigerated shelves within a freeze drying chamber. The evaporated moisture is collected on the walls of a cold trap that is at a temperature of at most -40°C. This can be a refrigerator coil or an acetone/CO_2 cold trap. To speed freeze drying the samples may sometimes be heated in the early stages but not enough to cause them to melt.

Various assemblies are illustrated. *Figure 9.6a* is a standard vacuum pump assembly mounted on a trolley. The mechanical pump is on the lower shelf and is suspended by anti-vibration mounts

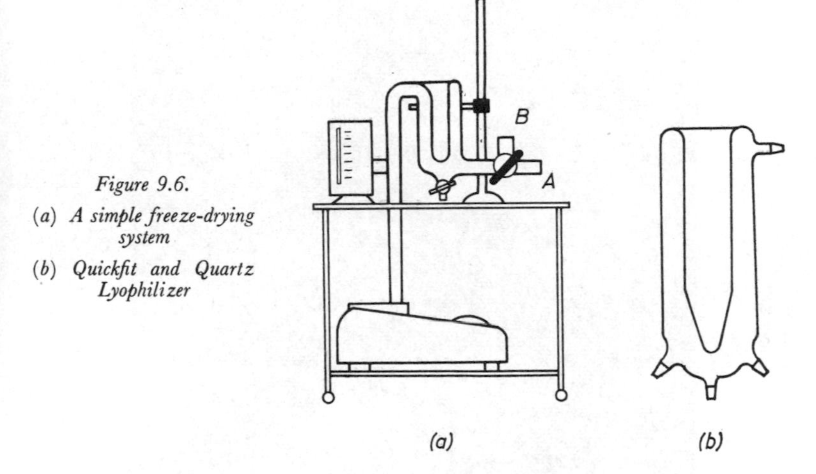

Figure 9.6.
(*a*) *A simple freeze-drying system*
(*b*) *Quickfit and Quartz Lyophilizer*

to minimize noise and vibration. Either 1 cm bore rubber pressure tubing or copper tubing connects the pump to a vacuum gauge, a moisture trap and a 3-way stopcock all of which are on the upper shelf and clamped on to laboratory scaffolding. The 3-way stopcock may be replaced by two diaphragm valves, these are less constrictive and are designed for attachment to 1 cm copper pipe. Except for a small bore connection to the gauge the tubing used should not be of smaller bore than that of the pump entry tube.

Flasks containing the pre-frozen samples can be attached at *A* either singly or two or more with the use of an adaptor. Where a greater volume is to be dried a second larger trap can be attached at *A* illustrated by the lyophilizer *Figure 9.6b* manufactured by Quickfit & Quartz which will hold four 1-litre flasks each containing 100 cm^3. On no account must moisture enter the pump, the traps must be kept efficient and must be able to hold all the water that is being evaporated. An ampoule manifold can be attached at *B* leaving *A* as a vacuum release.

Figure 9.7 shows a centrifugal freeze dryer and *Figure 9.8* shows a large capacity freeze dryer both manufactured by Edwards High Vacuum Ltd. A Pirani gauge is fitted as this is more suitable than a McLeod type. The latter are much cheaper and may be used provided that it is a partial pressure gauge and the vacuum system is intended to contain water vapour.

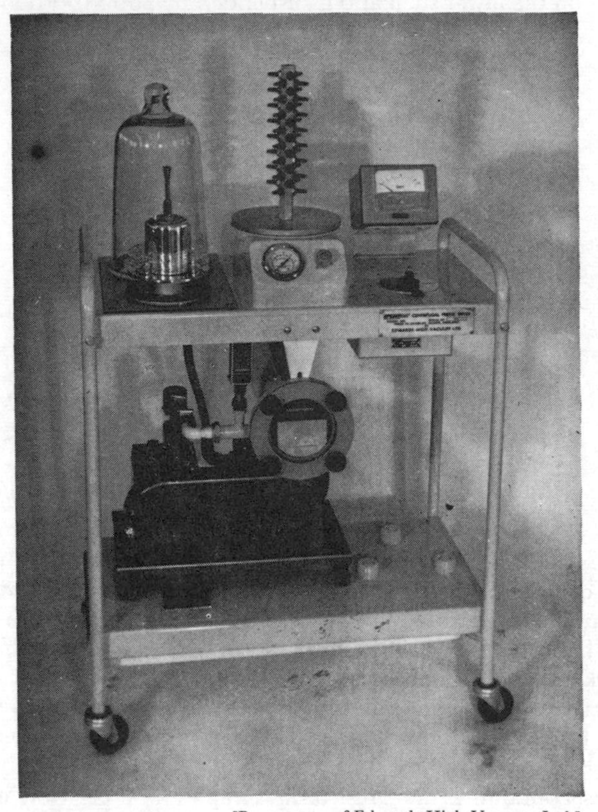

[By courtesy of Edwards High Vacuum Ltd]

Figure 9.7. Edwards centrifugal freeze dryer

The most important rule in vacuum work is to release the vacuum before switching off the vacuum pump (except with diffusion pumps) otherwise pump oil may be drawn into the traps and gauge. As a precaution against this happening accidentally, for instance because of power failure, a non-return valve can be fitted to the pump.

Fault Finding and Leak Testing

For efficient freeze drying the pressure must be 25 Nm^{-2} (0·2 torr) or less. This is readily obtainable in the pumps mentioned although

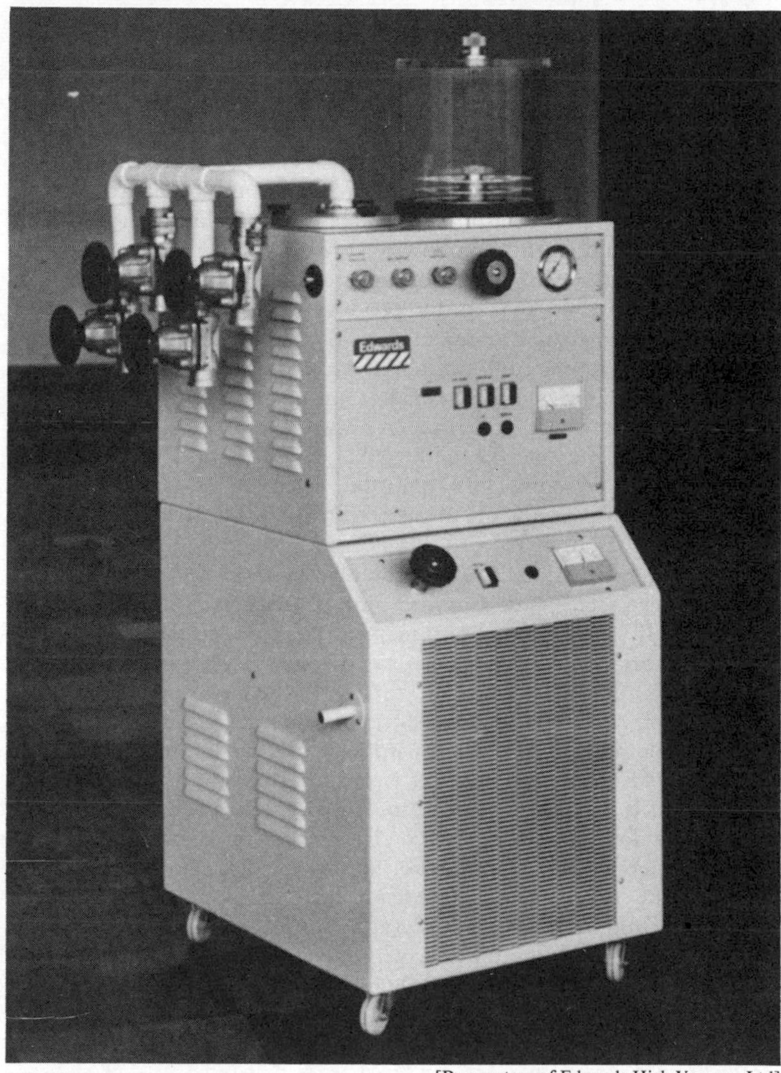

[By courtesy of Edwards High Vacuum Ltd]

Figure 9.8. Edwards model EF03 freeze dryer with mechanical refrigeration. A flask manifold is fitted to the condenser chamber lid

where large vessels are to be evacuated a pump must be chosen with a greater pumping speed. If the pressure is slow to drop or fails to reach its usual low pressure there are three possible causes.

(1) A leak has developed in the system.

(2) A highly volatile contaminant is outgassing from the oil or walls.

(3) A mechanical fault has developed in the pump. The latter is very rare.

Leak testing should not be haphazard but systematic. The various sections in a vacuum system should be isolated and pumped down individually, thus narrowing the area of search. Primarily a check should be made to ensure that the pump or pumps on their own are able to rapidly produce their maximum vacuum. Each section in turn is connected to the pump until the leaky one is found. Couplings are tightened and joints and taps examined and greased if necessary, then after pumping down, if the leak is not obvious, the section is removed and pressurized with compressed air. The leak may be felt or heard, but if it is not, the joints and other vulnerable points are 'painted' with soapy water or if the section is small enough it is immersed in water. A stream of bubbles will indicate the leak.

A leak detector that is primarily a pressure indicator is the high-frequency tester. This has an output frequency of 300 kc/s or more and if the probe is held near to the glass sections a discharge glow will appear in the tube if the pressure is between 1·3 Nm^{-2} (0·01 torr) and 1300 Nm^{-2} (10 torr). If the probe is placed near a crack or pinhole in the glass a stream of sparks will radiate from it.

These methods may not show up leaks that are imparing a higher vacuum system as produced by a vapour diffusion pump. Assuming the diffusion pump is able to function and produce a vacuum the search for leaks may be carried out using a modified Pirani or ionization gauge. These are connected into the system and a small jet of search gas is played around the likely spots for leaks. The gas, usually hydrogen or butane (Calor gas), has a higher thermal productivity and viscosity and will almost immediately cause a movement on the galvanometer if the gas enters the system.

Pinholes and cracks may be temporarily stopped with Apiezon 'Q'. This is a black, putty-like substance of low vapour pressure which is pressed over the affected part.

Where there is no apparent leak, contamination of the oil or walls can be suspected. Water in the oil of a mechanical pump has already been mentioned and is easily removed if the pump is fitted with a gas ballast valve. If not, the oil must be drained, the pump flushed with a little clean oil, and then refilled with the correct grade oil. Emulsified oil and water can be separated by centrifugation.

In high vacuum systems efficiency is often impared by the outgassing of molecules adsorbed or absorbed on to the walls. This effect is enhanced if the inner walls are not clean and have traces of grease, oil, flux or even greasy fingermarks. Desorption is carried out by pumping down the system and then heating the evacuated vessel or apparatus. This can be carried out with electric heating tapes, careful heating with burners and, where practicable, by the use of an oven. The pressure will be seen to rise on heating and then fall again to below the original ultimate vacuum.

THE POLAROGRAPH

Figure 9.9 shows a simple electrolytic circuit in which the potential across the cell may be adjusted by the variable resistance. If the voltage is gradually increased from zero and the current noted,

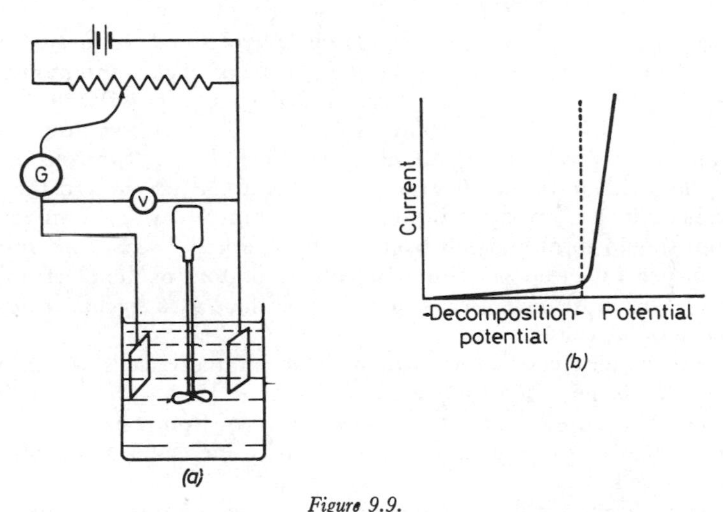

Figure 9.9.
(*a*) *Simple electrolytic system to demonstrate decomposition potential*
(*b*) *Current/voltage curve through an electrolyte*

a current-voltage curve can be traced and will take the form of the graph in *Figure 9.9b*. The current stays low and constant until, at a 'breakthrough' voltage electrolysis begins and the current rises proportionately with the increase in voltage. The potential at which the sharp inflection in the curve occurs is specific for the positive ion in the electrolyte and is called the *decomposition potential* of the ion. Every ion has a different decomposition potential and an analytical system can be developed based on the behaviour of electrolytes.

In 1922 Heyrovsky designed the dropping mercury electrode which is particularly suitable for this purpose. This was used as part of an electrolytic cell in which one electrode, usually the cathode consists of a succession of fine mercury drops falling regularly through a fine capillary tube. Using this system it was usually possible to construct a reproducible current voltage curve for any electrolyte, and by examining the curve to state both the identity and the concentration of the electrolyte. In 1924, Heyrovsky and Shikata automated this system to record the current voltage curve automatically. The machine, called *the polarograph*, has made this analytical system much more useful as the manipulation is simple and results speedily obtained.

The Polarographic Circuit

A simple circuit for manual polarography is shown in *Figure 9.10*. The dropping mercury electrode consists of a hard glass thermometer capillary of about 0·04 mm bore, sealed to a piece of wider bore tube about 10 cm long. This is attached by a rubber tube to a glass reservoir of 1 cm bore, and about 70 cm long. If the electrode is filled with pure mercury to a height of about 60 cm and the tip immersed in a strong solution of potassium chloride, a mercury drop should form and fall from the tip every 3–5 sec. The anode is formed by the pool of mercury at the bottom of the electrolytic cell. This must have a relatively large surface area so that it never becomes polarized.

The battery consists of two small accumulator cells which are capable of giving a steady e.m.f. of up to 4 V.

The two potentiometers allow an e.m.f. of from zero to 3 V to be applied to the cell and to be gradually increased or decreased as required.

The current, rarely more than 50 mA, is measured by the galvanometer. This is usually fitted with an adjustable shunt so that the sensitivity of the instrument may be varied to cope with a wide

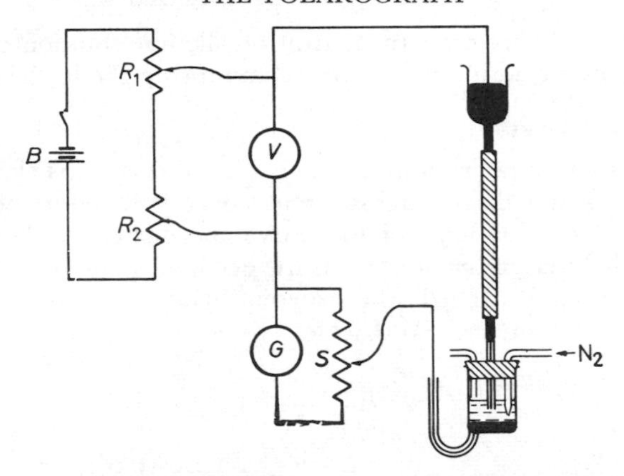

Figure 9.10. Circuit for a simple manual polarograph
R_1 — 1,000 Ω radio type wire-wound variable resistance
R_2 — 10 Ω radio variable resistance S — Galvanometer variable shunt

range of substances and concentrations. The current through the system pulsates with the growth and fall of the mercury drops, and a heavily-damped galvanometer of long period is used to minimize the oscillations.

The Solution

The solution of substances to be analysed is very dilute, from 10^{-2} to 10^{-5} M being usual. In addition to the mixture to be analysed a much larger concentration of *supporting electrolyte* is present. This has a decomposition potential much higher than that of the 'unknown' materials. Potassium chloride is the most commonly used supporting electrolyte, although other salts of alkali metals or ammonium are sometimes employed. The supporting electrolyte carries the current through the bulk of the solution and ensures that the material to be determined, if charged, does not actually migrate to the dropping mercury electrode. As will be seen in the section on wave height, the determination of concentration is made easier in this way.

Interfering substances are usually removed by direct precipitation often without subsequent filtration, the precipitate being allowed to settle, subsequent examination being carried out on a portion of the supernatant which is withdrawn.

Oxygen is a special case which requires special treatment. In most polarographic determinations it is necessary to remove oxygen from the solution, usually by sweeping out with a stream of nitrogen or hydrogen. In neutral or alkaline solution oxygen is destroyed by the addition of a little sodium sulphite to the solution.

Polarographic Waves

If a solution of potassium chloride, free of oxygen, is electrolysed in a polarograph the voltage current curve will be of the form I in *Figure 9.11.* For most of the curve the current will be small. This is called the *residual current.* As the decomposition (or reduction) potential is approached the current rises sharply until the galvanometer pointer goes off scale.

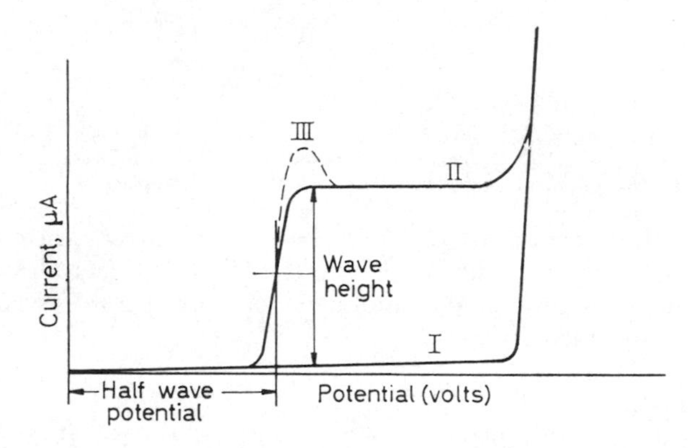

Figure 9.11. The polarographic wave

If a very small amount of zinc chloride is added and the experiment repeated the curve will be of the form II in *Figure 9.11.* Again only the residual current flows at first, but as the decomposition potential of the zinc ion is reached the current rises sharply. Instead of the rise continuing indefinitely, however, the curve flattens out into a step which shows no marked change until the decomposition potential of the potassium chloride is reached. The step is called a *polarographic wave* and the almost constant current flowing at the upper flattened section of the curve is called the *diffusion current* or expressed as the *wave height.* The wave height in this case is directly proportional to the concentration of zinc in solution, and the quantitative aspect of polarographic analysis is based on the direct relationship between wave height and concentration.

This direct relationship can be explained by examination of the conditions around the dropping mercury electrode as electrolysis proceeds. As the decomposition potential of the reducible substance is reached, electrolysis begins and the current rises sharply as the applied voltage increases. As electrolysis continues, however, the layer of solution in the immediate vicinity of the drop rapidly becomes exhausted of the reducible substance and a state called *concentration polarization* sets in. This can be offset in two ways:

(1) by diffusion of further amounts of reducible substance from the bulk of the solution to the vicinity of the electrode,

(2) by migration of a charged reducible substance from the bulk of the solution.

In the solution to be analysed the large excess of supporting electrolyte carries practically all the current through the solution and the migration of ions of reducible substance towards the electrode is negligible. The rate of electrolysis and the current flowing is controlled almost entirely by the rate of diffusion of reducible substance from the bulk of solution to the electrode. This is quite independent of applied e.m.f. but directly dependent on the concentration. The diffusion current remains constant as an expression of concentration until the decomposition potential of another constituent in the solution is reached and the current rises once more.

It has been stated that the decomposition potential is specific for the reducible substance. There is, however some variation in value depending on both electrode characteristics and concentration. In 1935, Heyrovsky and Ilkovic demonstrated that the potential at that point on the wave at which the current reaches one half of its diffusion or limiting value is not dependent on these factors and is much more characteristic of the reducible material. This is the *half-wave potential* of the substance. This factor is reported using the saturated calomel electrode as the reference standard.

Factors Which Affect Wave Height

Although the wave height or diffusion current is directly proportional to concentration there are several other factors which affect this figure. These have been summarized in the Ilkovic equation.

$$I_d = 605.n.D^{\frac{1}{2}}.C.m^{2/3}.t^{1/6}$$

Where I_d is the average diffusion current (the current pulsates slightly as the drops grow and fall), n the amount of electricity involved per mole of the reducible substance in Faradays, D the diffusion coefficient of the substance, C its concentration (mmol/dm^3),

m the rate of flow of mercury (mg/sec) and t the drop time (sec).

D depends on total salt concentration and on viscosity, the latter being dependent in turn on temperature. I_d is thus temperature dependent and it is usually found necessary to control the temperature in the cell with a thermostat to $\pm$ 0·5°C.

Both m and t are characteristics of the specific electrode in use. The factor m is almost entirely controlled by the 'head' of mercury above the tip of the electrode, and this is practically constant and hardly affected by temperature; m may be regarded as constant. The drop time t depends on the mercury pressure but also on the tension at the interface between the mercury and the solution. This depends on the potential and at first increases to a maximum with increase in potential, falling off again as the potential increases further. As only the one-sixth power of t is a factor in the equation the effect is small, but may be noticeable if two waves are separated by a wide difference in potential. The characteristics of the electrode represented by the product $m^{2/3}t^{1/6}$ are usually noted on the polarogram so that results obtained with different electrodes may be compared.

Maxima

Sometimes the point at which the wave flattens out is distorted by an extra peak called a *maximum* (*Figure 9.11*, III).

The causes of maxima may be irregularities of stirring effects at the surface of the drop, adsorption effects, or other causes of asymmetry. They are a nuisance as it becomes impossible to measure wave height accurately. Several *maximum suppressors* have been reported usually in the form of dyestuffs (e.g. methyl red) or colloids (e.g. gelatin) which return the wave to its normal shape. Only traces of suppressors are added as excess amounts may radically alter the shape and position of the wave.

Analysis of Mixtures

When several reducible substances are present in a mixture a polarogram of a series of waves may be traced (*Figure 9.12*). The concentration of each substance may be estimated from the height of its wave.

Advantages of Polarography

In working laboratories the polarograph is usually of the automatic recording variety in which both the increase in potential and the tracing of the polarogram are automatic after the initial

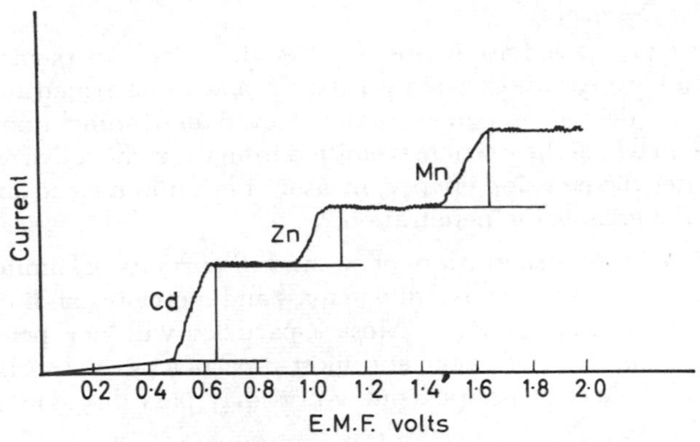

Figure 9.12. Polarogram of cadmium, zinc and manganese

adjustment of the machine. In the laboratory where the method is rarely needed, or in teaching laboratories where an expensive and sensitive machine is not required, the manual polarograph gives adequate results. In both cases the characteristics of this technique of microanalysis may be summarized as follows:

(*a*) The preparation of solutions is a simple matter.

(*b*) The amounts of substances to be analysed is very small. Solutions as dilute as 10^{-5} M and volumes as low as 5 mm^3 may be used although 2–20 cm^3 cells are more common.

(*c*) A permanent record of the analysis is available.

(*d*) Any reducible substance, whether ionized or not, may be analysed by polarography, e.g. dissolved oxygen, organic materials. The technique is thus important in biochemistry.

(*e*) The actual electrolytic decomposition of the solution is negligible. The determination may be repeated several times if necessary on the same sample, and the constituents of the sample may be recovered after analysis.

TECHNIQUES INVOLVING RADIOACTIVE SUBSTANCES

Over the last twenty years techniques in which radioactive materials are used have penetrated into every field of scientific endeavour and instruments for the detection and measurement of radioactivity have become standard laboratory equipment. This section will consist of a short introduction to the techniques and apparatus of radiochemistry.

Radioactive Reactions

Three types of radiation are considered. The two particles, α, and β and γ rays differ widely in their powers of penetration of matter, the degree of penetration achieved depending upon the energy with which the particle is emitted from the radioactive source. The greater the emission energy, measured in million electron volts (MeV), the greater the penetration.

The degree of penetration of α- and β-particles is limited by their mass; γ-rays have virtually no mass and therefore exhibit much greater penetrating power. Most α-particles will not penetrate through a thin sheet of paper and most β-particles are absorbed by a $\frac{1}{2}$ in. thick sheet of perspex but γ-rays may pass through several inches of lead.

In the absorption of radiation several different reaction mechanisms may be involved. When α- and β-rays penetrate a substance they collide with the atoms of the substance displacing electrons and producing ions. The α-particle is large (2 protons, 2 neutrons) and experiences many collisions in a short distance. The β-particle, which is very much smaller, experiences few collisions for much of its path until it has been slowed down sufficiently to allow collisions with surrounding atoms to occur more frequently. The energy of these particles is dissipated in the ionization of the media through which they pass. Gamma rays have no mass and their travel is limited only by their energy and the density of the medium through which they pass. Low energy γ-rays may be absorbed quickly while high energy rays might travel through the medium without absorption taking place. Some γ-rays are scattered, i.e. collision may cause a change of direction.

The Decay of Isotopes

Radioactivity is caused by changes in the atomic nucleus. If the nuclear protons and neutrons form an unstable combination radioactive decay results until the components of the nucleus have achieved a stable configuration. This may be accomplished through reactions of the following type:

(*a*) A negative β-particle is emitted from a neutron. The loss of negative charge transforms the neutron to a positively charged proton. An example of this process is the decay of radioactive phosphorus ${}^{32}_{15}\mathrm{P}$ to stable sulphur.

$$^{32}_{15}\mathrm{P} \rightarrow {}^{32}_{16}\mathrm{S} + \beta^-$$

(*b*) A positive β-particle (positron) is emitted transforming a proton to a neutron. The position has only a transient existence combining immediately with a negative electron the combination disintegrating to form two γ-rays which are emitted at 180 degrees to each other. Radioactive carbon decays to stable boron in this manner.

$$^{11}_{6}\mathrm{C} \rightarrow {}^{11}_{5}\mathrm{Bo} + \beta^{+}$$

(*c*) An α-particle is emitted. This method of decay is only found in atoms of high atomic weight (atomic number greater than 82).

(*d*) The emission of γ-rays. The loss of α- or β-particles leaves the nucleus in an excited state and γ-rays are ejected to restore the ground state. The γ-ray is identical with the photon, the electromagnetic quantum which is uncharged and travels at the speed of light. It originates within the nucleus but is otherwise similar to the x-ray formed in discharge tubes.

(*e*) The combination of a proton and an electron within the atom (called *K* electron capture) results in the transformation of the proton into a neutron with the rearrangement of the electrons in their orbital shells and the emission of x-rays. Radioactive argon is transformed to a stable isotope of chlorine in this manner.

$$^{37}_{18}\mathrm{A} \rightarrow {}^{37}_{17}\mathrm{Cl} + \text{x-rays}$$

Some isotopes may be described as pure α or β or γ emitters but many give a mixed radiation. Table 9.2 shows some of the isotopes commonly used in work with radioactive materials together with the type of radiation they emit.

TABLE 9.2. Characteristics of Some Commonly Employed Radioactive Isotopes

Element	*Isotope*	*Radiation*	*Half-life*
Carbon	^{14}C	beta	5700 years
Calcium	^{45}Ca	beta	180 days
Chlorine	^{36}Cl	beta, gamma	10^6 years
Cobalt	^{60}Co	beta, gamma	5·3 years
Hydrogen	^{3}H (tritium)	beta	12·1 years
Iodine	^{131}I	beta, gamma	8 days
Iron	^{59}Fe	beta, gamma	47 days
Potassium	^{42}K	beta, gamma	12·4 h
Phosphorus	^{32}P	beta	14·3 days
Sodium	^{22}Na	gamma	2·6 years
	^{24}Na	beta, gamma	14·8 h
Sulphur	^{35}S	beta	87·1 days

The decay of radioisotopes is a random occurrence and must be calculated in terms of probability. A decay constant is used which is the probable decay per unit time which equals:

$$\frac{\text{No. atoms decaying per unit time}}{\text{No. atoms originally present}}$$

The negative sign means that the number of decaying atoms is decreasing. The decay constant is expressed thus:

$$\gamma = \frac{dN/dt}{N} = \frac{-dN/N}{dt}$$

where dN = the atoms disintegrating during time dt. A graph indicating decay of a nucleide can be produced by measuring its activity over a period of time and plotting the results on semi-log paper (*Figure 9.13*).

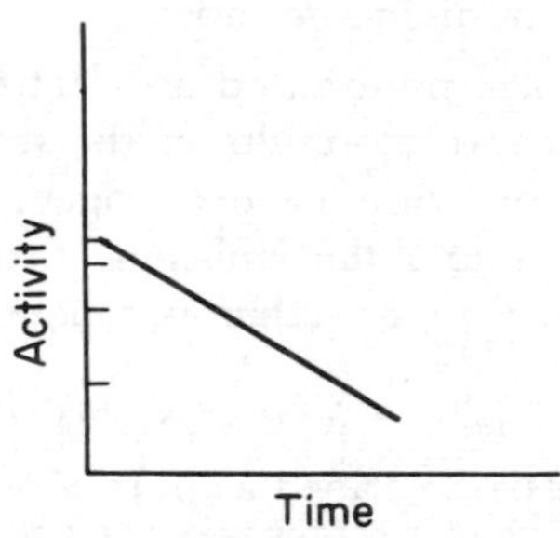

Figure 9.13. The decay of a nucleide (plotted on semi-log paper)

The following derived equation allows the extent of decay to be found if the decay constant is known,

$$\text{Log } N = \text{Log } N_0 - 0{\cdot}4343\gamma t$$

Using common logs, N = number of radioactive atoms, N_0 = number of radioactive atoms at time 0.

If N is taken as one, a proportion of the original activity is given. A common way of expressing decay of a nucleide is the half life $t_{\frac{1}{2}}$, that is the time for its activity to decrease by half

$$t_{\frac{1}{2}} = \frac{0{\cdot}693}{\gamma}$$

The knowledge of decay is important when designing experiments because considerable damage may be done to organs. Also it is necessary to dispose of waste after experiments. Safe storage must be arranged until radiation is below prescribed safety levels before discarding into public sewers. The choice of a suitable nucleide is relevant in this connection.

Radioactive Quantities

Before any quantity can be measured a scale of measurement must be devised, i.e. a unit of measurement must be defined. The unit of radioactivity is the *millicurie* (mc) based originally upon the rate of emission from radium. In one millicurie of radioactive material 37×10^6 atoms disintegrate per second, thus the measurement of disintegration gives the measurement of radioactivity for all materials.

Other units have been defined to state the dosage received by a body being acted upon by radioactive emanations. These are based on the dosage units devised for earlier work with x-rays.

The *röentgen* is defined as 'that quantity of x- or γ-radiation such that the associated corpuscular emission per 0·001293 g of air produces, in air, ions carrying one electrostatic unit of quantity of electricity of either sign'.

This complex definition (0·001293 g is the mass of 1 cm^3 of dry air at S.T.P.) is inadequate for modern work with radioactive materials especially as it applies only to x- and γ-rays.

The *rep* ('equivalent physical') was introduced to define the total quantity of all radiations which produces the same ionization in air as the röentgen.

In 1954 the International Commission on Radiation Protection define a third unit, the *rad*, which was stated to be a dose amounting to 100 ergs/g.

In dealing with problems of human protection against radiation damage a further unit the *rem* ('equivalent man') is used to correlate the damage caused by different sources of radiation.

The energy which the nuclear particles possess when they are ejected from nuclei is expressed in megaelectron volts. The electron volt is defined as the kinetic energy acquired by an electron falling through a potential difference of 1 V. It is equivalent to $1{\cdot}602 \times 10^{-12}$ ergs. Particles are commonly emitted from naturally occurring radioactive materials with an energy of from 3 to 8 MeV. The target of a 200 kV x-ray machine is struck by electrons with a maximum energy of 200 000 eV. Chemical bonds are of the order of a few electron volts in energy.

The Detection and Measurement of Radioactivity

The radiation emitted from radioactive substances has the ability to cause ionization in a medium through which it passes, that is, it is able to remove electrons from atoms and molecules. This phenomonon is used in the detection and measurement of radioactivity. The ionization caused by the radiation is detected and measured. Methods for doing this fall into two broad classes:

(1) Those based on the ionization phenomenon mentioned above and
(2) The ability of certain molecules to be excited to a high energy level when irradiated and to emit light when returning to their original energy level.

Methods included in (1) are the direct observation of ionization tracks with a photographic plate or cloud chamber and those based on ion collection. The cloud chamber is more appropriate to a physics laboratory and will not be considered further. The use of a photographic plate is employed in the very useful technique of autoradiography. Where mixtures containing radioactive components have been separated by paper or thin-layer chromatography the radioactive components may be detected by placing the flat chromatogram face-to-face with a suitable photographic plate. After some time the plate is developed and a 'picture' of the radioactive spots on the chromatogram is obtained. The technique is directly derived from the very earliest work in the field when Becquerel actually discovered radioactivity when he noticed the effect on a covered photographic plate of the invisible emanations from uranium salts. Autoradiography is also used as a histochemical technique, the radioactive tissucs in histological sections from organisms which have been treated with tracer materials being identified in a similar manner.

Ion collection methods are the most common in class (1). These are best described by considering a simple gas-filled chamber containing two electrodes connected to a simple circuit as illustrated in *Figure 9.14a.*

If radiation causes a molecule to split into an ion pair when positioned between the electrode plates at zero volts the ions would simply recombine. But if a low potential were applied they would tend to drift towards the respective electrodes with a few actually reaching them causing a current to flow. If the potential is increased, more and more ions are discharged at the electrodes. At approximately 100 volts all ions reach the electrodes and a saturation

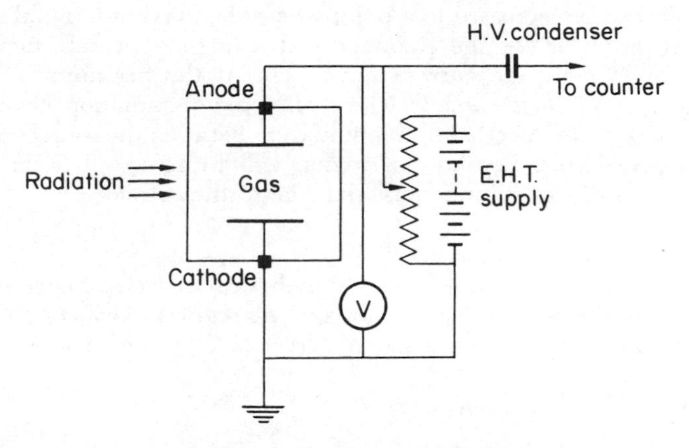

Figure 9.14a. Gas counting chamber

current is reached (*Figure 9.14b*). Continuing to increase the potential accelerates the ions causing them to collide with molecules that in turn split into more ion pairs. This is called *gas amplification.* As potential is increased more and more secondary ion pairs are produced and the current flow increases in proportion to the number of primary ion pairs caused by the incident radiation. Hence the counters designed to work in this region are called proportional counters. They are most useful as they can differentiate between sources of radiation. Alpha particles cause more ionization than lower energy beta particles, therefore a larger current flows for the former.

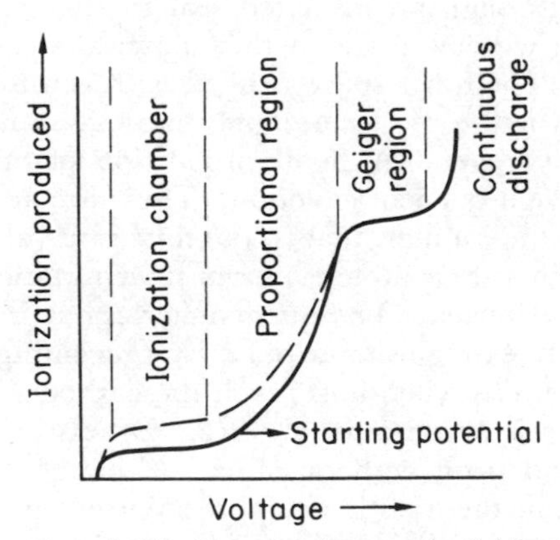

Figure 9.14b. Characteristic curve of a G.M. tube

Volts can be increased to a point where the maximum number of ions is produced for the chamber and a further increase in volts will not increase the current flow. This is the beginning of the Geiger region (*Figure 9.14b*) where all types of radiation have the same maximum effect and differentiation between the sources ends. To increase volts beyond this region will bring about a state of continuous discharge and cause damage to the counter.

Design of ion collection counters

No one counter will give the hypothetical curve of *Figure 9.14b* and various types are designed for specific regions. Geiger counters are of the simplest construction *Figure 9.15*. They consist of a

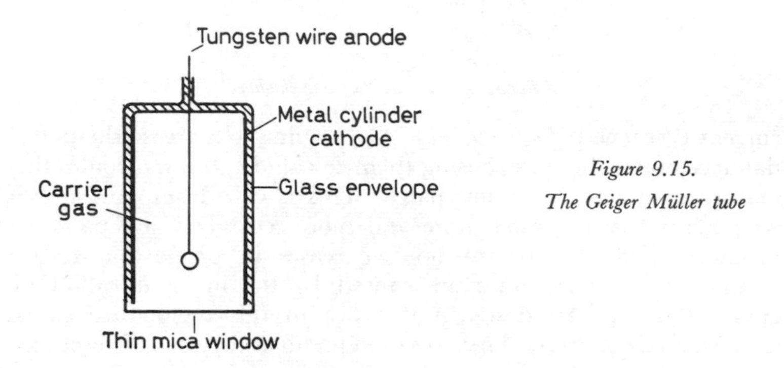

Figure 9.15.
The Geiger Müller tube

metal cylinder usually of brass and sometimes with a glass top. A very fine tungsten wire, the anode, runs along the centre axis of the cylinder through an insulated seal in the cap. Radiation enters through a window made of thin material such as mica and aluminium foil. The inner space is filled with counting gas at low pressure. Argon is the most commonly used counting gas, others are neon and hydrogen. A small proportion of other gases are added which have a quenching action. These are added to reduce the dead time of the counter, that is, the time interval during which the tube and circuit are completely insensitive and the second ionizing event is not detected. The quenching vapour is a polyatomic gas which quickly extinguishes a pulse by preventing the emission of secondary electrons when ions reach the cathode. A commonly used gas mixture is argon/ethanol 95:5. Others are helium with ethyl formate and neon with bromine. Although this effectively limits the dead time there is still an appreciable error when quantitative counts of above 5×10^3 disintegrations per second are attempted.

The proportional or gas amplification counter is very similar in construction to the Geiger counter. It is designed to give a maximum voltage gradient between cathode and anode. This requires the use of the finest wire for the anode that is compatible with the necessary mechanical strength. It is often shaped in a loop. The gas mixture, such as argon/methane 90 : 10 flows gently through the tube and has a greater quenching action than that used in the Geiger tube. The signal from the proportional counter will vary according to the energy of the radioactive type. The instrumentation must be able to separate and record counts at different pulse heights and have a sensitive amplifier (input less than 1 mV). The Geiger counter on the other hand needs a less sensitive amplifier and simpler instrumentation. A typical assembly of instruments for counting with G.M. tubes is shown in the block diagram *Figure 9.16*.

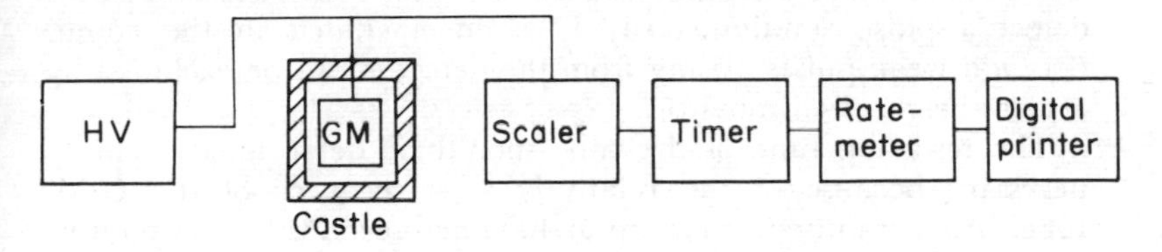

Figure 9.16. Apparatus for use with a G.M. tube

It needs a power supply ranging from zero to 2000 volts D.C., high stability not being necessary. The G.M. tube is housed in a lead castle. The sample chamber below a window counter usually has a sliding sample tray which may be positioned at varying distances from the window. The pulses are fed into an amplifier where the signal is also modified before being fed into the scaler. The scaler counts these pulses which are indicated on a series of cold cathode decade or digital counters. Additional refinements are a device for automatic shut-off when a pre-set count is reached, a digital print-out of results and a pulse height discriminator. The latter is of little use with a G.M. counter. Instead of a scaler a count rate meter can be used, this does not record individual pulses but averages the number of pulses and displays this average on an electrical meter usually in counts per minute.

An accurate timer is necessary to obtain counting rates. This may have a means of obtaining counts for pre-set time intervals and can be linked with the digital print-out. The power supply, amplifier, discriminator, scaler and timer, are often built into a single unit and called the scaler.

Before experimental work starts certain preliminaries must be carried out. First the Geiger plateau must be determined. This is done by plotting a graph of counting rate (C.P.M.) with voltage. Voltage is increased in 50 volt steps until the plateau is reached. If voltage is increased beyond the plateau, continuous discharge will occur and the G.M. tube eventually ruined. A suitable working voltage is approximately $\frac{1}{3}$ along this plateau.

Next the background count must be determined and subtracted from the total count. The background count is of the order of 30/70 C.P.M. and results from several sources. These may be natural sources such as cosmic rays and surrounding construction materials, or contamination and fall-out. The background is kept to a minimum by the lead shielding of the castle. It may also be reduced electronically by using (1) an anticoincidence shield of G.M. tubes surrounding the analytical tube. If both the tube and shield detect a pulse simultaneously it is not included in the count. (2) Odd weak pulses arising from the detector can be excluded by a pulse height discriminator.

The resolving time of the tube should be determined. This is necessary because of the relatively slow recovery of the G.M. tube. After a pulse it takes up to 400 microseconds for the positive ions to reach the cathode. If further ionizing radiations occur coinciding with this dead or resolving time of the tube they will be omitted from the count. Resolving time is found by producing a series of samples of known counts, counting them and drawing a graph of observed counting rate and theoretical number of counts expected. The difference is the resolving time. A coincidence correction is made using the formula:

$$N = \frac{n}{1 - nT}$$

where N = true count, n = observed count, T = resolving time.

Gas counters are sensitive to alpha and beta radiation but relatively insensitive to gamma rays.

Scintillation Counting Technique

This is the second class of detection referred to earlier and is the more important and commonly used analytical technique. Radiation is detected by the production of light photons when it falls on certain types of substances called phosphors or fluors. The radiation in this instance does not remove electrons as in the gas amplifiers but raises the energy of the molecule which on returning to its former state dissipates the energy as light. These minute flashes of light are detected by a photomultiplier tube (see page 236). The phosphors may be of solid or liquid form. For counting gamma radiation the solid phosphor is either well-shaped to contain a sample tube and the whole placed in a light tight housing or, more generally, a disc of phosphor rests on the P.M. tube which in turn is covered by a thin aluminium light tight shield. The latter system allows the samples to be easily and automatically placed against the phosphor without elaborate precautions to exclude light as the aluminium shield is transparent to the radiation. These counters are suitable for counting gamma sources such as iodine, iron etc.

Liquid scintillation counting is almost exclusively used for low energy beta counting and is the most important tool for biochemical work with H^3, C^{14} and S^{35}. The sample is solubulized in the liquid scintillant, efficiency is therefore very high due to the intimate contact of the radiation source and the scintillant. Preparation of the sample for liquid scintillating counting requires careful consideration and the reader is referred to the specialist literature, especially the reviews by the Radiochemical Centre.

Basically the samples consist of (*a*) a solvent such as toluene, xylene and 1:4 dioxane and also naphthalene in combination with another, (*b*) primary phosphors such as

p-terphenyl, 2,5-diphenyloxazole (P.P.O.),
2(4-biphenyl)-5-phenyl-1,3,4-oxadiazole (P.B.D.),
2(4′-*t*-butylphenyl)-5-(4″-biphenyl)1,3,4, oxadiazole (butyl P.B.D.)

that when irradiated emit light at a wavelength of approx. 360 nm. (*c*) secondary phosphors such as

1,4-bis-(5-phenyloxazol-2-yl) Benzene (P.O.P.O.P.)
1,4-bis-(4-methyl-5-phenyloxazol-2-yl) benzene (D.M-P.O.P.O.P.)

that convert the light emitted by the primary phosphor to a wavelength of 430 nm, this being the P.M. tubes area of maximum sensitivity. (*d*) The test sample itself. This may require a solubulizing agent such as hyamine or an emulsifier such as Triton X100.

There are numerous ‘cocktail’ mixtures of these solutions described in the literature, the object of which is to obtain maximum efficiency and a minimum of quenching. In liquid scintillation counting, quenching refers to any colour or chemical quenching within the sample. Common causes of quenching are coloured solutions and oxygen and water, so it is obvious that the extent of quenching must be determined.

Liquid scintillation counting equipment (*Spectrometer*)

The samples are dispensed into glass vials that in automatic machines are placed in an endless belt. The vials are positioned one after the other on to a lift that lowers them into the photocell chamber, made light tight by a sliding shield. In most instruments the sample and counting chamber are kept at a lower temperature, e.g. 5°C. This reduces photomultiplier noise levels and spurious counts. The counting chamber has two photomultiplier tubes horizontally opposed with the sample sitting between them. Only simultaneous pulses from the tubes are counted, thus eliminating almost all pulses due to noise. This is called coincidence counting, the opposite to the technique mentioned earlier for gas amplification counters.

The P.M. tubes are fed with a stable high D.C. voltage. The signal from the tubes is pre-amplified and then fed into the amplifiers of up to four counting channels. Each channel has two gates between which pulses will be counted. If the gates or discriminators are set at 0 and infinity respectively, pulses are counted across the whole energy spectrum. The setting of the first discriminator is called the window height and the difference between the discriminators is the window width. The radiation from a given source has energy (MeV) covering a certain band width or spectrum. For example, the maximum for carbon 14 is 0·156 but the most intense activity is up to 0·05 MeV. Individual isotopes vary considerably (see Table 9.2). It is possible to discriminate between pulses of different energy or pulse heights. The discriminators may be set to count pulses of an energy peak of an isotope. If samples contain a mixture of isotopes each channel may be given a window height and width setting to include a suitable energy peak of each isotope making it possible to count two or more isotopes simultaneously.

The pulses are counted on a scaler as before which usually has a timer. Automatic machines include provision for printing out results and some give partial or complete calculation of D.P.M. from an on-line computer.

It is usual to express results in distintegrations per minute. One micro curie $= 2{\cdot}220 \times 10^6$ D.P.M. No system is able to detect and record a pulse for each disintegration, therefore it is necessary to determine the efficiency.

$$\text{Efficiency} = \frac{\text{the observed counts per minute}}{\text{D.P.M.}}$$

Efficiency is affected by the limitations of the equipment and physical and chemical factors already mentioned. In liquid scintillation counting, quenching is a major factor effecting efficiency. Various techniques are used to determine efficiency, the following two methods are commonly used.

1. Internal standard method. After the sample has been counted a quantity of the same isotope of known D.P.M. is added. This is in a small volume, usually 0·1 cm³, so as not to dilute the sample. The additional isotope suffers almost the same degree of quenching. The sample is re-counted and the efficiency calculated as follows:

$$\frac{(\text{C.P.M. of sample} + \text{added isotope}) - (\text{C.P.M. of sample})}{\text{D.P.M. of added isotope}}$$

$$\text{D.P.M. of sample} = \frac{\text{C.P.M. of sample}}{\text{efficiency}}$$

2. The channels' ratio method. Two channels are set to cover between them the energy spectrum of an unquenched sample. An increase in quenching increases the proportion of counts in the channel detecting the lower energy band. By counting a set of standards of varying degree of quenching then plotting the channels' ratio (1/2) against the percentage efficiency a calibration curve is produced. Efficiency can be found for the experimental samples by first calculating their channels' ratio. This technique has the advantage of needing only one counting operation.

Chromatogram Scanners

Apparatus is available for detecting radioactive spots separated on chromatograms by the use of proportional or G.M. counters. In one type paper strip chromatograms are fed between two window counters at a speed that may be varied according to the activity present. Two dimensional, thin layer or paper chromatogram scanners scan the surface in a zig-zag fashion. The thin layer plate can only have a detector on one side, but spots once found may be scraped up and transferred to a vial and counted in a scintillation counter. Paper squares may be scanned both sides simultaneously and equipment has been developed which gives a computer replica of the chromatogram containing C.P.M. and D.P.M. of each spot.

Radioactive Hazards

Many of the early workers in the field of x-rays eventually suffered crippling and often fatal injury from prolonged exposure to the emanations from their primitive discharge tubes. The dangers of this work were not appreciated for some time, but today the work of radiologists and those who use radioactive materials is very strictly controlled by regulations based upon the recommendations of the International Commission on Radiological Protection which are issued from time to time. Several publications on this topic are noted in the Appendix.

The Uses of Radioactive Isotopes

Because the radioactive isotopes of an element are almost identical in their purely chemical properties to the non-radioactive isotopes and because radioactive materials are so easy to detect and measure, several completely new avenues of analytical work have been opened. Some of the more important will be discussed hereunder.

The use of radioactive *tracers* has been mentioned frequently in earlier chapters. The completeness of precipitation, partition, solubility, diffusion, or exchange may be tested by the addition of a small amount of radioactive material to a mass of isotopic non-radioactive material.

Isotope dilution analysis depends upon the determination of *specific activity*. The latter is defined in terms of curies/g. A mass M of a radioisotope which has a specific activity S is added to a mass M_x of an inactive isotope and equilibrium is established. The specific activity of the mixture (S_x) is determined and M may be determined by:

$$M_x = M\left(\frac{S}{S_x} - 1\right)$$

An example is the method of measuring blood volume. A very small amount of radioactive compound is injected, and after it has been allowed to circulate the specific activity of a small blood sample is determined. From the dilution of specific activity in the sample the total blood volume may readily be calculated.

In *radioactivation analysis* a sample and a standard known mass of the 'unknown' are irradiated together in the same radioactive field. The specific activities (curies/g) will be the same, so by comparing the activities of the 'known' and the 'unknown' the mass of the unknown constituent can be determined.

In biochemistry a vast amount of work has been done on *incorporation* and *turnover* of compounds in the living system, especially in research into intermediary metabolism. Substances ingested into the body cannot normally be detected or separated from similar materials stored or synthesized within the system. By the addition of small amounts of isotopically labelled foods to the diet the fate of compounds within the body can be easily determined. One of the earlier experiments in this field was the demonstration that the creatine of muscle is built from three amino acids, methionine, glycine, and arginine. Similarly it has been shown that the haem of haemoglobin is derived largely from glycine.

Turnover studies follow the fate of constituents within the body by measuring the rate of loss of ingested material. The fate of proteins has been studied by using labelled amino acids and it has been shown that whereas the proteins of the main organs must be replaced at about fortnightly intervals, those of skin and muscle are exchanged much more slowly while the collagen of bone and cartilage is almost inert.

MASS SPECTROMETRY

In Chapter 6 it was shown how qualitative and quantitative analysis could be performed by measuring the absorption and emission of a wide range of spectra of light, how pure compounds exhibited a spectral pattern only attributable to themselves. Another physical parameter that can be similarly utilized is that of mass. Mass spectrometry is a means of elucidating the structure of molecules and the accurate measurement of molecular weight up to 700. It has also been the means of detecting and measuring isotopic ratios.

When working with discharge tubes in 1886, Goldstein saw luminescent streams that came from the back of the cathode through a small hole. These were ionized particles not trapped on the cathode. In 1898, Wein showed that this stream of positive ions could be deflected by a magnetic or electric field. If a stream of positive ions composed of a mixture of different masses was exposed to a magnetic field, the components of least mass will be deflected most. The extent of deflection will also depend on the energy and velocity of the particles. A mass spectrograph compares the deflection of each beam and its intensity. Most instruments are based on the design of Dempster and Aston that measure mass to charge ratio m/e.

The apparatus consists of a chamber under a high vacuum, one end contains the ion source, the other the detector. In the middle is the deflector, either a magnetic or electric field for high resolution

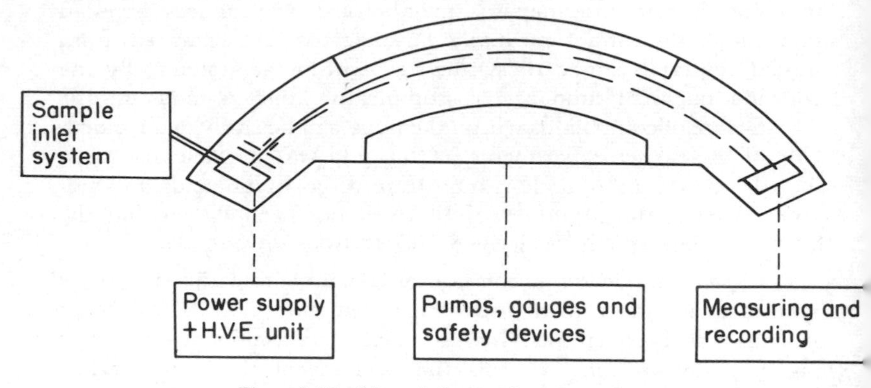

Figure 9.17. Schematic layout of a mass spectrometer

both may be used (*Figure 9.17*). The vapourized sample is ionized in an electron gun. The ionized particles are focused into a beam which passes through the deflector into the detector. The detector was originally a photographic plate but now the split beam is scanned by passing each separated beam through a slit onto a collector plate. This scanning is done by having a fixed slit and varying the current on the electro-magnet or by varying the accelerating voltage thereby passing each beam in turn on to the detector. The resultant voltage change at the detector is amplified and displayed on a meter or oscilloscope. The spectrum may be recorded on a pen or U.V. recorder or photographed. The instrument requires to work in a high vacuum between 10^{-3} and 10^{-5} Nm^{-2}(10^{-5} and 10^{-8} torr). A backing pump and two or more diffusion pumps are usually used (see this Chapter). The pumps are protected by liquid nitrogen traps. Operation requires a long sequence of events including pumping down procedure, baking parts to reduce background from absorbed volatiles, many valve adjustments, and beam detector and recorder adjustments.

There are several systems for admitting the sample. Gas samples are leaked in at a constant rate via a sintered disc. The sample is stored in glass reservoirs which in turn are connected to a diffusion pump. A liquid sample can be volatilized in a temperature controlled chamber. They may also be introduced through a sinter that is covered by molten gallium. This is done using a syringe to inject below the gallium and allows rapid introduction of the sample. A pressure drop of 1 to 10^{-3} Nm^{-2} (10^{-2} to 10^{-5} torr) is arranged across the sinter between sample reservoir and ionization chamber.

Solid samples may be pyrolysed at the end of a probe placed in the electron beam.

The value of the mass spectrometer has been enhanced by the development of gas chromatography and it is now possible to examine specific peaks as they elute from a G.L.C. column. The problem has been to introduce the sample separated from the carrier gas. Several methods are used, the most common is a tube with walls of sintered glass. The sample, in a carrier gas of helium, passes through the tube and the helium diffuses out thus enriching the sample which then passes into the ion chamber. A sintered metal disc has been used and also a P.T.F.E. tube at a temperature of 250°C for diffusing out the carrier. Other methods involve small gaps between jets or slits such that the lighter carrier gas molecules diffuse away from the sample stream into a vacuum chamber. The enriched sample continues into the ion chamber. As peaks elute very rapidly from a G.L.C. column the mass spectrometer must be capable of very fast scanning of the spectrum.

The mass spectrum is usually recorded on a multi-channel recorder. Each trace is set at a different attenuation thereby making it possible to measure both small and large peaks. As the resolution is good these peaks are very narrow. For visual examination a line diagram is drawn as in *Figure 9.18* making comparison of spectra easy. Each peak corresponds to one ion species. If one electron only has been removed from the sample molecule the ion has a mass of the parent molecule and will be found at the highest significant *m/e* position. A close smaller peak will be positioned near to the parent peak due to the presence of some isotopes in the molecule. The normal abundance of isotopic carbon, nitrogen and chlorine is 1·1%, 0·4% and 33% respectively so it will be seen that the size of the secondary peaks will vary with molecular structure. The peak heights are expressed as a percentage of the largest peak which is

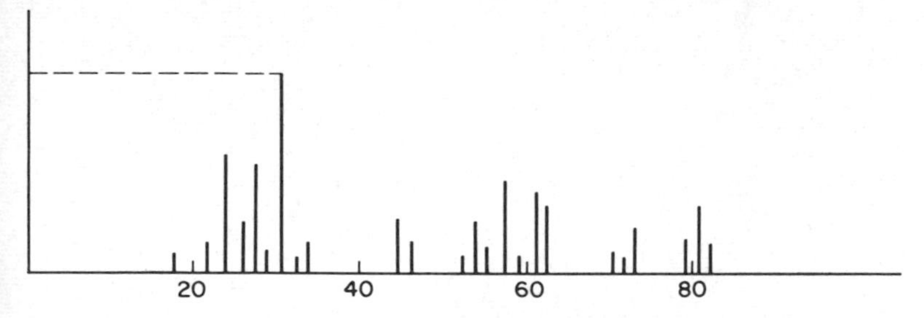

Figure 9.18. Stylized mass spectrogram

called the base peak. The other lines are fragments of the molecule or are contaminants. Libraries of mass spectra have now been accumulated and are available for reference in qualitative analysis. Some examples of peaks are as follows:

Fragment		*Peak*
Air	H_2O	18
	N_2	28
	O_2	32
	Ar	40
	CO_2	44
Ether		29
		31
		45
		59
		74
Bromine isotopes		79–81
Organic acids: a large peak at		60
Ketones: a large peak at		43

CHAPTER 10

BIOCHEMICAL LABORATORY APPARATUS AND METHODS

The apparatus and techniques so far described are common to both general chemistry and biochemistry. The apparatus described now is essential to the biochemist although it may find some applications in the general chemistry laboratory.

The equipment falls into two general groups:

1. Apparatus for the control of experimental conditions, environment and storage.
2. Apparatus for the preparation of biological material, its separation and extraction.

Apparatus in group 1 requires temperature control.

OVENS AND INCUBATORS

These are usually rectangular in section and vary considerably in size depending on individual requirements. The inner shell may be made of stainless steel, aluminium, copper, enamelled steel or asbestos boards. The shelves are perforated to allow movement of air either by convection or forced draught. The insulation layer is of asbestos, glass wool or fibre matting, or cork at least 5 cm thick. For high temperatures refractory materials are used such as fire-clay bricks. The outer casing is usually of enamelled or rust-proofed steel. A ventilation grid is supplied fitted with a shutter. The door or double doors close on to gaskets, an inner one of asbestos and an outer one of rubber, plastic material or asbestos. A glass inner door is often provided and is especially useful for incubators. Some ovens have a glass window in the door itself. Heating elements are situated on the outside of the inner shell around the sides and back and underneath the floor. If the shell is insulated on the outside, resistance wire may be wound completely round it giving intimate heating conditions. Temperature control is often a bimetal thermostat operating a relay or hot-wire switch. Expansion capsules and gas-filled sensing probes connected to bellows operating electrical contacts are also used.

Ambient to 100°C *Range*

These are generally called incubators and are required for accurately controlled conditions from about 25°C to 50°C. They invariably have an inner glass door for observation, this minimizes loss of heat. If the inner door is opened for a short while, the temperature recovery time may be 15 min. Good lagging and thermal capacity of the walls is essential. This is obtained in some models by a water-jacket forming the inner shell itself. They are made of copper sheet containing a drain cock at the bottom rear, and have an opening in the top for filling and a floating level indicator. The electrical heating controls are sited on one side or at the top so that in the event of a leak, damage to the controls is minimized. They should be filled with water; topping up will be necessary about every 6 months depending on siting and temperature of incubator.

Incubators using the more common lagging material of asbestos and glass wool and of improved design are now capable of giving equally good conditions. Although the insulation layer is at least 6·5 cm thick they have the advantage of lightness, and may be stacked where several are required. Controls are positioned underneath. The temperature scale has arbitrary divisions and should be calibrated, the time for equilibrium at several temperatures being assessed.

Occasionally biochemical work has to be performed at constant temperature which requires continual manipulation and the use of large apparatus such as chromatography columns, fraction collectors and shakers. In this case a room is usually converted into an incubator, and should be sited preferably in the centre of the building, without windows and having one door. In general, walls will not require further insulation but the ceiling may if the room is on the top floor. Entrance should be through an ante-room forming a warm-air lock. Heaters are ranged low on the walls and air-flow over them should not be obstructed. The heaters may be hidden behind a false wall which has a gap top and bottom (*Figure 10.1*). Shelves should be erected away from the walls and be constructed of perforated sheet materials or wire mesh. The local electricity authority or a heating contractor will estimate the amount of heat required, but as a guide a room 2·5 m cube will need a minimum of 4 kW and should have two thermostat controls in series to keep temperature variation to a minimum.

Glass house problems are similar for the plant biochemist but twice as much heat is required as good ventilation is essential for plant growth and heat loss is greater especially under winter conditions. All electrical installation must be waterproof, power

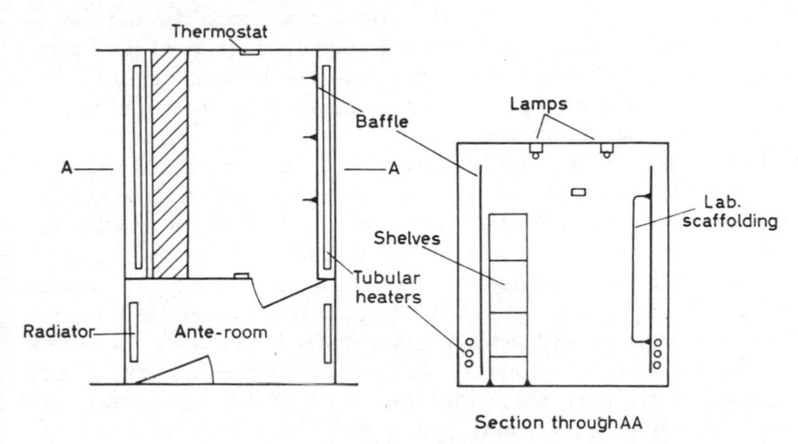

Figure 10.1. Schematic diagram of a temperature controlled room

points should be sited high up or hung from the roof. Temperature fluctuations may be considerable due to a high concentration of heaters. This is minimized by using several methods of heating such as soil warming cables and fan heaters. The reader is referred to the specialist horticultural literature for further information.

Humidity ovens—These are able to give any temperature from 0–100°C with variable humidity conditions. A fan re-circulates the air which is heated by an element in the air duct. The lower temperatures are obtained by a refrigeration plant and humidity is balanced by drying agents and a small electrical boiler. They may be used for the biochemical study of plants, algae and fungi, etc., lamps for irradiation being incorporated in the oven.

Temperature regulating mechanisms are discussed later in the chapter.

Ambient to 300°C *Range*

Ovens in this range have the same basic structure and vary in size from 30 cm cube to 60 × 60 × 45 cm, larger types usually being custom built. The interior liner is made of aluminium, stainless steel or enamelled steel. The design depends on particular requirements. Ovens in this range are suitable for digestions, hydrolysis, sterilization, curing, drying and many other applications. For drying purposes good ventilation is required, and vents are situated at the top and the bottom of the door. A forced draught may be provided by a fan in the back. *Chromatography ovens* have a chimney in the top and a fan under the bottom to direct the unpleasant vapours of the solvents often used in chromatography into a fume exhaust system. These ovens are provided with a frame or rack that pulls out to facilitate the hanging up and removal of chromatograms.

Ovens used for sterilization may have a time switch incorporated in the mains input. The thermostat is set to the desired temperature, usually 160°C, and the timer adjusted to cut off the heat after the necessary warming up and sterilizing time.

Some domestic appliances are admirably suited to certain laboratory applications. For drying apparatus a cheap wall mounting plate warmer with sliding glass doors is available and some clothes dryers will be found suitable. The domestic cooker could be used more, it is about half the price of the laboratory oven and incorporates a timer and several useful hot plates. Temperature fluctuation is about 0·2–0·5°C, the domestic types probably being greater than this.

Vacuum Ovens

These operate up to 250°C and are either rectangular or cylindrical in section, the latter is preferred from a design point of view. They are made of aluminium and vary from 30–50 cm in diameter and are up to 60 cm long. Perforated shelves are fitted and the door has an observation window. A vacuum gauge, usually reading in either inches or millimetres of mercury, and two air entry ports with valves are supplied. If a water pump is to be used for evacuation, a trap or non-return valve is advisable. The heater of Nichrome wire or tape is wound round the outside of the oven and covered with insulation. The oven and controls are housed in an aluminium or fibre-glass case. They are most useful for drying purposes and are often operated as heated vacuum desiccators. Care should be taken to prevent desiccants from contacting the walls as they may corrode the aluminium.

300–1500°C *Range*

Ovens in this range are called *furnaces*. The designs fall into two types, those in which the heaters are fixed to the inner wall of the furnace and therefore exposed to the prevailing internal atmosphere, and those where the heating elements are on the outside of the refractory. The latter are called *muffle* furnaces and are suitable where corrosive vapours are involved. The smallest furnaces are those designed to hold one crucible and are used mainly for ashing. Some are gas heated, these sit on a Bunsen burner of the Meker design. They have the disadvantage of poor temperature control. They are constructed of fire-clay or silica. Small electrical crucible furnaces are also available, all these are vertical types.

Horizontal muffle furnaces vary in size from 12·5 cm cube to 25 × 25 × 45 cm and are made of refractory brick, the doors are of cast metal, usually aluminium, holding a refractory brick. They open either by hinging down or up with the aid of a counter weight. Some are held by a rod so that the door may be clamped in the open position. The outside case is made of asbestolite, a hard non-corrosive material. At the back of the furnace is a chimney with a vane for controlling the draught and a temperature fuse which prevents the furnace over-heating. The highest temperature in a muffle is 1,100°C; if the furnace temperature is to be higher than this, heating elements are placed internally. The heating elements are of Nichrome wire but where a temperature of over 1,250°C is required special elements have to be fitted. These are made of silicon carbide or molybdenum disilicide.

All these furnaces are suitable for glass annealing purposes but custom built ovens to fit a particular space are more usual. Pottery kilns are a useful substitute, these rise to a temperature of about 1,200°C. All furnaces can be supplied with temperature controls and time switches for automatic control allowing them to be used overnight so that both the room and the oven may cool before morning.

Carius Furnace

This is a specially designed oven to accommodate Carius tubes (see Chapter 5). These lie in aluminium tubes which are heated by external electric elements or by gas burners. The whole is protected by an asbestolite case. In one design the tubes slope downwards to the rear and if a Carius tube cracks and explodes the contents are collected safely in a fibreglass lined compartment.

All of this equipment uses resistance heating but other forms of heating may have application and are used occasionally for special purposes. Induction heaters are suitable for heating and melting metals, e.g. alloy baths, metal rods or tubes. The object to be heated is placed within a copper coil through which an a.c. current is passed. This induces eddy currents in the surfaces of the metal, thereby generating heat.

Dieletric heating is most suitable for plastics and other non-conducting material. It is similar to a capacitor. An electrode is placed either side of the material (a dieletric) and a high voltage, high frequency a.c. applied. Heat is generated by increased molecular movement. Another means of heating is to focus micro waves onto an object, this method has a radiation hazard.

All these methods have the advantage that heat is applied immediately to the object. There is no delay as it is heated from within. Temperature is easily controlled, quickly applied and there is less likelihood of overheating and burning.

Temperatures Below Ambient

Cold storage and cold laboratories are essential for many biochemical operations. Two temperatures are commonly required for refrigerated rooms, 2 to 4°C and −10 to −15°C. Particularly sensitive biological material may require very low temperature storage such as −70°C or −190°C.

The −10°C cold laboratory is generally situated within the +2°C room for the convenience of workers and economy of insulation and power consumption.

The +2°C room sould be fitted as a normal laboratory with one or two services omitted and with extra provision for storage. At least one wall should be devoted to racks of shelves with additional peninsula racks if the space is available. A workbench of teak or metal should be supplied containing a sink with cold water taps. Due to the usually low ceilings, about 2·5 m maximum, one wall should be free of benches and shelves so that tall apparatus such as chromatography columns may be accommodated. Lighting should be as efficient as in a normal laboratory. Fluorescent fittings are most suitable as they throw less shadow, and are best left turned on during working hours. Power sockets should be of the waterproof type.

The lighting in the sub-zero room should be a tungsten lamp housed in a waterproof fitting. The other requirements are as for the other room except the water supply.

The refrigerator compressors should be housed in a separate room to eliminate the noise, and ample ventilation is required. The heat exchangers and fans may be either fitted high on the walls or in a false roof, with the air circulated through shutters. An external temperature indicator, preferably with a chart recorder, should be positioned where temperature changes due to refrigeration plant failure will be quickly noticed.

An alarm should be fitted that can be sounded from both rooms, either by push-buttons or cord operated pull-switches. The hazard of light failure may be overcome by permanently lit neon bulbs supplied from a separate fused circuit and without a switch.

The great fault with cold rooms is that they rapidly become dumping grounds and this should be resisted at all costs. Refrigerators are used where cold room facilities are not available or as supplementary storage in the laboratory. From the large range of domestic refrigerators there is a good chance of finding one that will fit under existing benches. These have a capacity of up to 1·5 m^3 at +2°C to 4°C and a small tray maintained at sub-zero temperatures. A survey of these refrigerators giving results of tests for safety and efficiency has been published in *Which?* the journal of the Consumers' Association, and will be found useful in making a choice.

Larger refrigerators are available and may be constructed to fit into recesses or other suitable spaces. Cold cabinets with a capacity of 6 m^3 are made for blood storage.

All these units are now supplied with sealed compressors which give reasonably trouble-free service. A variable thermostat is often supplied so that the temperature may be varied from 2–10°C. The internal finish and fittings are of galvanized, enamelled or plastic-coated steel. The doors can be hinged on either side and those

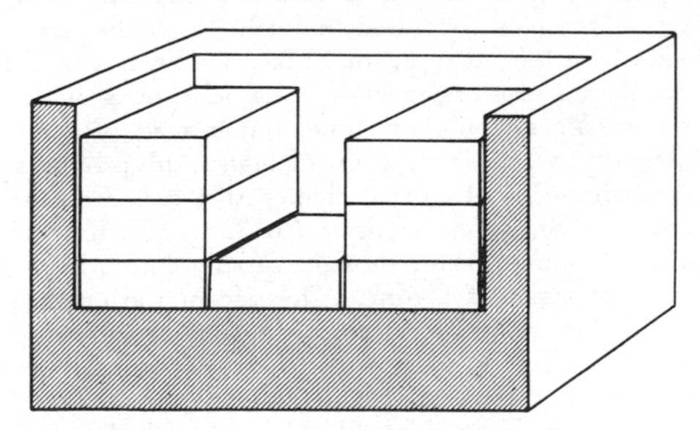

Figure 10.2. A tray storage system for a low temperature cabinet

with magnetic catches open and shut with the minimum of vibration.

For sub-zero temperatures down to −15°C the 'ice-cream' and frozen food type of refrigerators are suitable. The majority of these open at the top and, therefore, cannot be housed under a bench. Storage in this type presents a problem, as to remove containers from the bottom, other stores and bottles have to be lifted out. This can be most irksome and it is worth sacrificing some storage space to have a sliding tray system as illustrated in *Figure 10.2*.

For temperatures around −80°C refrigerators may be obtained that have two refrigerator units in a cascade arrangement. That is the evaporator of one takes heat from the condenser of the other. This type of machine has of necessity a thick layer of insulation and is therefore very bulky compared with the amount of cold storage space it gives.

For the preservation of tissues, whole organs and cells such as micro-organisms and sperm, the liquid nitrogen refrigerator is used. These are large stainless steel dewar vessels containing special holders and trays. These vessels require topping up two or three times per week with liquid nitrogen depending on the use they receive. Before most items can be stored they must have their temperature gradually reduced to −180°C. In some instances this must be done at a strictly controlled rate, for example, sperm. It is preferable to do so in any case to avoid violent boiling of the nitrogen.

Ice is always needed in the laboratory and machines are available to freeze large or small trays of water, some with divisions for the production of cubes. An ice crusher is a useful accessory, this has a spiked shaft which is rotated manually, ice cubes are forced past the spikes by a hinged ram, the broken ice being collected directly into a beaker or tray underneath. Another type is similar in design to the domestic mincing machine, but instead of a screw feed a ram forces the ice through the cutting blades, this produces powdered ice. A small electrically driven ice crusher called the Ice-O-matic is supplied by Shandon Scientific Co. Ltd., and this breaks down ice cubes or solid carbon dioxide rapidly into a plastic drawer that has a capacity of 1 dm^3. The size of the crushed granules can be varied from ½ cm to 1½ cm. Machines are now available that supply crushed ice directly, water from a cistern is frozen on a special evaporator, the ice particles fall into a storage bin. When this becomes full the production of ice is automatically stopped. The ice remains dry and does not pack into a solid lump. This innovation from America, primarily intended for hotel use, is a worth while investment saving time and labour as it obviates the need for crushing and the production of ice cubes. Their capacity varies from 15 to 90 kg/24 h.

Freezing Mixtures

These are used to produce temperature conditions varying from 0°C to −50°C and are made by mixing various salts with crushed ice and water. The most common mixture is 3 parts of crushed ice and 1 part of sodium chloride (technical grade or agricultural salt), this will give temperatures of from −5°C to −20°C depending on how well it is made. The salt should be added slowly to the ice with vigorous mixing; a minimum of tap water is then added to produce a slush so that the vessel to be cooled will be in intimate contact with the mixture. Made in this way a temperature of −15°C is assured. Other salt mixtures are tabled below:

Crushed ice or snow	*Salt*		*Temp.*
3 parts	Sodium sulphate $10H_2O$	1 part	−1°C
3 parts	Potassium chloride	1 part	−10°C
3 parts	Sodium bromide $2H_2O$	1 part	−28°C
3 parts	Calcium chloride $6H_2O$	3 parts	−50°C

Other freezing mixtures:

Sodium sulphate 1 part
Sulphuric acid dilute 1 part −15°C

Sodium nitrate 1 part
Nitric acid 1 part −19°C

Another freezing agent is solid carbon dioxide. When added to ethanol, methanol or acetone, temperatures of below −70°C are obtained. The solvent must be cooled slowly by gradual addition of the carbon dioxide otherwise the evolution of gas may cause the solvent to overflow. When cold the maximum quantity of crushed carbon dioxide is added.

For lower temperatures liquid air or nitrogen may be used. These are delivered in special Dewar vessels. They should be handled with care as contact with the skin will cause severe burns. Gloves should be worn when pouring. There is also a potential fire hazard as material which normally only smoulders when ignited, will burn fiercely if soaked in liquid air or oxygen.

A convenient and quickly available means of cooling is the ice bag stored in the refrigerator. This is placed in or on the object to be cooled. Braided Products Ltd., of Montreal, produce a sealed plastic bag called the Ice-Pak which contains a gel. It is pliable and if punctured will not leak. As well as an ice bag it may also be used for heating by first placing it in an oven or hot water.

COLD BATHS

If the nature of the work in a laboratory demands the continuous use of ice baths it will be found more economical and more convenient to use a refrigerated bath. (*Figure 10.3.*) Thermostatically controlled baths with heaters and refrigeration units vary in design. One type has a comparatively small bath with a capacity of 10–20 dm^3 and a working diameter of 30–40 cm. The temperature may be controlled to fine limits by a contact thermometer (see page 395) and stirrer to −40°C the variation being less than 0·1°C. A pump allows the bath contents to be circulated through external apparatus, such as jacketed columns. Both bench and floor-standing models are available.

For temperatures down to −80°C, two refrigerator units are used, one backing the other as in the deep-freeze refrigerator. The evaporator of the refrigerating unit is usually immersed in the bath fluid, thus giving efficient heat exchange. A larger bath volume is often useful for immersing large flasks, etc. The Grant refrigerated bath has a large tank volume and is mobile. The bath fluid is circulated by a paddle and the temperature is controlled by a bimetal thermostat operating an immersion heater and a refrigerator which is mounted underneath the bath. A circulating pump is also provided.

Existing water baths may be converted to cold baths by immersing the evaporator of a refrigerating unit. These are obtainable with the evaporator coupled to a flexible tube. *Figure 10.4a* shows the unit supplied by Townson and Mercer for cooling baths, and also the thermostatic refrigerator type TK1 supplied by Camlab (*Figure 10.4b*) which may be controlled by a contact thermometer. For temperatures below 0°C the bath fluid is either ethylene glycol, white spirit or alcohol.

Finally, the Peltier cooling modules supplied by Messrs. Delarve Ltd., may find some applications in a laboratory. These are panels of thermo-electric couples that rely on the effect discovered by

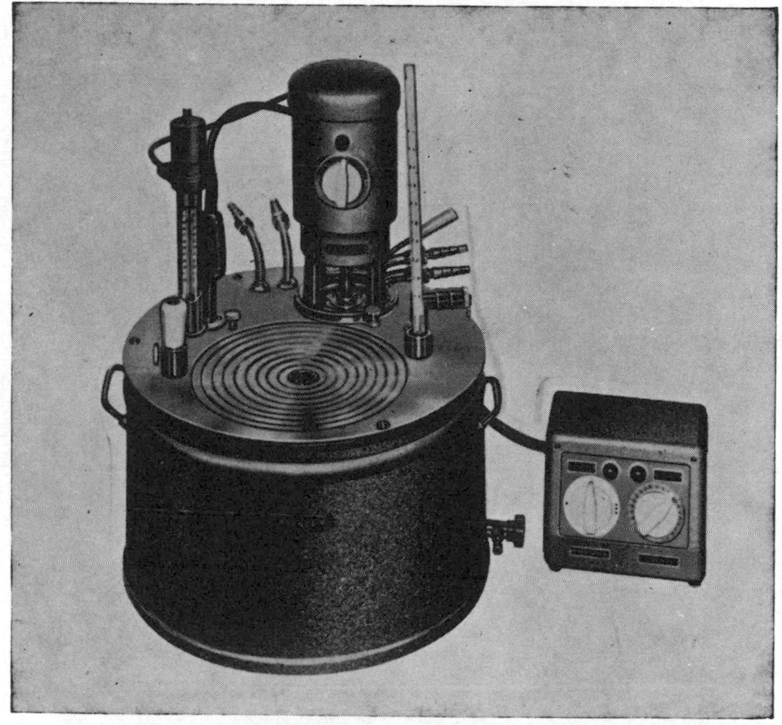

[By courtesy of Shandon Scientific Co. Ltd]

Figure 10.3. A refrigerated bath

Peltier in 1834, that when a d.c. current flows across a thermocouple junction, heat is absorbed to maintain the energy level between the semiconductor lobes. They require a d.c. source of not more than 3·5 V but with high currents of up to 30 A. The temperature difference across the device can rise to 60°C depending on the voltage used.

WATER BATHS

Controlled temperature conditions are often required by the biochemist for experimental work such as the study of enzyme reactions and metabolic processes, and for this reason the thermostatically controlled water bath is essential. Metabolic shaking bath incubators have efficient circulation and temperature control; they are able to contain racks which hold flasks, tubes or beakers in

[By courtesy of Townson and Mercer Ltd]

Figure 10.4. (a) Townson and Mercer refrigeration unit

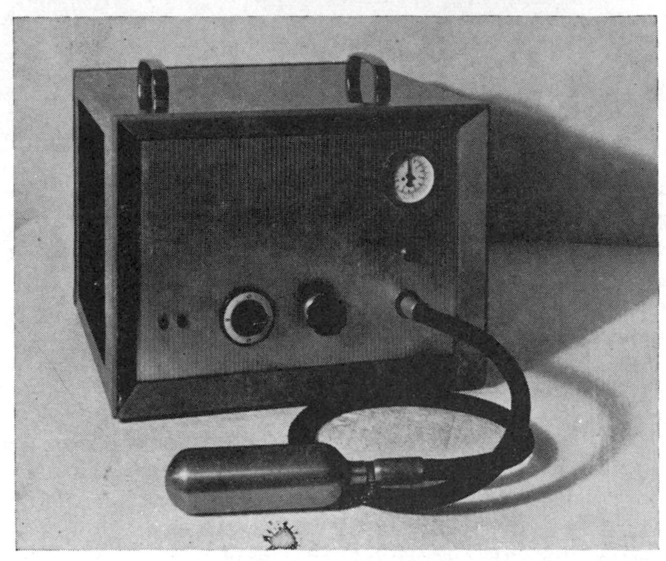

[By courtesy of Camlab Ltd]

Figure 10.4. (b) TKl refrigerator unit

the water. These racks are fixed to a shaking mechanism which imparts a reciprocating motion, and is variable from approximately 10–200 strokes/min, the amplitude may also be variable. One model made by the New Brunswick Scientific Co., has a shaking platform that has a circular or 'gyro-rotary' motion of a half inch diameter circle. The speed may be varied from 85–285 rev/min. The flask contents are swirled up the sides giving excellent aeration conditions. Some models are supplied with a hood whereby the atmosphere around the reaction vessels may be changed. A gas seal is effected by having the bottom of the hood immersed in the bath water. Two models are illustrated in *Figure 10.5a* and *b*.

Figure 10.5.
(*a*) *Metabolic shaking bath*

[By courtesy of H. Mickle]

Several self-contained units are now available that can turn any vessel into a thermostatically controlled water bath with an accuracy of at least $\pm 0{\cdot}05\,^{\circ}C$ with a bath volume of 20 dm^3. The thermostat, immersion heater and stirring mechanism are so compactly arranged that even a 1 dm^3 beaker could be used as a bath. In the Thermomix II the bulb of a contact thermometer is in the stream of pumped liquid, and in the Techne a bimetal thermostat dips into the bath close to the stirrer. On some models a pump serves the dual purpose of agitating the bath liquid and circulating the liquid through an external apparatus. This type of apparatus is ideally suited for use with refractometers, viscometers, constant temperature cell housings on absorptiometers and for the temperature control of chromatography.

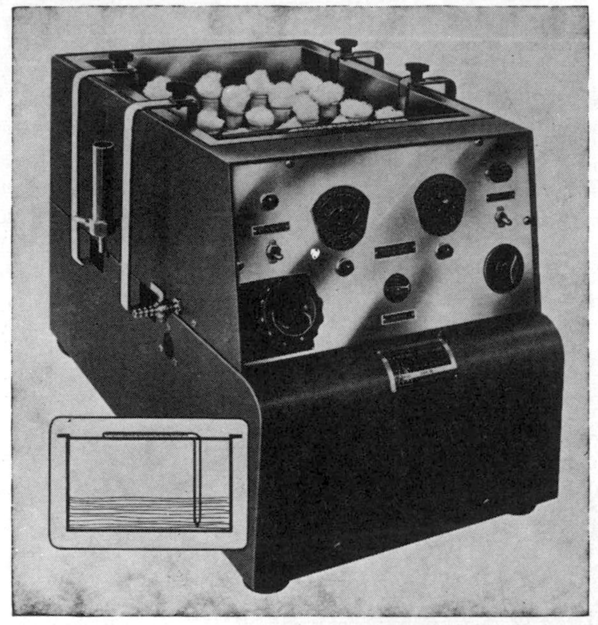

[By courtesy of V. A. Howe Ltd]

Figure 10.5. (*b*) *Metabolic shaking bath*

For manometry and constant volume respirometry an accurate constant temperature bath is required with the minimum of temperature fluctuation and, more important still, no temperature variation. To obtain the latter, bath stirring efficiency must be high. These baths are provided with a mechanism for shaking respirometers and are called after Warburg, a pioneer in this field. The respirometers have to be stopped at intervals and depending on the bath model this may be done individually or as a group, and the shaking motion can be either oscillatory in a straight path or in an arc, or semi-rotary. The baths are of two general types, rectangular and circular (*Figure 10.6a* and *b*). The latter takes up less room, sits on a bench and can be placed in a corner. The respirometer holder is so designed that by rotating it the manometers can be positioned before the operator for reading.

In the Braun circular water bath circulation is by a central paddle. The heater is two electrodes dipping into the water, which has sodium chloride as an electrolyte, with an a.c. supply connected across them. Temperature control is by contact thermometer.

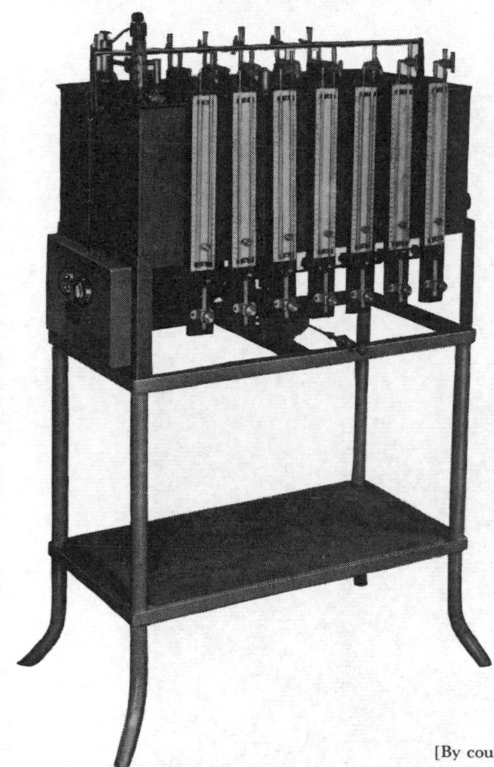

Figure 10.6. Warburg bath
(a) Rectangular (Hoslab)

[By courtesy of Hoslab Ltd]

The bath itself is made of perspex and for photo-chemical work the reaction flasks can be illuminated from outside, one model having tungsten lamps under each flask.

The rectangular baths require floor space and the operator has to move round the bath to take readings. The manometers may be stopped singly or as a group of seven on one side of the bath. Stirring is either by large paddles or by pumps and the temperature is controlled by contact thermometer or mercury toluene regulator. This type of bath is more easily adapted as a shaking metabolic bath. The Hoslab apparatus is very suited to general purpose work as well as respirometry, the mechanism is simple and readily accessible for maintenance purposes.

For temperatures from 100–200°C an oil bath is used. These are simply steel baths with an energy regulator or bimetal thermostat operating a 2–4 kW heater, and a more powerful stirrer to circulate the viscous oil.

[By courtesy of Shandon Scientific Ltd]

Figure 10.6. Warburg bath (b) Circular (Braun)

TEMPERATURE MEASUREMENT AND CONTROL MECHANISMS

Most of the temperature measurement in the laboratory is done with a liquid-in-glass thermometer, and these cover the range of −120°C to 500°C with the use of a variety of liquids such as alcohol, pentane and toluene, and mercury under pressure. The principle of expanding mercury is used in two types of thermostats. The first is the contact thermometer which is fast becoming the preferred method within its working range. The rise and fall of the mercury makes and breaks an electrical circuit operating a relay which in turn switches the power to the heaters. On no account should the mains be connected to the thermometer or a relay of high power consumption used. The current through the thermometer should only be a few milliamps to prevent sparking and oxidation of the wire. The thermometer is also hydrogen filled for this reason. This small current operates a vacuum hot wire switch or an electronic relay, varieties of which are able to switch currents up to 30A at 600 V.

One wire in the contact thermometer is sealed into the capillary below the lower scale, and another is inside the capillary and poised above the mercury thread. A nut which cannot rotate is attached to the other end of this wire. Through the nut is threaded a long piece of studding which has a piece of iron at its upper end. This studding can be turned from outside the glass tube by a magnet, which moves the nut and the wire up or down. The end of the wire indicates the control temperature, but as this is difficult to see a second scale adjacent to the nut is used, the position of which gives an indication of the temperature, a reference thermometer being used to obtain an accurate setting.

The second device is the liquid expansion thermoregulator (*Figure 10.7*). This may be filled with mercury or, more commonly, toluene which is trapped by mercury. The regulator is immersed in the bath up to the side arm. The toluene expands with the rise in temperature pushing the mercury up capillary *C*, this is connected electrically by a wire fused into the capillary wall to a pool of mercury in *A*, which in turn is connected to one terminal by a length of Nichrome wire. Another Nichrome wire from the other terminal is positioned in the capillary above the mercury, and is adjustable in depth. When the mercury reaches this wire the terminals are shorted, and a relay is operated which turns off the current to the heaters. The reservoir *R* allows the thermoregulator to be used over a wide range. For higher temperatures

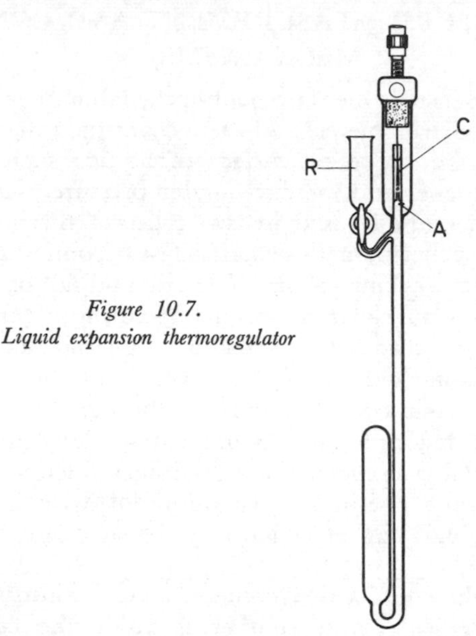

Figure 10.7.
Liquid expansion thermoregulator

the adjustable wire is retracted and the reservoir tap opened. As the wires are not shorted, the temperature rises, and when this is within a degree of that desired, the tap is closed and the final adjustment made by screwing down the centre wire. For lower temperatures the tap is opened and the adjustable wire is made to contact the mercury.

The toluene used should be sulphur free or redistilled over mercury, otherwise the mercury will become 'dirty' and have a scum, and will cause sticking in the capillary.

One type of relay used with these thermostats is the hot wire vacuum switch (*Figure 10.8*). In this, a coil of wire which is part of the subsidiary circuit through the regulator is under tension and controls one of a pair of mains contacts. When the electrical circuit is completed by a regulator, the wire heats and expands causing the mains contacts to open. In the electronic relays a very low current is allowed through the regulator, the passage of which causes a larger current to flow in a hot-wire switch or sensitive magnetic relay.

A most useful device for measuring temperature changes and controlling them is the pyrometer. This employs a thermocouple as a sensing element, it is very small and may be a distance apart from the meter and controls. Temperatures can be read in difficult

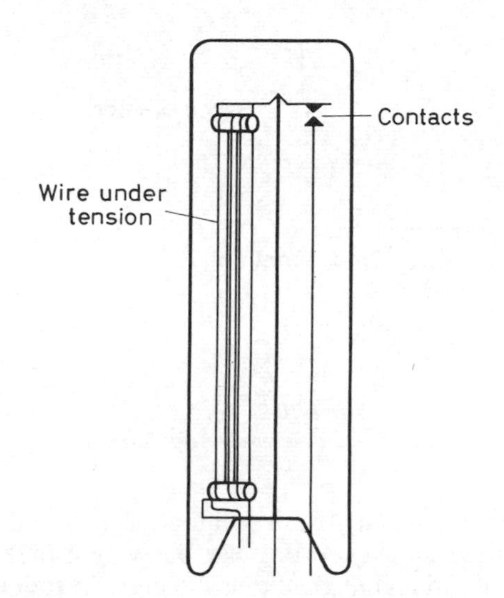

Figure 10.8.
The hot wire vacuum switch

positions such as the centre of ovens or vacuum vessels and even in a substance undergoing treatment.

If two wires of dissimilar metals are joined together at each end and one end is heated, an electric current will flow. This arrangement is called a *thermocouple;* the e.m.f. generated depends on the metals or alloys used and the temperature difference between the junctions. If one junction temperature is kept constant and the other allowed to rise (*Figure 10.9*), it will be found that the e.m.f. flowing in the circuit increases in a near linear fashion with temperature.

To make a thermocouple the wires are cleaned and twisted together at their ends, and then soldered or brazed together. For practical purposes the millivolt meter used can be regarded as constituting the reference junction provided connections in the meter and to the meter are at the same temperature. Ideally the wires from the variable temperature junction should continue to the meter or reference junction, but this may be too expensive or impractical. In this case, connections may be made with wires having similar thermoelectric properties.

Table 10.1 gives a list of metals commonly used with their maximum working temperatures. Thermocouples can also be used for measuring low temperatures below the range of liquid expansion thermometers. The reference junction will still be the

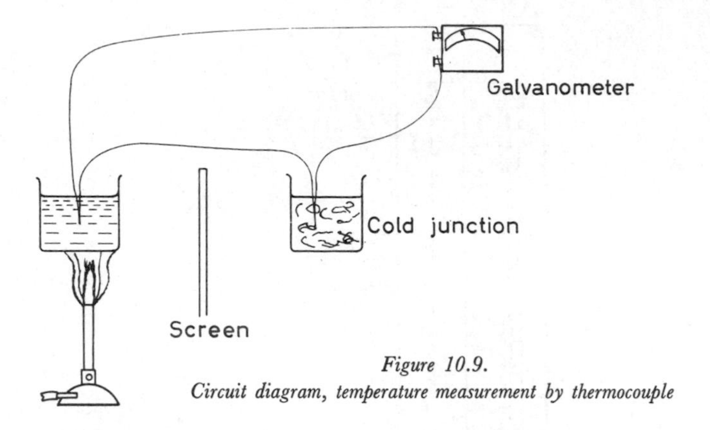

Figure 10.9.
Circuit diagram, temperature measurement by thermocouple

meter connections, a melting ice bath or a boiling pure liquid. The two latter are impractical for day to day use but where high temperatures are being measured a change of a few degrees in room temperature represents a small percentage error at say 1000°C. For greater accuracy various devices are employed to compensate for ambient temperature changes. One is a bimetal strip acting on the hairspring of the meter, and in others the thermocouple is connected into a potentiometer circuit in which resistances are

TABLE 10.1. Metals Used in the Construction of Thermocouples

Couple	*Max. temp.*	*Connecting wires*
Copper/Constantan (Eureka)	400	Same
Chromel/Constantan	800	Same
Iron/Constantan	850	Same
Chromel/Alumel	1100	Same
Platinum Rhodium 10%/Platinum	1400	Copper/Copper Nickel alloy
Platinum Rhodium 13%/Platinum	1400	Copper/Copper Nickel alloy
Platinum Rhodium 20%/Platinum Rhodium 40%	1900	—

Alloy	*Composition*
Constantan or Eureka	Nickel 40%, Copper 60%
Chromel	Nickel 90%, Chromium 10%
Alumel	Nickel 94%, Aluminium, Silicon, Manganese Mixture 6%

incorporated that change in resistance with temperature and thereby alter the e.m.f. fed to the meter.

The thermomocouple pyrometer is used as a temperature control device for high temperature furnaces. The switching mechanism may be mechanical with the meter pointer acting as a contact or positioned between index levers and involved in a feeler or chopper mechanism. The more popular and accurate controllers have an electronic system for switching which may be one of a variety of principles such as a photo-electric device or potentiometer. As with liquid expansion thermostats an intermediate relay operates the supply to the heaters.

Resistance Thermometers and Controllers

When a metal is heated, its resistance to electrical flow increases in a near linear fashion with temperature. This fact is made use of in a *resistance thermometer*. The metals most commonly used are platinum, nickel and copper, these are either wound on a former of insulating material such as porcelain or other refractory, or the wire is coated with shellac and close wound. The first method is used for temperatures above 100°C the bare wire is wound into grooves or notches, the whole being enclosed in a metal sheath or pocket. It is essential that the wire and former are kept perfectly dry as dampness will cause tracking and therefore changes in resistance. The diameter of wire used is usually less than 0·25 mm. The approximate resistance of wires at room temperature is given in Table 10.2.

The shellacked wires may be wound on any suitable former even on one of the connecting wires. The thermometer is then protected from mechanical damage by coating with epoxy resins or covering with thin glass or plastic tubing. Thermometers of this type have been constructed small enough to be implanted in animal and plant tissue.

TABLE 10.2. Characteristics of Wires Used in the Construction of Resistance Thermometers

Metal		*Ω/ft*	*Temp. coeff. of resistivity*
Copper	0·0125 cm	0·415	0·004
Platinum	0·0125 cm	2·41	0·003
Nickel	0·0125 cm	1·88	0·006

The thermometers are wired into one arm of a Wheatstone bridge circuit (*Figure 10.10*) the galvanometer, or chart recorder being calibrated in degrees. R_4 is the thermometer, L is a loop of the same length as the connecting wires to R_4 and laid adjacent to them. This compensates for any resistance change that may take place in the leads to R_4, for example, when the thermometer is being used in a growing plant some distance from the recorder.

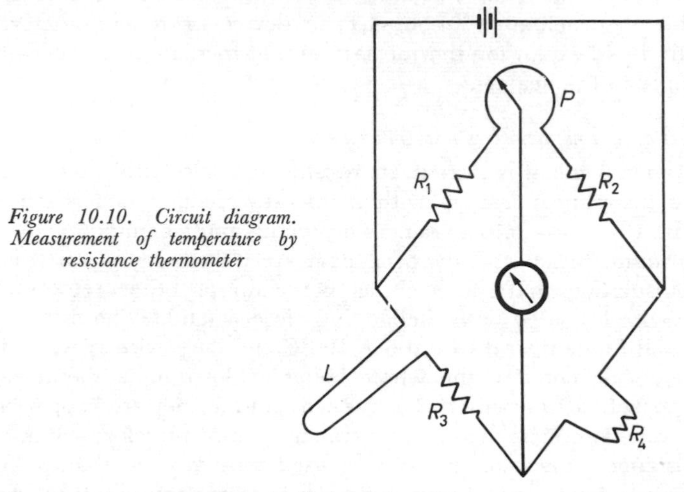

Figure 10.10. Circuit diagram. Measurement of temperature by resistance thermometer

The control devices are similar to the thermocouple pyrometer, the bridge circuit being used instead of the potentiometer.

The advantage of resistance thermometers over thermocouples is that no reference junction is required; the resistance characteristics remain constant.

Bimetal Thermostats

A less sensitive control is based on the fact that metals have differing coefficients of expansion. If a strip of Invar is rivetted to a strip of brass and then heated, it will bend. If one end is fixed, the movement at the other can be made to operate an indicating needle or a set of electrical contacts. In practice, the metals are fused together by careful heat treatment and then wound in a helix about a rod. The rod and bimetal are joined at the temperature sensing end. A change in temperature causes the rod to turn axially and operate electrical contacts. This type of thermostat does not need an intermediate relay for low wattage heating but it is advisable to use one as this prolongs its working life indefinitely.

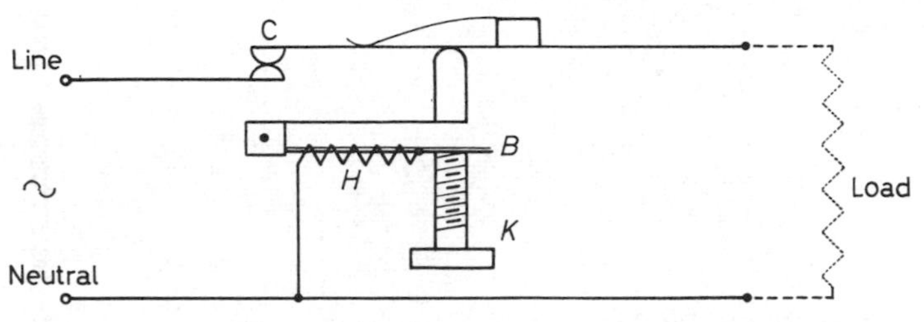

Figure 10.11. The bimetal thermostat

The Energy Regulator

This is the most common form of temperature regulating device, being used for controlling all types of heating equipment, baths, ovens, furnaces, heating mantles and tapes, etc.

Figure 10.11 illustrates the principle of operation. In the position shown current is flowing through the load and the heater element *H* that is adjacent to the bimetal strip *B*. As the latter heats up, it bends and opens contacts *C* thereby cutting off the supply to both the load and the heater. The bimetal cools, the contacts close and the cycle starts all over again. The ratio of the time that the contacts are open to the time they are closed is varied by operating control knob *K*. When the ambient temperature changes, so will that of the apparatus being controlled. This will only be a few degrees and may not be of any consequence but when finer control is needed a true thermostat or controller may be used in conjunction with the regulator. This is the standard practice with high temperature furnaces.

Liquid or Gas Expansion Thermostats

In this type of thermostat the expanding agent is in a closed system and a rise in temperature increases the internal pressure and operates a Bourdon tube or a metal bellows attached to a microswitch. Messrs. Gallenkamp use such a device called a 'Compenstat' in many of their ovens (*Figure 10.12*). In this a bellows operates a 15 A microswitch, subsidiary bellows compensate for ambient temperature changes.

This principle is generally used for low temperature control either by operating a valve which controls gas flow or a bellows that tips the mercury switch. The latter is the method often used for cold room control.

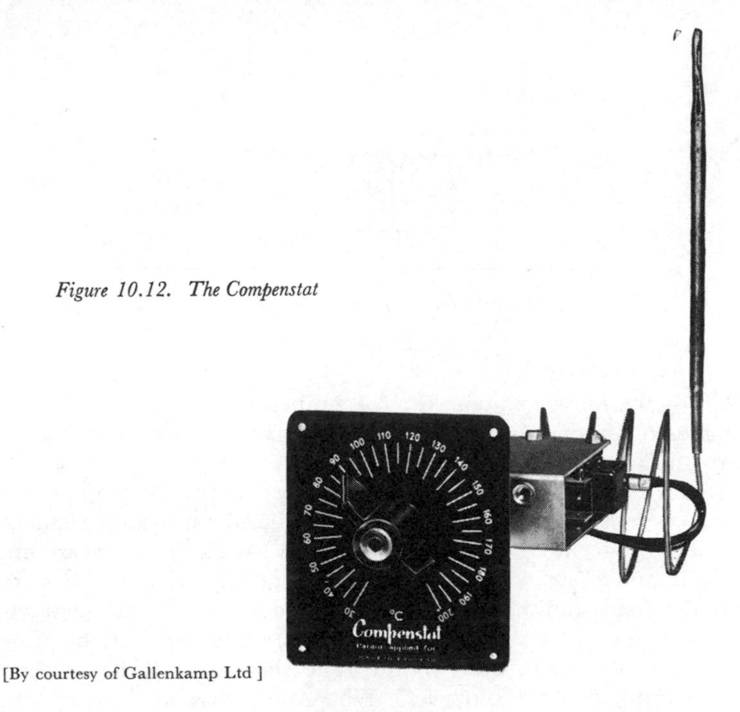

Figure 10.12. The Compenstat

[By courtesy of Gallenkamp Ltd]

APPARATUS AND METHODS FOR THE PREPARATION OF BIOLOGICAL MATERIAL

In this chapter it is assumed the starting material has been harvested or dissected. Biochemical material may be grouped by the sizes of sample needed, from the largest, the whole organ such as the liver or a leaf, down to sub-cellular particles. The whole organs are often required to be sliced. Slicing of animal tissue is performed in several ways. With practice thin slices may be cut with a 'cut throat' razor or a safety razor in a holder. The tissue is placed on a piece of filter paper which is made damp with saline, and then put on a glass slide or upturned petri dish. The slices are cut at an angle of 45 degrees, the thickness should be less than 1 mm. For enzymic experiments the slices must be 0·2 mm or less. This is more easily done by lightly gripping the tissue between two pieces of frosted glass and using half a safety razor blade under the upper glass the slices being cut with a slow to and fro movement.

The apparatus in *Figure 10.13* is a squat glass jar, the glass stopper of

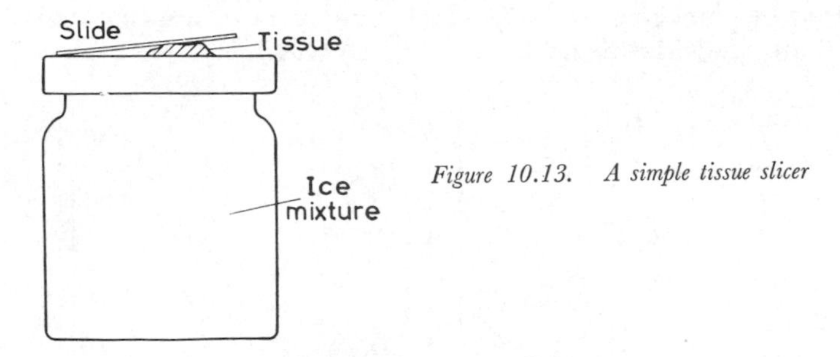

Figure 10.13. A simple tissue slicer

which has been ground with emery powder (about 200 mesh), this constitutes the lower piece of frosted glass. A frosted microscope slide is the upper glass. The tissue is cooled by filling the jar with iced water or ice/salt mixture. Stainless safety razor blades will be found ideal, these are broken in half and attached at one end to a home-made handle.

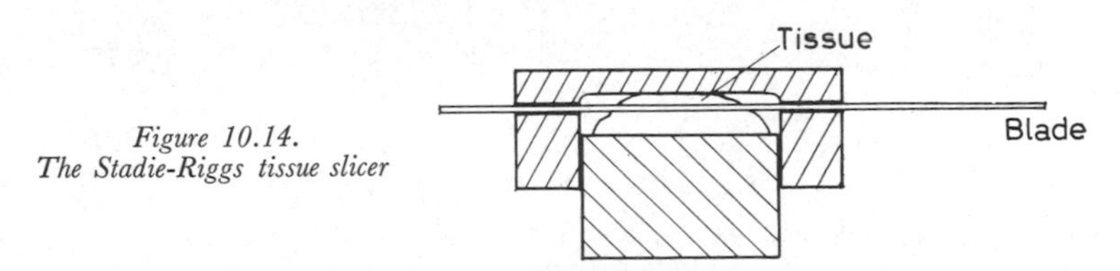

Figure 10.14.
The Stadie-Riggs tissue slicer

A commercially available tissue slicer has been designed by Stadie and Riggs* the principle of which is illustrated in section in *Figure 10.14.* This gives very uniform slices. In all these methods the tissue and knife should be moistened with a suitable saline or buffer solution.

A mechanical tissue slicer or chopper has been designed by McIlwain and Buddle†. In this apparatus the slices are obtained by a chopping action. A safety razor blade is clamped in an arm that is spring-loaded and actuated by a revolving cam. The specimen, on a carriage, is moved transversely after every stroke of the knife by a linked ratchet system. The distance moved can be varied from 0·026 to 0·3 mm. A similar apparatus available from Messrs. Hoslab cuts slices from 0·1 mm to 1 mm thick.

* *J. biol. Chem.* **154** (1944) 687.

† *Biochem. J.* **53** (1953) 412.

Often a higher degree of fragmentation is required prior to separation of cells and cellular components. With wet tissue this is done by homogenization. An early homogenizer was that designed by Potter and Elvehjem* (*Figure 10.15a*) which is a test tube with a

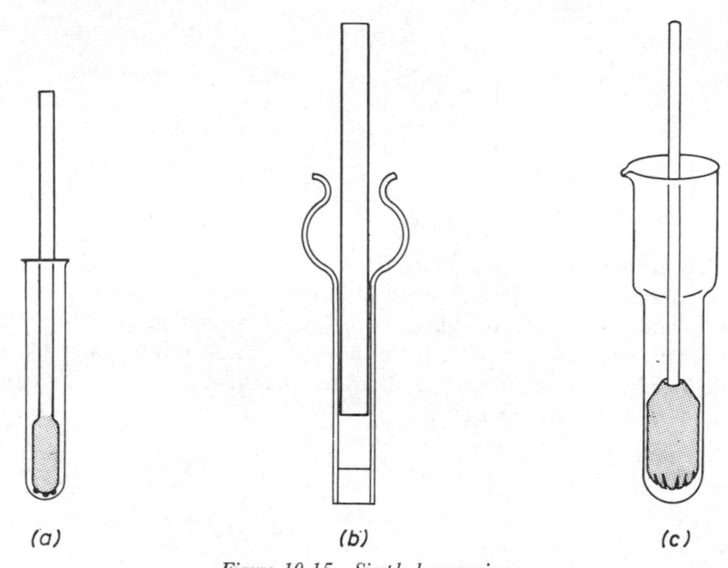

Figure 10.15. Simple homogenizers

(*a*) *Potter and Elvehjem* (*b*) *Precision bore tube homogenizer* (*c*) *P.T.F.E. pestle homogenizer*

close fitting glass pestle, the end of which has teeth. The tissue with some saline or buffer solution is placed in the tube, the pestle driven by an electric motor is introduced into the tube, which is then slowly moved up and down. The tissue is first macerated by the teeth and then finely ground as it passes the pestle. Homogenization is complete in about 2 min.

The softer tissues such as liver and brain are more easily broken down, so the teeth may be dispensed with if desired. The modified apparatus (*Figure 10.15b*) is made of precision bore glass tubing one end being plugged and heat-sealed or cemented with Araldite. In *Figure 10.15c* the pestle is made of P.T.F.E. with a stainless steel shaft. It is possible to construct this to produce a given maximum particle size.

* *J. biol. Chem.* **114** (1936) 495.

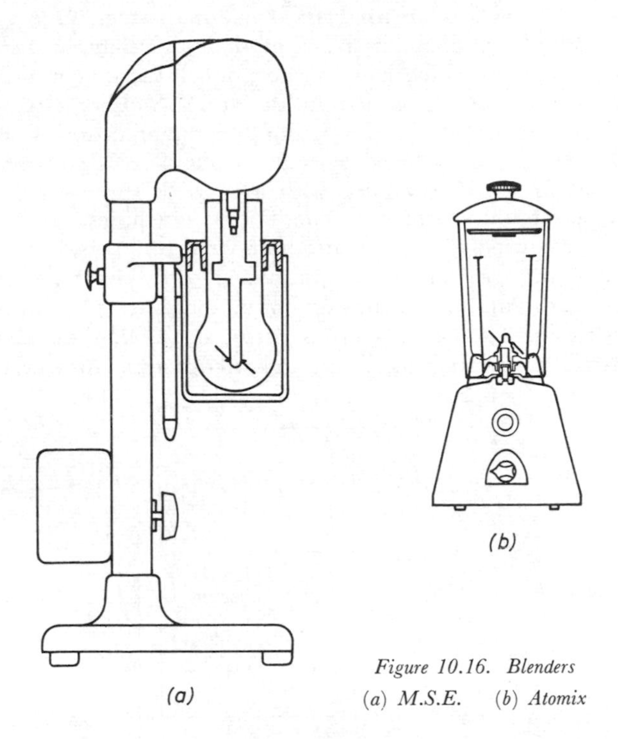

Figure 10.16. Blenders
(*a*) *M.S.E.* (*b*) *Atomix*

The coupling to the motor should not be rigid as the tube is manipulated by hand. The motor itself may be fixed vertically to a wall where the coupling could be a short piece of heavy walled rubber tubing, or fixed to a bench with a flexible drive mechanism. The pestle must be in the tube before the motor is turned on.

Homogenates are also produced by machines that use high-speed cutting knives. These are of two types. In the first, the chopped tissue is placed in flat sided glass beakers and the blades are introduced through the top of the vessel (*Figure 10.16a*) and in the second type the blade assembly is fixed into the bottom of the vessel (*Figure' 10.16b*), and driven from underneath. The former can deal with volumes down to 5 or 10 cm^3 and is comparable with the Potter and Elvehjem type in this respect. The second type can deal with volumes of 200 cm^3 to 5 dm^3. All the vessels are made liquid tight whilst the homogenization is in progress. The blades must be thoroughly cleaned by dismantling and washing with the aid of detergents between experiments if contamination is to be avoided. The vessels are of glass or stainless steel. One type has the lid rigidly clamped on a gasket seal and may be used with infectious material which is fed in and out of the machine continuously through tubes provided in the lid. Another vessel has a built-in heat exchanger coils so that the temperature of the fluid may be controlled. The speeds of cutting knives can be varied from 5000 to 20 000 rev/min. with a two- or three-speed switch, or a variable transformer. The larger machines are used for preparative purposes or for pre-homogenization before using a method giving greater disintegration. These machines are not designed for continuous running over long periods, and where large portions of tough or fibrous tissue are used, these should be pre-treated by passing them through a domestic mincer. Roots

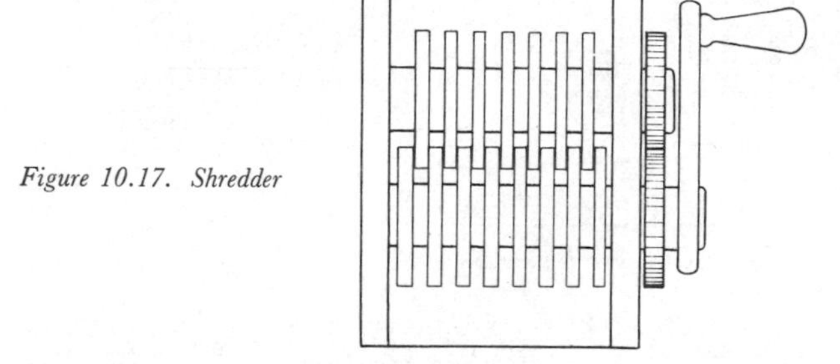

Figure 10.17. Shredder

or tubers may be chopped in a domestic potato chipper. Leafy material may be shredded before further reduction. A suitable machine for this has been designed. Basically it is two sets of revolving discs that mesh with each other. Leaves are fed between them producing 1 mm strips (*Figure 10.17*).

An apparatus combining the two methods of homogenization is available (the Kontes Disintegrinder) and has stainless steel blades mounted below the P.T.F.E. plug of a motor-driven pestle.

Plant tissues may be frozen in saline or buffer at −20°C and then ground in a pestle and mortar. The suspension or *brei* is then filtered, or more often, centrifuged (see page 413).

The reduction of dry material is performed in a mill. The most common type is the familiar pestle and mortar. These range from small 3 cm diameter agate mortars to 30 cm diameter stoneware mortars. Apart from the tediousness of hand grinding it is difficult to obtain reproducible specimens in this way and mechanical mortars and pestles are to be preferred. These are called *end runner mills*. The mortars are revolved on a motor-driven table, the pestle being suspended over it and pivoted in the centre with a counter balance arm attached to the top. The weight of this arm holds the pestle against the side of the mortar, the pressure applied being varied in some models by a sliding weight. End runner mills are available in the same sizes and materials as hand mortars and pestles. Refinements are the enclosure of the machine so as to contain dust and prevent dirt entering, electric timers and variable speed controls.

For the grinding of large quantities of dry material *ball mills* may be used. These are of two types. In one a closed jar containing two or more balls is clamped in an eccentrically rotating table. The balls of different sizes roll round the jar in a planetary motion. In the second type jars containing balls are placed horizontally on moving rollers. The rollers arc adjustable to accommodate a variety of bottle sizes. The balls are made of porcelain or stainless steel.

Plant and fibrous tissue wet or dry is easily reduced by a cutting mill. In this the material is fed into a hopper which directs it into blades revolving at high speed. It is then graded as it passes through a sieve into a collecting bag or bottle.

Disintegrators

Some materials, especially uni-cellular suspensions (e.g. bacteria, yeasts and fungi), require more drastic measures to rupture them. For these various disintegrators have been designed. These fall into three types:

1. Ultrasonic methods.
2. Forcing a suspension under high pressure through a narrow orifice.
3. High-frequency shaking with microscopic beads.

These methods are mainly used to obtain cell contents and precede fractionation processes to obtain mitochondria, nucleoli and chloroplasts, etc.

Supersonic vibrations have been found to disrupt some types of cells. These inaudible sound waves are generated electronically and then transmitted mechanically to a cell suspension by a probe or blade. A phenomenon called *cavitation* takes place. Cavities are minute bubbles, caused by the oscillations and extremely high pressures are generated by their formation and collapse. These pressures disrupt the cells, usually within 15 min. The frequency used varies but is of the order of 10 000 c/sec with an amplitude of about 10 μm. The principle is also applied for cleaning purposes (page 20). The main criticism of the method appears to be the generation of heat over a period of time which may inactivate sensitive enzymes.

The ultrasonic disintegrator manufactured by M.S.E. makes use of a special cell allowing refrigerated fluid to be circulated. This machine dissipates 150 watts. The probes, which may be of titanium, shed fine particles into solution and after long runs of 15 min or more they are seen as a fine grey suspension.

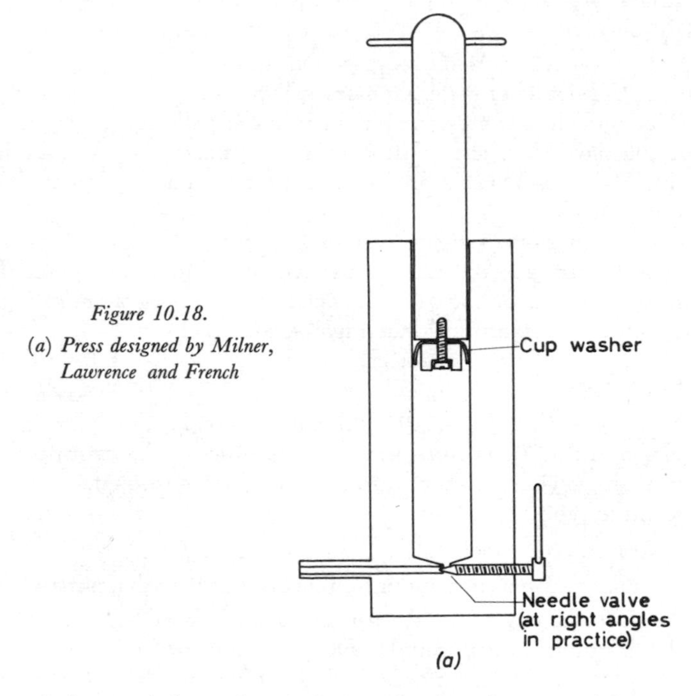

Figure 10.18.
(*a*) *Press designed by Milner, Lawrence and French*

Several 'presses' have been designed whereby suspensions are subjected to the combined effect of a shearing action and the velocity and expansion after release from pressure. One type designed by Milner, Lawrence and French* is a simple cylinder and piston with the exit orifice controlled with a needle valve

* *Science, N.Y.* **111** (1950) 663.

(*Figure 10.18a*). Several modifications have appeared in the literature. *Figure 10.18b* is an adjustable spring-loaded needle valve designed by A. J. Cox of Twyford Laboratories to give a predetermined minimum pressure at which the valve will open.* After filling with suspension the whole assembly is placed in an hydraulic

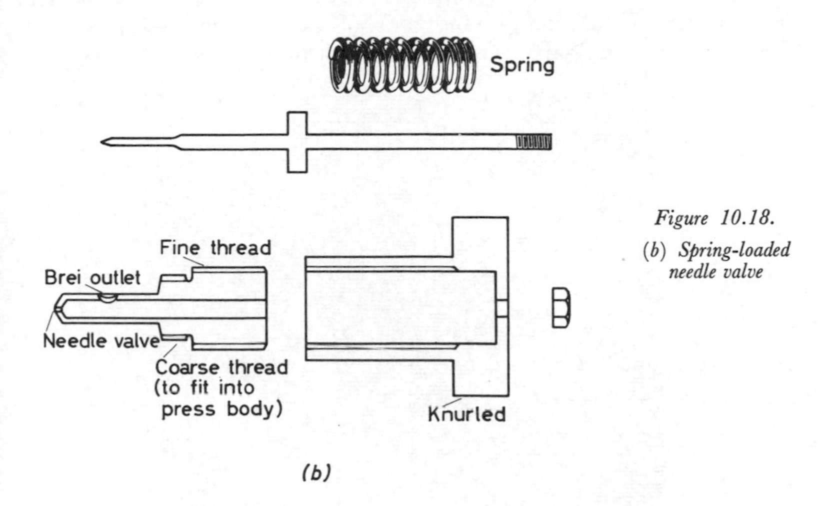

Figure 10.18. (*b*) *Spring-loaded needle valve*

10 000 kg press.* A receiver is placed beneath the exit nozzle and the press is operated to give the desired pressure, which may be up to 140 MNm^{-2} (20 000 lb/in^2). The cylinder is usually constructed from 7·5 cm round stainless steel bar with a hole of $2\frac{1}{2}$–3 cm diameter drilled in the centre. The heavy walls are needed not so much to withstand the pressure (a thickness of 3 mm would do), as to keep the solution cool, as heat is generated by the pressure and at the orifice by the work done. Many workers cool the assembly in a refrigerator before use.

Ribi *et al.* designed a similar press, the variation being that the orifice is continuously cooled by a stream of carbon dioxide.

For processing larger volumes of suspension the A.P.V. Co. market the Manton-Gaulin homogenizer. This can cope with 50 dm^3 per hour. A single action piston pump forces the slurry through an orifice closed off by a spring loaded pin. Simple valving allows recirculation of the material. It is possible to modify the valve body to allow cooling by pumped refrigerated fluid.

* Both these apparatuses are obtainable from Iver Scientific Instruments Ltd., Iver, Bucks.

Another type designed by Emmanuel and Chaikoff* nominally for animal tissues, has a fixed orifice size while it is in use (*Figure 10.19*) and a built-in cooling jacket. The piston-head has a protruding precision ground rod that passes through the orifice and projects into the receiver. The movement of this rod tends to prevent blockages and its diameter governs the size of the orifice. A number of interchangeable rods of varying size are provided. The assembly is again placed in a hydraulic press.

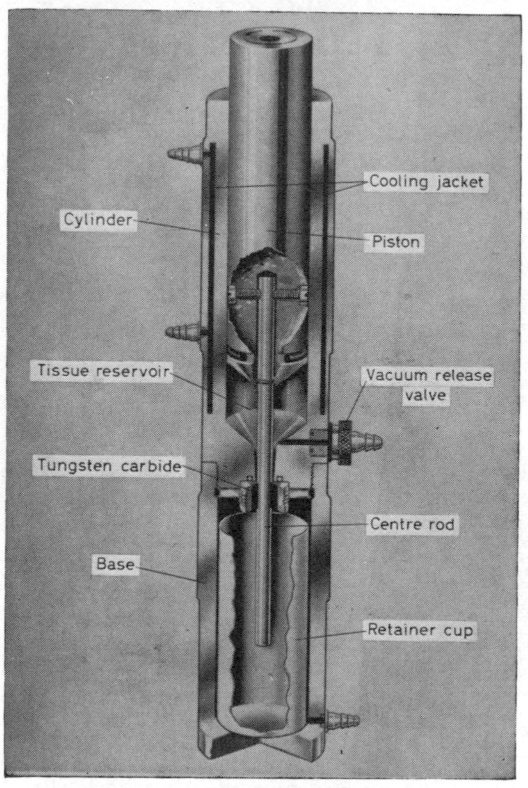

[By courtesy of V. A. Howe and Co]

Figure 10.19. Press designed by Emmanuel and Chaihoff

In the third type designed by Hughes† (*Figure 10.20*) a cylinder is made in two halves and the gap between these halves, after they are bolted together, forms the orifice. The material is frozen

* *Biochim. biophys. Acta* **24** (1957) 254.
† D. E. Hughes. *Brit. J. exp. Path.* 32 (1951) 97.

in the cylinder and then forced between the metal faces by a piston that is repeatedly struck by a fly press.

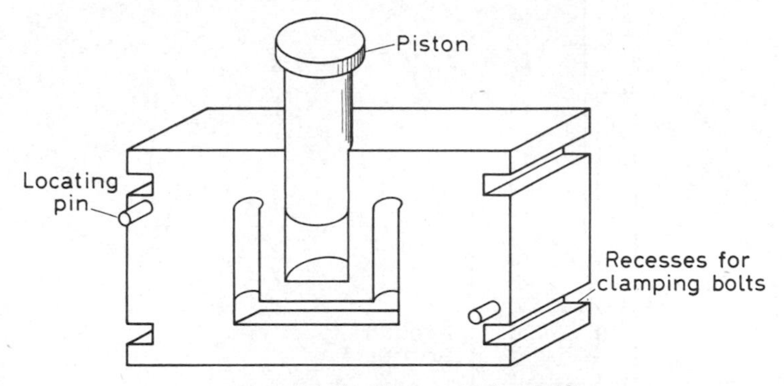

Figure 10.20. Hughes press (one half)

Working on a similar principle to the Hughes press, Professor Edebo has designed an apparatus (*Figure 10.21*) whereby the sample can be immediately reprocessed without dismantling. It is a double cylinder connected by a central orifice. A floating piston is placed in one end, the sample introduced into the other and a second piston inserted on top. The whole is frozen in an alcohol/CO_2 bath. A connecting rod is placed on the second piston and with the aid of a hydraulic press, the sample is forced into the other chamber. The connecting rod is removed, the press upended, the rod placed in the other end and the sample forced back again. By this means and by ensuring the sample remains frozen, it can be treated as many times as is found necessary.

Shaker disintegrators rely on the abrasive action of fine glass balls or beads shaken with the material in a liquid suspension. The beads are 0·2 mm in diameter called *ballotini* and are supplied by the English Glass Co. and firms that produce glass sinters. The volume of beads required has to be found by experiment, the suspensions being examined microscopically before and after treatment. The shaking machine designed by Mickle* for this type of disintegration has glass cups clamped at the ends of two steel arms, the other ends of which are clamped rigidly. These arms, like a tuning fork, have a natural frequency of vibration, this vibration is maintained and the amplitude controlled electrically.

The other type of shaker has the arm holding the vessel attached to a ball-bearing. The drive spindle hole in the centre of the

* *Jl. R. microsc. Soc.* **68** (1948) 10.

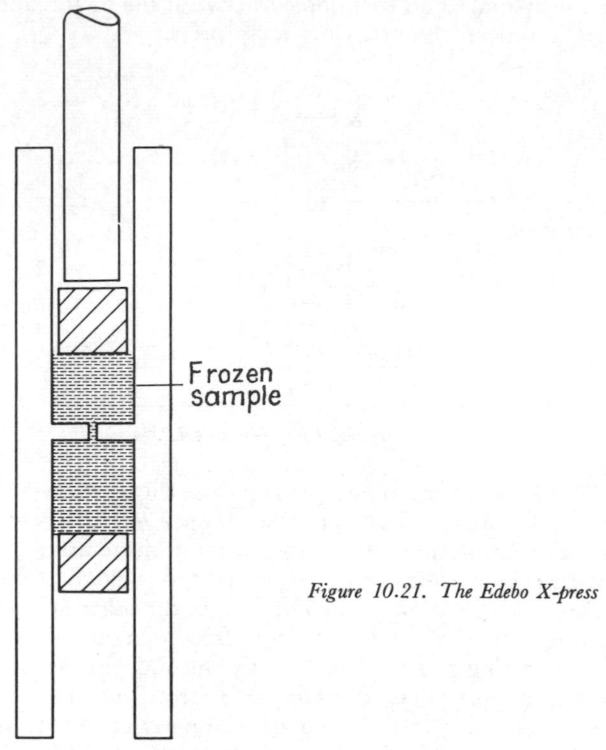

Figure 10.21. The Edebo X-press

bearing is at an angle of less than 90 degrees to the bearing faces, so that if the arm is prevented from rotating, and the spindle revolved, a figure of eight shaking motion is imparted to the sample vessel. The Mixer Mills supplied by Glen Creston Ltd. use this principle and can cope with volumes up to 5 cm^3. Shockmann, Kolb and Toennies* designed a shaker head to fit the International Equipment Co.'s centrifuge Model PR2 employing the same principle. The arms are in the form of a cross and hold four stainless steel or glass vials and may have a volume of 50 cm^3 each. The shaker head is prevented from rotating by a strong spring. The maximum spindle speed is 2600 rev/min but this depends on the amount of vibration imparted to the centrifuge. Hints on the reduction of vibration are given by Shockmann.†

* *Biochim. biophys. Acta* **24** (1957) 203.
† *Biochim. biophys. Acta* **59** (1962) 234.

In another type designed by Rehacek, Beran and Bicik, a disc is rotated either vertically or horizontally in a suspension of ballotoni. The rate of disintegration was found to be proportional to the peripheral velocity of the disc.

After disintegration the glass beads are separated from the cell fragments by centrifugation and washing. The heavy larger fragments are more difficult to separate and require repeated washing and centrifugation. It may be possible to overcome this difficulty by performing the disintegration dry using ground sodium chloride instead of glass beads. After disruption water is added, the sodium chloride goes into solution and the particles recovered by centrifugation.

A variation on this technique is the use of synthetic Zeolite in place of ballotini. Anhydrous Zeolite type 4A (Union Carbide Corp.) is mixed two to one with the microbial paste in a mortar. This is ground for 3 min with a pestle. This treatment effectively disrupts many micro-organisms, especially yeasts, from which good preparations of cell walls can be made. The mechanism of rupture is unknown, it certainly is not a solely abrasive action. It is thought that the great affinity of Zeolites for water when in intimate contact with the cells may be the major cause of disruption.

Unicellular material can be ruptured by grinding with powdered glass in a mortar and pestle. The powdered glass can be prepared by grinding broken Pyrex for 24 h in a ball mill using porcelain balls. Powdered glass is dangerous, when handling the dry material, a face mask must be worn.

Providing time is no objection and where only simple apparatus is at hand, many cells rupture simply by repeated freezing and thawing.

SEPARATION METHODS

After homogenization and disintegration the suspension is a mixture of particles of many sizes, including organized fragments such as mitochondria and chloroplasts, macro-molecules, colloids, soluble proteins and protein fragments. Whereas the chemist turns to filter funnels and papers as a first step, the biochemist prefers to centrifuge as the material is more easily recoverable from a tube than filter paper, and differential separation can take place from the same vessel without recourse to several grades of filters.

Centrifugation

Suspensions left to stand will sediment and a simple suspension such as barium sulphate in water quickly settles. If it is in a graduated cylinder the sharp dividing line of suspension and clear liquid is easily seen as it moves downward under the gravitional pull of the earth. Precisely the same happens in centrifugation but at an accelerated rate. Use is made of the centrifugal force that acts whenever an object is turned or made to describe an arc. This force increases with speed of rotation and also depends on the distance of the moved object from the fulcrum or centre of curvature. The *relative centrifugal force* (*RCF*) is a measure of the strength of pull acting on an object. *RCF* is the number of times the force is greater than gravity and is found by the following formula:

$$RCF = \frac{(\text{rev/min})^2 \times \text{radius (in cm)}}{89\,500}$$

The simple centrifuge has a rotor head driven by a variable speed electric motor. The rotor is designed to hold centrifuge buckets swinging in trunnions set symmetrically opposite each other. The suspension is placed in a tube of suitable size which in turn is placed in a centrifuge bucket. The rotor must be perfectly balanced and, before loading, the buckets are balanced in pairs and placed opposite each other in the rotor. When only one tube is required a counterpoise containing liquid of a similar density is used, for although two cups may balance by weight their volumes may be quite different and the centre of gravity of the rotor will alter. This will cause excessive vibration when running. When the tubes are in position and the protective cover in place, the motor is started and the speed is increased over several minutes to the maximum required. The speed used and the time of centrifugation depends on the sediment and the capabilities of the centrifuge. After the desired time the motor is turned off and the rotor allowed to coast to a halt. On no account attempt to stop it by hand. If it slows down too quickly the sediment may be disturbed. Some machines have a brake that is worked electrically.

Rotors

Two types are normally used, the first, already mentioned, is called a *swing-out head*, the tubes lying in a horizontal plane when revolving. The second is the angle head. This is usually made of aluminium alloy and the tubes go into symmetrically placed holes that are at an angle of 10–60 degrees from the vertical. As the tubes are at an angle they either have to be tightly sealed or only partially filled to prevent spillage. The advantage of this type of head is that it shortens the distance the particles have to move towards the walls, and therefore speeds centrifugation.

Rotors designed for high speed running and experiencing forces in excess of 50 000 *g*, have a finite life due to deterioration of the metal. Manufacturers recommend derating rotors after a period of time or amount of use. High speed rotors must be treated with care and examined at regular intervals for damage such as corrosion spots, dents and scratches and hair line cracks. *Figure 10.22* illustrates a damaged rotor bucket which exploded at 100 000 *g*. Investigation showed the rotor had been used over twice its permitted life span without derating. Another bucket from the same rotor had a hair line crack that was associated with a corrosion spot inside the bucket. Rotors can now be obtained made from titanium. These can withstand much longer use and higher speeds.

Figure 10.22. A damaged rotor

Tubes

Centrifuge tubes and pots are to be had in a variety of sizes, shapes and materials. Pyrex glass is used extensively, and if well annealed and free from blemishes, pots of this material can withstand 3000 × g and with thick walled tubes of about 10 cm^3 capacity, 10 000 × g is possible.

The microchemical analyst has found conical tubes of 0·5 cm^3 to 5 cm^3 capacity extremely useful as light precipitates are collected as a small pellet. Both conical and round-bottom tubes are available with volumes up to 100 cm^3 with and without graduations. The largest glass vessels in common use are the 500 cm^3 bottles which are used extensively by the blood transfusion service. The preferred technique for blood separation is to collect specimens in 600 cm^3 sterile plastic bags which are then placed in a special angle head capable of producing 4000 × g, twice the force that may be used with glass bottles of the same size. Vessels having the same capacities as the glass ones are available in plastic materials for use at speeds above 4000 rev/min. The materials used are polythene, cellulose nitrate, polypropylene, polycarbonate and nylon. These are designed to fit snugly in the head and are able to expand under pressure so that the strain is taken by the rotor walls. They vary in their ability to withstand common solvents and in their transparency, so that the choice of plastic depends on the solution and suspension. Stainless steel is also widely used especially when centrifuging materials that dissolve plastics and centrifuging at speeds not permissible for glassware. The sealing closures for use when using the angle head are made of stainless steel or alloy and are designed to seal the tubes by expanding an O-ring on to the inside walls of the tube. For high-speed work it is wise to completely fill plastic tubes as they may collapse. If filling is impossible use the steel tubes.

Centrifuges

A large number of centrifuges are now on the market. Choice should be governed by the work that is required of them, service facilities and of course, available capital. The biochemist will almost certainly require a refrigerated machine to keep the rotors cool. Even so, it must be remembered that the temperature inside the rotor may be 5°C above the recorded bowl temperature at high speeds, i.e. 10 000 rev/min or above. Preparative centrifuges are available that run at speeds up to 75 000 rev/min giving a maximum of 500 000 × g. Although higher speeds are obtainable no rotor is as yet available that can withstand the tremendous forces acting on the periphery. The cost rises proportionally with speed. There are several machines suitable for most biochemical purposes in the 20 000–30 000 × g range coping with volumes up

Figure 10.23. A high-speed centrifuge

to 1500 cm^3, having automatic controls that maintain refrigeration, that bring the speed to maximum, time the run and reduce the speed by applying the brake.

Continuous Flow Rotors

A boon to the biochemist are the several continuous flow centrifugation systems on the market, making it possible to spin off large volumes and collect a slight precipitate in the minimum number of vessels. It is especially useful for recovering protein precipitates after semi-scale fractionation and separating bacteria from large volumes of media, thus taking away the drudgery of numerous spins with bucket centrifuges.

The Servall system has a manifold above the rotor leading the fluid in and out of normal centrifuge cups. The IEC system passes the liquid through a plastic tube spinning at high speed.

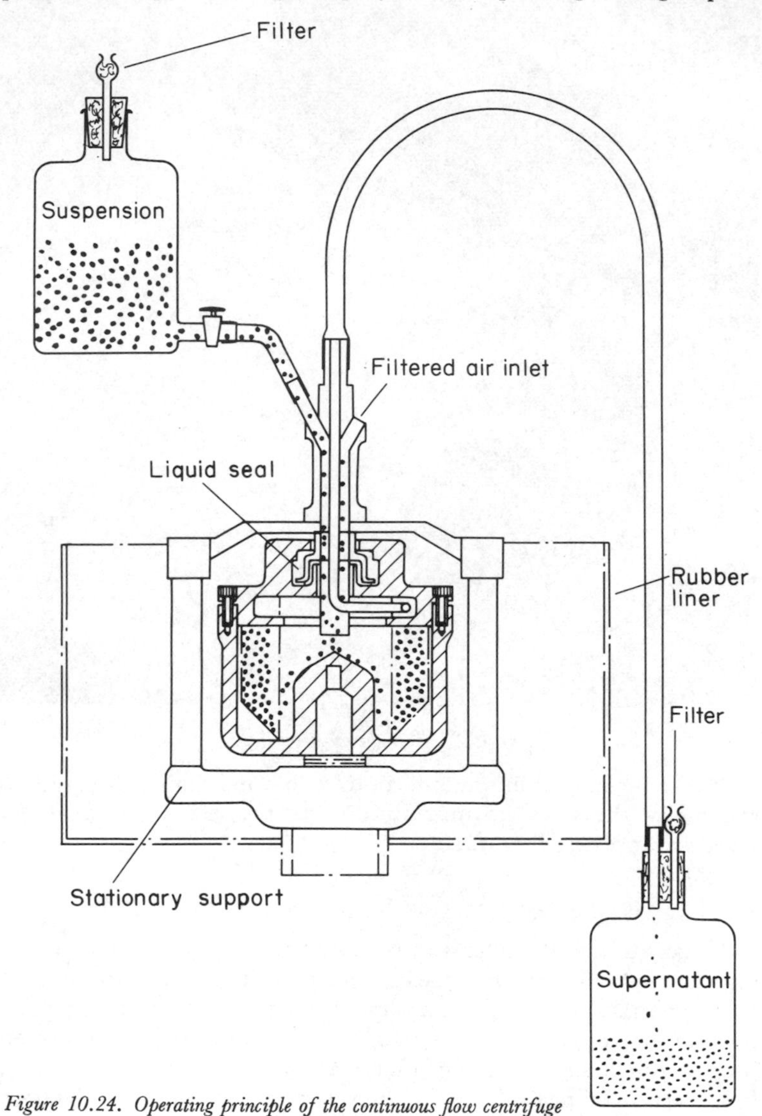

Figure 10.24. Operating principle of the continuous flow centrifuge

The sediment collects on the walls of the tube and gives some further fractionation of particles, the heaviest settling at the beginning of the tube and the lightest at the end. M.S.E. have a spinning bowl, the fluid enters in at the centre, the precipitate collects on the outer walls and the supernatant overflows into a channel, and is then transferred to a receiver via an inserted stationary tube. The Sharples Centrifuge has a spinning bowl that is a long tapering tube, the fluid is fed into the bottom, travels up the tube and out at the top, and is collected in a cowl which directs it into the receiver. This machine is the only one that is supplied in various bowl sizes. In continuous flow types the degree of separation depends on the flow rate, which is variable, and the nature of the material in suspension.

The ultra-centrifuge may also be used analytically being a method of determining molecular weights of proteins and showing the number of components in a protein mixture. Very stringent conditions are required for the work as the liquid under investigation must not be affected by disturbing influences such as convection currents. These are controlled by accurate thermostatic apparatus to within $\pm$ 0·1°C, and the use of cells, the walls of which are radial, allowing free movement of the molecules.

The exception is the basket centrifuge. This has a drum rotor with a perforated wall. This in turn spins in a stationary drum that has a drain outlet at the bottom. The rotor is lined with filter paper covering the perforations. The suspension for filtration is then poured into the spinning rotor. The filtrate is retained by the outer drum and collected in a receiver attached to the drain. For fine precipitates, a suspension of filter aid may first be put through forming a cylindrical cake. This is similar to pressure filtration on a Buchner funnel (page 29) and similar rules apply (page 30). That is, the rotor is revolved at 500 rev/min first, then as a thicker filter cake is formed the rotor speed is increased.

Gradient or Zonal Centrifugation

In normal centrifugation a mixture is suspended in a solution which is then centrifuged so that the heavier component sediments to the bottom of the tube and the lighter component stays suspended. Unfortunately, the heavier material remains contaminated with the lighter and the practice was then to pour off the supernatant, resuspend in fresh liquid and centrifuge again. By repeating this process several times a high degree of purity of the heavier material could be obtained. This process could be speeded up if the mixture was suspended in a liquid having the same density as the heavier component. On centrifugation the heavier would sediment and the lighter would rise to the surface, thus purifying both components at one go. This is ideal for a simple mixture of two substances but commonly in biochemistry one works with a multi-component mixture of compounds. Separation may be possible by centrifuging through a density gradient. A centrifuge tube can be prepared having a near linear density gradient of, for example, 1·2 g per cm^3 at the bottom, to 1·0 at the top. If the sample is then carefully layered on the top, on centrifugation the several components will sediment into zones of comparable density. After preliminary experiments, gradients within narrower limits may be selected for easier separation of specific components.

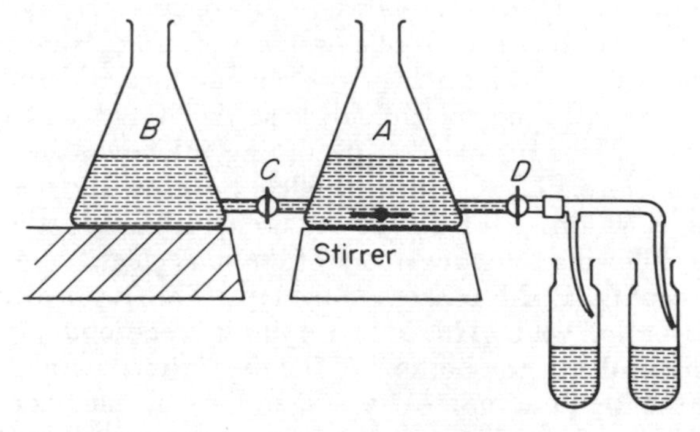

Figure 10.25. Simple apparatus for producing a gradient for zonal centrifugation

A gradient may simply be produced using the assembly in *Figure 10.25*. With taps *C* and *D* closed, equal volumes of solutions are poured into the flasks. *A* contains a high concentration of a suitable solute and *B* a weaker solution or pure solvent. Stirring is commenced in *A*, tap *C* is opened then tap *D* allowing a gentle flow through a capillary tube into the centrifuge cups. A gradual dilution of *A* continues while the cups are filling. It is common to prepare two tubes at once, the tubes then need the minimum of balancing after the sample is layered on the top with a pipette. After centrifuging the tubes or cups may be photographed to record the number of bands present before attempts are made to remove the various zones. This is a tricky operation when glass or stainless cups have been used. If only a few bands are present they may be sucked off using a water vacuum pump or pipette. It is difficult not to disturb and mix lower layers. Another way is to introduce a long large bore syringe needle or similar tube to the bottom of the cup. Then insert down this tube a finer needle that is attached to a peristaltic pump. The contents are slowly pumped into a fraction collector. Wherever possible gradients are run in cellulose nitrate tubes as this material is easily pierced with a needle. Apparatus is available to clamp these tubes and to insert a needle in the base. The contents can then be withdrawn and fractionated.

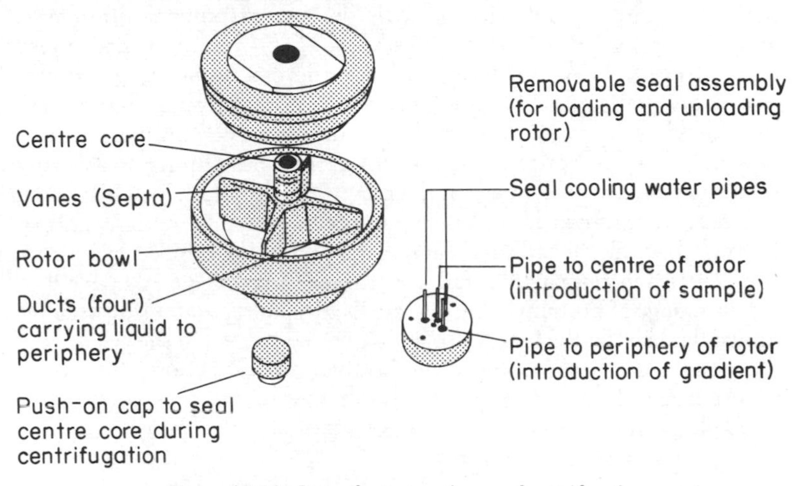

Figure 10.26. Rotor for preparative zonal centrifugation

Rotors have been developed for larger scale preparations (*Figure 10.26*) using density gradients and used for zonal centrifugation. The rotor is a bowl having four vanes radiating from the centre. These vanes have a channel or hole along them. There is also a separate set of holes into each segment at the centre. The rotor is spun at between 3–5000 rev/min. A gradient is then introduced, the less dense portion first, to the periphery of the bowl via the channels in the vanes. The rotor fills until liquid seeps from the other set of holes at the centre. Pumping is stopped and the sample is introduced by syringe or pump onto the top of the gradient. Connections to the rotor are removed. The centrifuge lid is then closed and the speed increased to the desired level. Rotors are available for speeds of 48 000 rev/min and 170 000 *g*. At the end of the running time, the speed is reduced to 2000 rev/min and the centrifuge opened. The tube connections are remade and a dense solution is pumped in via the vanes pushing out the gradient which is then fractionated. This technique has been very successful in making protein, R.N.A. and virus preparations.

Ultra-filtration and Dialysis

The ordinary filtration methods are discussed in Chapter 2 but the biochemist has the problem of separating or concentrating substances that are actually dissolved or are colloidal in nature. The finest filter paper has an approximate pore size of 500 nm and will not retain colloids except with the use of powder filter aids, and this is acceptable if the filtrate only is required and there is no adsorption by the filter aid. For the examination of proteins and other molecules in biological fluids by analytical techniques such as electrophoresis and the various forms of chromatography it is necessary to concentrate the solutions and remove interfering salts and small molecular fragments. One method, gel filtration, has been mentioned already on page 306. In this method gels or porous glass microbeads are chosen that have the desired exclusion. Concentration may also be effected by adding dry gel powder to a solution, e.g. of protein. The gel swells absorbing water and smaller molecules leaving the larger molecules concentrated in the remaining solution. Another common method is ultrafiltration through a semi-permeable membrane. The pore size of the membrane used depends on the molecular size to be retained. The range available is from 500 nm to less than 5 nm:

300 nm will retain bacteria

200 nm will retain bacterial fragments and pyogens

35–100 nm will retain viruses and bacteriophages

10–20 nm will retain proteins of about 100 000 mol. wt., e.g. albumen

5 nm will retain substances of about 25 000 mol. wt.

less than 5 nm will retain substances of about 10 000 mol. wt.

Selective membranes are available giving a specific molecular weight cut off from 500 000 to 300 000.

The membranes are made from specially treated cellulose and cellulose esters, P.V.C. nylon and P.T.F.E. and must be chosen according to the experimental conditions pertaining such as degree of heat or acidity used. They have the advantage over the Seitz type filter in that they are very thin, do not absorb and are true filters, i.e. do not rely on depth for effectiveness.

Ultrafiltration can only be performed under increased pressure so the apparatus must necessarily be sturdy. The membrane is supported on a grid or sintered disc which may be of stainless steel, bronze, plastic or a carbon block. The membrane may also be clamped between two blocks having spiral grooves cut in them, thus giving a long channel bisected by a membrane. The membrane and its support fit into a funnel and a cell body clamped on top.

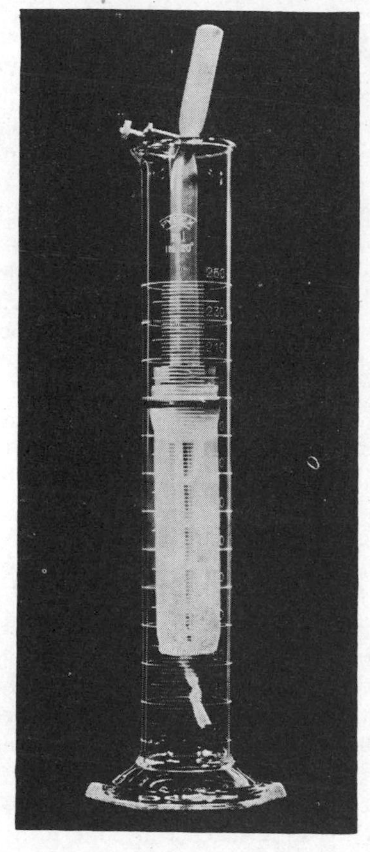

Figure 10.27.
Pressure dialysis L.K.B. dialyser

[By courtesy of L.K.B. Ltd]

The top will have a port and pressure release valve. The filter assembly is mounted over a receiver, the sample introduced through the port and a pressure applied by attaching the port to a pressure pump or compressed gas cylinder. Do not exceed the recommended safe working pressure, this is usually about 10 or 20×10^5 Nm^{-2} (200 or 300 p.s.i.). This means of filtration or concentration is very slow, for being static the membrane pores tend to block as a layer of large molecules forms, often gel like, on the surface. Also the higher the concentration at the surface, the higher will be the osmotic pressure that has to be overcome. In fact another name for this technique is reverse osmosis, a term used in the field of water purification. One must remember that solutions, often used biochemically such as sucrose and ammonium sulphate, may exert pressures of over 500 p.s.i.

An improvement on the static system is to use a magnetic stirrer. The Millipore ultrafiltration apparatus uses a magnetically driven agitator. Another method is to re-circulate the solution over the surface of the membrane. This is less liable to lyse cells or denature proteins. For large volumes it is possible to connect together a bank of filter units in parallel and re-circulate solutions on both sides of the membrane with a peristaltic pump. This method has been used for the concentration of virus particles. Amicon Ltd supply a filtration unit utilizing a narrow channel over the membrane. Movement through this channel prevents settling or gelling. This type is particularly useful for concentrating eluates directly from a chromatography column and before passing into an analyser such as a u.v. absorbance monitor.

Small filter units are available to cope with small volumes. Some are designed to fit on to a syringe. It is possible by this means to reduce, for example, 5 cm^3 of C.S.F. to 0·1 cm^3 which may then be examined electrophoretically.

There is very little distinction between ultrafiltration and dialysis now except that it is the gentlest treatment that can be given to a solution and also the cheapest.

Graham in the 1860's showed that proteins were large molecules by diffusion experiments using parchment as a semi-permeable membrane. Salts and mono- and di-saccharides diffuse through the membrane into distilled water on the other side until a pressure equilibrium exists. If the water is changed continuously the diffusible matter is completely removed.

The membranes used are parchment, collodion films and cellophane. The latter is obtained as both sheet and tube and is most commonly used. A coil of tubing is soaked in distilled water for 10 min and then knotted at one end. A test is then made for pin-holes by distending the sac by blowing or filling with water. Fill the tube with the sample using a funnel. If a large concentration of salt is present leave a space of 1 or 2 volumes above the solution before knotting the other end. This is to prevent the sac bursting as water diffuses in quicker than the salts diffuse out. The sac is then suspended in water or weak buffer, the speed of dialysis will increase if the water is continuously changed and agitated.

Dialysis can be assisted by a pressure of 1 atm using the L.K.B. apparatus (*Figure 10.27*) or the collodion shell holder. In the first the vacuum is applied inside the tube, in the second, it is on the outside.

An extension of the technique is electro-dialysis in which it is possible to separate protein molecules of differing size. In this method a d.c. current is passed through several cells separated by membranes. Electrolytes decompose and are liberated at the electrodes. The larger molecules are either neutral or possess a negative or positive charge and will attempt to migrate to the anode and cathode respectively. By careful choice of pore size of the membranes protein mixtures may be separated. A suitable apparatus is illustrated in *Figure 7.13*, several sizes of sample cell are provided and the electrode compartments have cooling coils.

The desalting apparatus described in Chapter 7 provides a quick method of removing electrolytes. It is especially useful in the preparation of samples for chromatography.

Another electrolytic method is that of electrofocusing. This technique has been perfected by L.K.B. and depends on a group of chemicals called ampholytes which are mixtures of aliphatic polyamino-polycarboxylic acids. A solution of these in a column has the ability when placed in a d.c. field to form a pH gradient. The ampholytes are incorporated in a linear density gradient of non-ionic material such as sucrose. The density gradient has a stabilizing effect and prevents convection currents. The gradient is formed as described earlier and is pumped into a glass column not unlike a Davis double surface condenser set on end. This gives thermostatic control of the enclosed annular space. The sample protein mixture may be layered in at any point or mixed with the gradient. A d.c. current is applied and the proteins migrate to a point where the pH equals the isoelectric point (see page 450). After separation the liquid column is gently passed through a flow analyser onto a fraction collector. The separated bands may then be dialysed to remove ampholytes and gradient solute.

Many preparative separation techniques are based on analytical methods described elsewhere in this book. Various forms of chromatography have been developed to supply pure samples for further analysis. Paper chromatography (page 296) using 3 MM or similar thick paper can be loaded heavily with sample and after a single dimensional run separated spots can be eluted. Similarly thick layers of T.L.C. media can be spread on large plates and run in the near horizontal position. The desired area is scooped out and absorbed components extracted.

Highly efficient liquid/solid and liquid/liquid chromatography column techniques have now been devised utilizing high pressures. Excellent and rapid repeatable separations are obtained using columns packed with uniform and very fine particles. High pressures are used up to 7×10^6 Nm^{-2} (1000 p.s.i.) to obtain high flow rates and prevent normal diffusion. As results are so reproducible it is possible to programme repetitive runs combining the desired fractions each time, thus accumulating a usable amount.

Similar apparatus is now available to continuously trap a peak from repeated G.L.C. runs. In this case the carrier and compound being gases must be condensed in a cooled trap.

Distillation and extraction techniques are frequently used especially vacuum fractional distillation methods and the multistage liquid/liquid extraction methods (page 39).

CHAPTER 11

AN INTRODUCTION TO BIOCHEMICAL COMPOUNDS

LIPIDS

The lipids are fats and vegetable oils, and compounds that are closely associated with them or in combination with them.

Lipids are usually classified in three groups: simple lipids, compound lipids and derived lipids.

They all have certain characteristics in common, being insoluble in water and soluble in 'fat solvents' including chloroform, benzene, ether, acetone and boiling alcohol. All are esters, or are able to form esters, with fatty acids.

The alcohol most commonly involved is glycerol (glycerine). This is a trihydric alcohol and was discovered by Scheele. Another is cholesterol, an aromatic alcohol. The acids involved are called 'fatty acids'. These are usually straight chain acids having an even number of carbon atoms, and may be saturated or unsaturated. One or two exceptions have side chains.

Saturated fatty acids—have the general formula $C_nH_{2n}O_2$, the simplest of the series being formic acid, followed by acetic acid. Table 11.1 shows some of their formulae and sources. I.U.P.A.C. terminology has not been used as trivial names are more often encountered.

TABLE 11.1

Fatty acid	R. COOH	*Source*
Butyric acid	C_3H_7	Butter
Caproic acid	C_5H_{11}	Milk, coconut
Caprylic acid	C_7H_{15}	Milk, coconut
Capric acid	C_9H_{19}	Milk, coconut
Lauric acid	$C_{11}H_{23}$	Laurel oil, spermaceti
Myrystic acid	$C_{13}H_{27}$	Nutmeg
Palmitic acid	$C_{15}H_{31}$	Common in animals
Stearic acid	$C_{17}H_{35}$	and plants
Arachidic acid	$C_{19}H_{39}$	Peanut oil
Cerotic acid	$C_{25}H_{51}$	Wool fat

Unsaturated fatty acids—contain one or more double bonds. Some of the common acids are listed below (Table 11.2).

TABLE 11.2 Some of the Unsaturated Fatty Acids

Name	*Empirical formula*	*No. of double bonds*	*Source*
Palmitoleic acid	$C_{16}H_{30}O_2$	1	Animal and plant fats
Oleic acid	$C_{18}H_{34}O_2$	1	Animal and plant fats
Linoleic acid	$C_{18}H_{32}O_2$	2	Vegetable oils (linseed)
Linolenic acid	$C_{18}H_{30}O_2$	3	Vegetable oils (linseed)
Arachidonic acid	$C_{20}H_{32}O_2$	4	Animal fats

The last three are also called essential fatty acids as they have been shown to be required in the diet of animals for normal growth and maintenance of health.

Ricinoleic acid has one double bond and a hydroxyl group branch at C 13. This is found in castor oil. Some cyclic fatty acids have been found. The unsaturated fatty acids are more reactive and their degree of unsaturation is found by reaction with iodine or bromine.

Simple Lipids

Fats are triglycerides, that is esters of fatty acids with glycerol. One molecule of glycerol reacts with three molecules of fatty acids, the general formula being

$$\begin{array}{l} CH_2O\cdot CO—R_1 \\ | \\ CH\cdot O\cdot CO—R_2 \\ | \\ CH_2O\cdot CO—R_3 \end{array}$$

The R groups are usually different in natural fats, except in beef and mutton fat which have large quantities of tristearin and tripalmitin.

Waxes are also esters of fatty acids but combined usually with a long chain alcohol such as cetyl alcohol which is found esterified with palmitic acid in spermaceti (whale fat).

General Properties of Simple Lipids

All are greasy compounds and in a pure state are mostly colourless and odourless. On hydrolysis they yield fatty acids and glycerol; this is brought about by the action of an enzyme lipase which is secreted by the pancreas. Exposure to air and light will also hydrolyse and oxidise fats. This has the effect of making edible fats 'rancid'.

Unsaturated hydrocarbons may be saturated by hydrogenation with the aid of a catalyst. Thus unpalatable vegetable oils are hydrogenated commercially to produce lard and margarine which have higher melting points.

Fats have many functions in animals and plants, they are an energy source, having twice the calorific value of carbohydrates or proteins. This energy may be stored in tissue, and forms insulation and protective layers for organs.

Compound lipids are esters of fatty acids that also contain another group or groups in the molecule.

Phospholipids are esters of glycerol but with one of the fatty acids replaced by phosphoric acid.

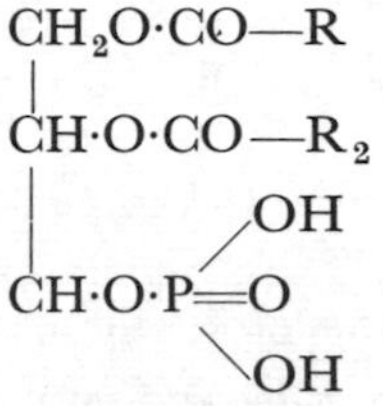

$CH_2O{\cdot}CO—R$
$|$
$CH{\cdot}O{\cdot}CO—R_2$
$|$
$CH{\cdot}O{\cdot}P(=O)(OH)_2$

Phosphatidic acid

An amine is attached to one of the free hydroxyl groups. Cephalin has an ethanolamine group and lecithin a choline group. These compounds are widely distributed in tissues, high concentrations being found in the brain. Also found in the brain and nerve tissue is sphingomyelin, a compound of fatty acids, phosphoric acid, choline and sphingosine.

Glycolipids. As the name suggests these are lipids that contain a carbohydrate, usually galactose. Cerebrosides are among these compounds and are found in the brain. On hydrolysis they yield a carbohydrate, sphingosine, and a fatty acid.

DERIVED LIPIDS

The most important group is the steroids. These include such substances as cholesterol, bile acids, sex hormones and the D vitamins. All are derived from a cyclic nucleus, the skeleton structure is indicated below.

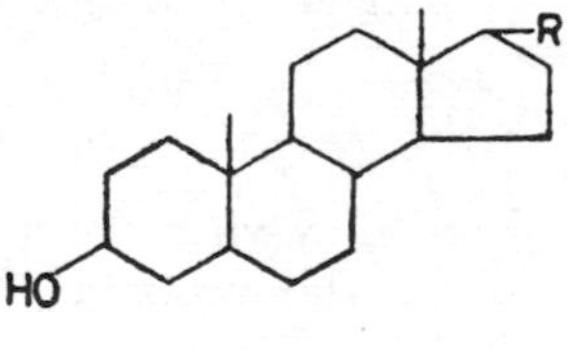

The various sterols have a different group at R and one or more double bonds. Cholesterol is found in mammals in the brain, blood, bile and egg yolk. Some gallstones are almost pure cholesterol.

Ergosterol is found in fungi and yeasts and can be extracted from ergot. Ergosterol and dihydrocholesterol may be converted to the D vitamins by the action of ultra-violet light.

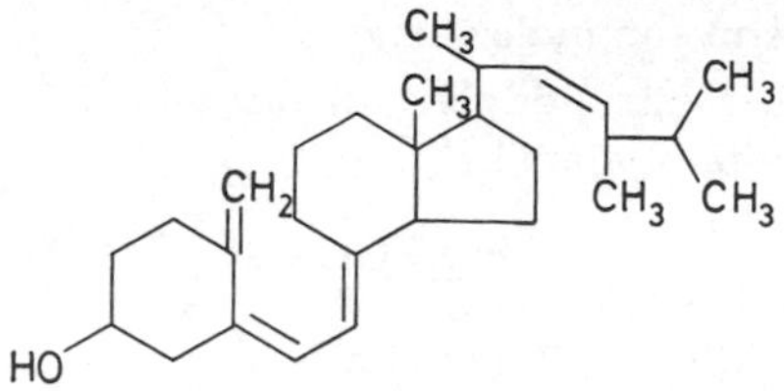

Calciferol (Vitamin D_2)

Other plant sterols are sitosterol, stigmasterol and spinasterols.

Bile acids are found in bile and are derived from cholesterol. They occur free and as salts of the acid. These acids, cholic acid, desoxycholic acid and lithocholic acid are found in combination with glycine and taurine. These salts act as soaps and emulsify fats present in the intestine.

The sex hormones, oestrogens and androgens and the hormones secreted by the adrenals, e.g. cortisone, are all sterols and are believed to be derived from cholesterol.

General Reactions

Fats are emulsified by soaps.

If a few drops of fresh, neutral olive oil are added to water and mixed, the oil will be seen to separate out. If a little sodium carbonate is added, followed by a trace of oleic acid the oil will be seen to emulsify due to the production of the soap sodium oleate. Bile salts will have the same effect.

Acrolein Test for Glycerol

To anhydrous potassium bisulphate in a test-tube add a drop of glycerol and heat. A pungent irritating odour of acrolein will be given off.

Saponification

When fats are hydrolysed they yield the individual components. If an alkali is used to do this, the alcohol is liberated and also salts of the fatty acids. The latter are called soaps if they are formed from long chain fatty acids such as palmitic, oleic and stearic acids. The soaps are soluble if the hydrolysis is carried out with sodium or potassium hydroxide. This process is called saponification and may be used as a measure of the fatty acid content of a fat. The saponification value or number of a fat is the number of milligrams of KOH required to 'neutralize' 1 g of fat. This value is obtained by back titration.

Method. Dissolve a known weight of the fat, about 1 g in 2·5 cm^3 of alcohol. Add to this 10 cm^3 of 0·5 M alcoholic KOH solution and reflux (Chapter 2), for 30 min. Similarly treat 2·5 cm^3 of alcohol and 10 cm^3 alcoholic 0·5 M KOH.

After cooling, titrate each flask against 0·5 M HCl using phenolphthalein as an indicator. The difference in titre is equivalent to the amount of fatty acid present. 0·5 M KOH = 28·05 g/dm^3 Calculate the milligrammes KOH used by 1 g of fat.

An estimate of the degree of unsaturation of the fatty acids is made by finding the amount of iodine that is taken up at the double bond position. This is called the iodine value or number and is the number of grammes of iodine taken up by 100 g of fat.

Determination of Iodine Absorption Value

Wijs method. Weigh about 0·3 g of fat into a small weighing dish and place it in a 250 cm^3 glass stoppered conical flask. Add 10 cm^3 of carbon tetrachloride to this and a similar flask for a blank. To both flasks add equal quantities of Wij's reagent, about 25 cm^3, using a burette. Stopper, mix well and leave in the dark for at least an hour. Titrate both flasks with standard 0·1 M sodium thiosulphate solution after adding 15 cm^3 of 10 per cent potassium iodide solution and about 100 cm^3 of distilled water. Use starch for the end-point and shake continuously whilst titrating to ensure that the iodine in the carbon tetrachloride layer is transferred to the aqueous layer. The weight of iodine absorbed by 100 g of fat is:

$$\frac{100 \times \text{difference in titration} \times \text{the thiosulphate factor}}{\text{Weight of fat used}}$$

1 cm^3 of 0·1 M sodium thiosulphate = 0·0127 g of iodine.

Wijs reagent: Dissolve 7·8 g iodine trichloride and 8·5 g of iodine in glacial acetic acid in separate flasks by warming. When in solution cool and transfer both to a 1 dm^3 volumetric flask and make up to volume with glacial acetic acid.

The reactions and analysis of the sterols and other derived and compound lipids will be found in the specialized literature. Cholesterol is present both free and as the ester in tissues. Total cholesterol is obtained therefore by saponification. Free cholesterol is precipitated by digitonin as the digitonide. This is then dissolved and a colour reaction applied, which is the Liebermann-Burchard reaction and is not specific for cholesterol.

Method. To a few crystals of cholesterol in a test-tube add 1 cm^3 chloroform. When solution is complete add 2 cm^3 of acetic anhydride and 2 drops of concentrated sulphuric acid. A purple colour develops changing to a deep green.

Modern methods of analysis have made it possible to separate many lipids and their products of hydrolysis. Chromatographic techniques have been used both for analytical and preparative purposes. The variation in solubility of lipids in different solvents also serves as a means of separation. Craig (Chapter 2) has developed a 'counter current' technique employing two immiscible solvents. The apparatus is designed so that many 'separator' extractions are performed simultaneously and in sequence. Great advances have been made in the use of vapour phase chromatography for analysis of complex mixtures of lipids using minute quantities. Thin layer chromatography has been found to be a very useful technique in lipid research.

CARBOHYDRATES

These important compounds are found in all living matter. They are composed of carbon, hydrogen and oxygen, the latter two being in the same proportions as water, although, in fact, they are not hydrates but aldehyde or ketone alcohols.

The simplest sugar containing the aldehyde group is hydroxyacetaldehyde

$$\begin{array}{l} CHO \\ | \\ CH_2OH \end{array}$$

and all sugars containing this group are called aldoses. The simplest ketone sugar is di-hydroxy acetone

$$\begin{array}{l} CH_2OH \\ | \\ CO \\ | \\ CH_2OH \end{array}$$

Classification

There are three general types, monosaccharides, disaccharides and polysaccharides.

Monosaccharides are further classified according to the number of carbon atoms in the molecule.

Glycerose

$$\begin{array}{l} CHO \\ | \\ CHOH \\ | \\ CH_2OH \end{array}$$

and di-hydroxy acetone

$$\begin{array}{l} CH_2OH \\ | \\ CO \\ | \\ CH_2OH \end{array}$$

are both trioses. Tetroses have 4 carbon atoms, pentoses 5, hexoses 6, heptoses 7, etc.

The most important biologically are pentoses and hexoses and these only will be considered with the exception of the triose, glycerose. All monosaccharides have one or more asymmetric carbon atoms and show optical activity. Glycerose has been chosen as an arbitrary standard, the dextro form being

$$\begin{array}{c} CHO \\ | \\ H—C—OH \\ | \\ CH_2OH \end{array}$$

and laevo form

$$\begin{array}{c} CHO \\ | \\ HO—C—H \\ | \\ CH_2OH \end{array}$$

This is their *spatial configuration*, the D and L forms respectively, referring to the position of the H and OH groups at the carbon atom next to the one that has the alcohol group attached to it.

Example:

$$\begin{array}{c} CHO \\ | \\ H—C—OH \\ | \\ HOCH \\ | \\ HCOH \\ | \\ HCOH \\ | \\ CH_2OH \end{array} \qquad \begin{array}{c} CHO \\ | \\ HOCH \\ | \\ HCOH \\ | \\ HOCH \\ | \\ HOCH \\ | \\ CH_2OH \end{array}$$

D Glucose L Glucose

This capital D or L does not mean they are necessarily dextro-rotary or laevo-rotary, this being indicated by the signs (+) or (−). For example:

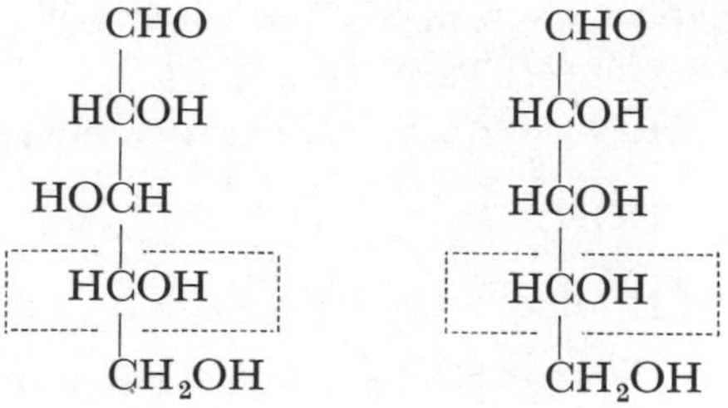

CHO
|
HCOH
|
HOCH
|
HCOH
|
CH_2OH
D (+) Xylose

CHO
|
HCOH
|
HCOH
|
HCOH
|
CH_2OH
D (−) Ribose

Most naturally occurring sugars are of the D configuration.

Pentoses

The most important is D ribose which occurs in plants and in a combined form with protein and is a constituent of nucleic acid and of coenzymes I and II (DPN, TPN). It exists in the ring form of a furan.

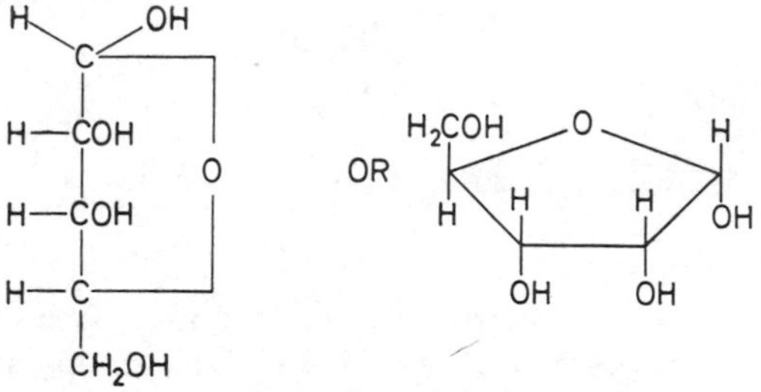

H, OH
C
H—COH
H—COH
H—C
CH_2OH
O

OR

H_2COH O H
H H H OH
OH OH

Two other important pentoses are D xylose and L arabinose

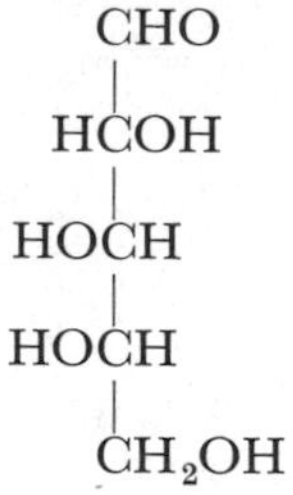

CHO
|
HCOH
|
HOCH L—Arabinose
|
HOCH
|
CH_2OH

hese are obtained from tree gums by acid hydrolysis.

Hexoses

The most important of these is glucose and is found unbound in blood, plants, grapes and honey and is the constituent of many polysaccharides. It is, in fact, obtained commercially from starch.

Glucose may be found in the ring form of a pyran

CHO
HCOH
HOCH
HCOH
HCOH
CH_2OH

D Glucose

CH_2OH
H H H
OH OH H OH
H OH

Pyranose form

Glucose is involved in one of the earliest known biochemical reactions, that of fermentation. Yeast (saccharomyces cerevisiae) is capable of converting 90 per cent of the sugar into CO_2 and alcohol.

$$C_6H_{12}O_6 \rightarrow 2CO_2 + 2C_2H_5OH$$

The final products are obtained after a series of enzyme catalysed reactions.

D *Fructose* is a keto hexose and is laevo-rotary, an old name being laevulose. It occurs in fruits, honey and in mammalian foetal blood and semen. With glucose it is a component of the disaccharide sucrose (cane sugar) and it is obtained by its hydrolysis.

CH_2OH
CO
HOCH
HCOH
HCOH
CH_2OH

D *Galactose*. This sugar in combination with glucose is synthesized into the disaccharide lactose in mammals. It is also found in vegetables in combination with other compounds.

$$
\begin{array}{c}
\text{CHO} \\
| \\
\text{HCOH} \\
| \\
\text{HOCH} \\
| \\
\text{HOCH} \\
| \\
\text{HCOH} \\
| \\
\text{CH}_2\text{OH}
\end{array}
$$

D *Mannose*. This is found chiefly bound in polysaccharides of vegetable origin called mannosans. It is obtained by hydrolysis of these and certain gums. It occurs in man attached to a number of proteins.

$$
\begin{array}{c}
\text{CHO} \\
| \\
\text{HOCH} \\
| \\
\text{HOCH} \\
| \\
\text{HCOH} \\
| \\
\text{HCOH} \\
| \\
\text{CH}_2\text{OH}
\end{array}
$$

THE DISACCHARIDES

The most important are lactose, maltose and sucrose.

Lactose is found in mammalian milk and occasionally will be present in urine of lactating women and anyone who has consumed a large quantity of milk in their diet. Lactose is not fermented by brewers' yeast but is by bacteria of the genus *lactobacillus* and also by *streptococcus lactis*, the end products being lactic acid and alcohol. It is produced commercially from the waste products of cheesemaking.

Maltose is a combination of two molecules of glucose. It is formed by hydrolysing starch with the enzyme amylase (diastase) and is produced during the malting or germinating of grain. Yeast contains an enzyme maltase, which hydrolyses maltose to glucose.

Sucrose, or cane sugar, is the sugar found extensively in plants, especially sugar cane and sugar beet. It is formed of glucose and fructose and is hydrolysed by the enzyme invertase, found in yeast. Sucrose is dextro-rotary but on hydrolysis becomes laevo-rotary, and the enzymic reaction may be followed in a polarimeter. This phenomenon is called inversion and the resulting mixture of glucose and fructose is called 'invert sugar'.

POLYSACCHARIDES

These carbohydrates are very complex and have high molecular weights of the order of 50 000. They have certain similarities to proteins in that they are in chain form and are insoluble, or form colloidal suspensions. The structural units are simple sugars, usually glucose. The chain may be either branched or unbranched depending on the polysaccharide. They occur extensively in the vegetable kingdom in fruits, roots and cereals.

Starch $(C_6H_{10}O_5)_x$

Starch is the form in which carbohydrate is stored in plants. It is made up of glucose units, these being grouped in two forms called amylose and amylopectin in the ratio 1:3. Amylose is in the form of a long helical chain and amylopectin has a highly branched chain structure. 'Soluble' starch is produced by heating in dilute acid solution. The amylose and amylopectin may then be fractionally precipitated by alcohol. The group of enzymes called amylases found in saliva and secreted by the pancreas break down the starch components to dextrin and then maltose. Maltose is then hydrolysed to glucose by the enzyme maltase.

Glycogen

This performs a similar task to starch in that it is a storage carbohydrate, but differs in that it is only found in animals. It is made up of glucose units in a highly branched molecule. It is stored in the liver and is also present in muscular tissue, where it may rapidly be hydrolysed to glucose to supply energy when the muscles are used. It is soluble in cold water and has a high resistance to attack by alkali, but is readily hydrolysed by acids and amylase.

Cellulose

This is the most common polysaccharide being the main constituent of plant cell walls. It is made up of glucose units in a long chain molecule. Cellulose is insoluble in water and dilute acids or alkalies and, in fact, is highly resistant to any chemical action. It is only soluble in special solvents, one being Schweitzer's reagent (ammoniacal copper hydroxide) from which it is immediately precipitated by acid. It is soluble in zinc chloride and hydrochloric acid (1 : 2 by weight) and sodium hydroxide and carbon disulphide. It can be hydrolysed by concentrated sulphuric acid to glucose. After the initial treatment with strong acid the solution should be diluted with water and boiled.

Many modern fibres, papers and transparent sheet materials are produced from cellulose derivatives, such as rayon, celluloid, cotton wool and filter paper. The latter is almost pure cellulose.

Although man eats large quantities of cellulose in his diet only a minute quantity is digested, and even this is believed to be due solely to the action of certain intestinal bacteria. Herbivorous animals are capable of digesting large quantities of cellulose as the flora of the intestine and rumen hydrolyse it by the action of the enzyme cellulase to the disaccharide cellobiose. This has a parallel in the carnivorous animals in the break-down of starch to maltose. Cellobiose is then hydrolysed to glucose by the enzyme cellobiase.

The soil is full of organisms capable of utilizing cellulose, such as species of cellvibria, cytophagia and clostridia and moulds. A well-cared for heap of vegetable waste seeded with soil may be completely decomposed within three months.

Inulin

This is a polysaccharide found in plants such as dahlias and dandelions, and is made up of fructose units.

Chitin

This is an interesting compound as it has a structural function like cellulose, as it forms the shells of crustaceans. It is made up of glucosamine units.

Other important sugar compounds are the mucopolysaccharides. These are found in tissues often acting as a lubricant, such as in the synovial fluid. They are usually viscous and are complex in structure, being composed of sugars together with hexosamines and uronic acids. The latter are the oxidation products of a sugar.

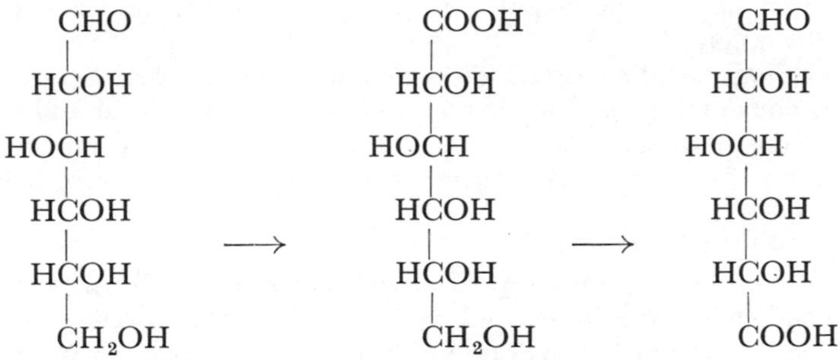

$$
\begin{array}{c}
\mathrm{CHO}\\
|\\
\mathrm{HCOH}\\
|\\
\mathrm{HOCH}\\
|\\
\mathrm{HCOH}\\
|\\
\mathrm{HCOH}\\
|\\
\mathrm{CH_2OH}\\
\text{Glucose}
\end{array}
\longrightarrow
\begin{array}{c}
\mathrm{COOH}\\
|\\
\mathrm{HCOH}\\
|\\
\mathrm{HOCH}\\
|\\
\mathrm{HCOH}\\
|\\
\mathrm{HCOH}\\
|\\
\mathrm{CH_2OH}\\
\text{Gluconic Acid}
\end{array}
\longrightarrow
\begin{array}{c}
\mathrm{CHO}\\
|\\
\mathrm{HCOH}\\
|\\
\mathrm{HOCH}\\
|\\
\mathrm{HCOH}\\
|\\
\mathrm{HCOH}\\
|\\
\mathrm{COOH}\\
\text{Glucuronic acid}
\end{array}
$$

Examples of mucopolysaccharides are the tree gums, such as gum arabic and gum tragacanth, hyaluronic acid found in the vitreous humour of the eye, and heparin in the blood. These compounds are often associated with proteins.

GENERAL REACTIONS AND QUALITATIVE TESTS

Many tests and their modifications are used for carbohydrates. Those given here will be sufficient to identify those commonly met.

In acid solution all carbohydrates will give a positive *Molisch reaction.* The carbohydrate is converted to a furfural derivative

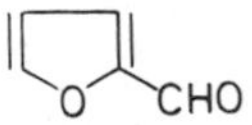

O CHO

and this will give a coloured solution with certain phenolic compounds such as naphthol, indole, resorcinol, and thymol.

Method

To 1–2 cm^3 of sugar solution in a $7{\cdot}5 \times 1$ cm test-tube add a few drops of a 5 per cent solution of naphthol in ethanol, mix, then carefully pour in about 1 cm^3 of concentrated sulphuric acid. A purple ring will develop at the interface. This reaction is not specific and aldehydes and certain acids will give a positive, so will combined carbohydrates such as glucoproteins and polysaccharides.

All monosaccharides and many disaccharides are reducing agents in alkaline solution. They are capable of reducing metallic oxides of Cu, Ag, Hg, Fe and potassium ferricyanide to ferrocyanide. Two common reagents are used, Fehling's solution and Benedict's modification.

Fehling's solution deteriorates and must be checked before use. It is not suitable for detecting sugar in urine as uric acid and other reducing compounds will give a positive reaction and ammonium salts interfere. It is more sensitive than Benedict's modification.

Benedict's test for reducing sugars

To 5 cm^3 of Benedict's qualitative reagent add 8 drops (approx. 0·5 cm^3) of sugar solution and mix. Boil for 2 min, a red or yellow precipitate of cuprous oxide will appear. The variation in colour is due to the particle size.

Benedict's qualitative reagent

Dissolve with heat 173 g sodium citrate and 100 g sodium carbonate anhydrous in 800 cm^3 of water. Cool. Dissolve separately 17·3 g copper sulphate in 100 cm^3 of water. Add the latter to the former slowly whilst stirring.

Barfoed's test for monosaccharides

This reduction test is performed in acid solution. To 5 cm^3 of Barfoed's solution add 8 drops (0·5 cm^3) of sugar solution. Boil for 30 sec. The yellow or red copper oxide precipitate will form. If not immediately apparent, stand for 15 min before assuming a negative test. Chlorides interfere with this reaction.

Barfoed's reagent

Dissolve 13·3 g copper acetate in 200 cm^3 of water and add 1·8 cm^3 of glacial acetic acid.

Tests for Pentoses

Bial's test

To 5 cm^3 of Bial's reagent add 2–3 cm^3 of sugar solution, and gently warm. When bubbles rise to the surface cool under the tap, a blue-green colour or precipitate appears.

Bial's reagent

Dissolve 1·5 g orcinol in 500 cm^3 concentrated hydrochloric acid. Add 20 drops of a 10 per cent ferric chloride solution.

Tauber's benzidine test

To 1 cm^3 of a 4 per cent benzidine solution in glacial acetic acid add 2 drops of the sugar solution. Quickly bring to the boil and immediately cool under the tap. A violet colour develops. N.B. CAUTION. Benzidine is highly carcinogenic. This test is given only for historic interest. Benzidine is now virtually unobtainable because of this hazard.

Seliwanoff's test for fructose (and all ketoses)

To 5 cm^3 of Seliwanoff's reagent add 5 drops of sugar solution. Bring to the boiling point. A red colour develops. Prolonged heating will hydrolyse disaccharides and other monosaccharides will eventually give a colour.

Seliwanoff's reagent

0·05 per cent resorcinol in concentrated hydrochloric acid. 50 per cent v/v hydrochloric acid is also used.

Osazone Formation

The monosaccharides and the reducing disaccharides form osazones when reacted with phenylhydrazine. It is possible by examination of these crystals to identify some of the common sugars. The easiest to prepare are glucosazone, lactosazone and maltosazone. Fructose and mannose give the same osazone as glucose due to similarities in molecular structure.

Phenylhydrazine reaction

To 5 cm^3 of sugar solution add a spatula end of both phenylhydrazine and sodium acetate ($3H_2O$) or a 1–2 mixture by weight. Dissolve and then place in a boiling water bath for 30 min. Allow to cool slowly. Yellow crystals will form, compare these under a microscope using a low power objective with known osazones prepared simultaneously.

Carbohydrate Fermentation

Another simple distinguishing test depends on the ability of yeasts to ferment some sugars, producing generally alcohol and carbon dioxide. The appearance of the latter indicates a positive reaction. Lactose is not fermented making it possible to distinguish it from glucose in urine.

Fermentation Test

To 10 cm^3 of sugar solution in a 25 cm^3 screw-capped bottle add a small piece of compressed bakers' yeast. Make a suspension by agitation then drop in an inverted small test-tube or ignition tube. Screw on the cap and by inverting the bottle fill the test-tube with suspension. Set the bottle upright in an incubator or warm bath at 37–40°C. If fermentation sets in gas will be seen to collect in the test-tube.

The fermentation of sugars by micro-organisms is used as a means of identifying and typing them.

Iodine test for starch

To a few cm³ of starch solution add a few drops of dilute iodine solution. A deep blue colour will develop. If this test is performed on glycogen a red colour develops. As starch is hydrolysed the iodine test gives colours ranging from blue through red to yellow, becoming colourless when hydrolysis is complete. Starch is used as a sensitive test for free iodine in solution improving the end-point of iodimetric titrations.

Cellulose and inulin do not produce a colour with iodine.

Quantitative Analysis of Carbohydrates

There are several excellent methods for reducing sugars based on the reduction of cupric ions or ferricyanide. The cuprous or ferrous compound is then reacted with iodine and the excess iodine found by titration with sodium thiosulphate. The test titration is then subtracted from the blank titration. One of these methods for blood glucose will be found on page 462.

In general biochemical carbohydrate analysis, it is preferable to use different methods as proteins and other substances encountered have reducing properties. Methods have been devised that are based on Bial's reaction and the orcinol reaction. One of the most useful methods involves the anthrone reaction. This depends on the production of furfural derivatives. Many variations have appeared in the literature.

Anthrone reaction (J. Kahan*)

2 cm³ of deproteinized solution is placed in a boiling tube ($15 \times 2\frac{1}{2}$ cm T.T.) and 0·5 cm³ anthrone reagent added. From a burette 6 cm³ of concentrated sulphuric acid is carefully run into the tube to form a lower layer. Mix and place in a boiling water bath for 3 min. Cool and leave 20 min then read at 625 nm in a spectrophotometer.

---

* *Archs. Biochem. Biophys.* (1953–54) 408.

It is important to ensure the same treatment of each tube or colour intensities will vary. Any precipitated protein should be removed by centrifugation before commencing test as cellulose fibres from filter paper may cause considerable error.

Anthrone reagent: 2 per cent anthrone in ethyl acetate.

Another simple method is that devised by Dubois *et al.**

Method. 1 or 2 cm^3 of sugar solution containing up to 70μg of sugar are placed in a 15 mm test-tube or absorptiometer tube. Add 1 cm^3 of 5 per cent phenol in distilled water. 5 cm^3 of concentrated sulphuric acid is then run in rapidly from a fast flowing pipette or burette ensuring immediate mixing. Stand 10 min, then shake the tubes and immerse them in a bath at 30°C for 20 min. A yellow orange colour develops. Measure absorption in a spectrophotometer at 490 nm for hexoses and 480 nm for pentoses and uronic acids. As in the previous method, traces of cellulose will give false readings.

Like proteins polysaccharides may be precipitated by salts and alcohols, e.g. starch is precipitated by 50 per cent saturation with ammonium sulphate and glycogen by 100 per cent saturation.

PROTEINS

Proteins are the most important class of biological compounds and are very complex in structure. In recent years, great advances have been made in the knowledge of their structure and function due to the discovery and improvement of physical methods of analysis.

Proteins are organic compounds made up of carbon, hydrogen, oxygen and nitrogen and occasionally sulphur.

The molecular weights are large and vary from several thousand to several million, special techniques being employed to ascertain them. When completely hydrolysed by boiling with strong hydrochloric acid overnight, proteins are broken down to amino acids. Some amino acids may also be partially decomposed. These are the building blocks that all proteins are made of and there are some 30 known amino acids occurring in nature.

Amino Acids

Naturally occurring amino acids are all α amino acids, and have the general formula

$$\begin{array}{c} \text{H} \\ | \\ \text{R—C—COOH} \\ | \\ \text{NH}_2 \end{array}$$

* *Analyt. Chem.* **28** (1956) 350.

two exceptions being the heterocyclic imino acids, proline and hydroxyproline. With the exception of glycine all are capable of forming isomers and are optically active, but with rare exceptions all naturally occurring amino acids are of the L configuration. The exceptions are found in certain micro-organisms and their products.

The amino acids are most easily classified in these groups, neutral, acid and basic, the neutral group being further sub-divided.

Neutral Amino Acids

A. Aliphatic amino acids

1. Glycine (Amino acetic acid)

This is the simplest and in fact the first discovered

$$\begin{array}{ccc} & \mathrm{NH_2} & \\ & | & \\ \mathrm{H}— & \mathrm{C} & —\mathrm{COOH} \\ & | & \\ & \mathrm{H} & \end{array}$$

2. Alanine (α aminopropionic acid)

$$\begin{array}{ccc} & \mathrm{NH_2} & \\ & | & \\ \mathrm{CH_3}— & \mathrm{C} & —\mathrm{COOH} \\ & | & \\ & \mathrm{H} & \end{array}$$

3. Serine (α amino β hydroxy propionic acid)

$$\begin{array}{ccc} & \mathrm{NH_2} & \\ & | & \\ \mathrm{HO—CH_2}— & \mathrm{C} & —\mathrm{COOH} \\ & | & \\ & \mathrm{H} & \end{array}$$

4. Threonine (α amino β hydroxy butyric acid)

$$\begin{array}{cccc} & \mathrm{H} & \mathrm{NH_2} & \\ & | & | & \\ \mathrm{CH_3}— & \mathrm{C}— & \mathrm{C} & —\mathrm{COOH} \\ & | & | & \\ & \mathrm{OH} & \mathrm{H} & \end{array}$$

5. Valine (α amino isovaleric acid)

$$\begin{array}{l} CH_3 \quad\quad NH_2 \\ \quad \diagdown \quad\quad\quad | \\ \quad\quad CH-C-COOH \\ \quad \diagup \quad\quad\quad | \\ CH_3 \quad\quad\;\; H \end{array}$$

6. Leucine (α amino isocaproic acid)

$$\begin{array}{l} CH_3 \quad\quad\quad\quad\quad NH_2 \\ \quad \diagdown \quad\quad\quad\quad\quad\quad | \\ \quad\quad CH-CH_2-C-COOH \\ \quad \diagup \quad\quad\quad\quad\quad\quad | \\ CH_3 \quad\quad\quad\quad\quad\;\; H \end{array}$$

7. Iso-leucine (α amino γ methyl valeric acid)

$$\begin{array}{l} CH_3-CH_2 \quad\quad NH_2 \\ \quad\quad\quad \diagdown \quad\quad\quad | \\ \quad\quad\quad\quad CH-C-COOH \\ \quad\quad\quad \diagup \quad\quad\quad | \\ \quad\quad\quad CH_3 \quad\quad H \end{array}$$

Aromatic Amino Acids

8. Phenyl alanine (α amino β phenyl propionic acid)

$$\begin{array}{c} \quad\quad\quad\quad NH_2 \\ \quad\quad\quad\quad | \\ C_6H_5-CH_2-C-COOH \\ \quad\quad\quad\quad | \\ \quad\quad\quad\quad H \end{array}$$

9. Tyrosine (α amino β hydroxyphenyl-propionic acid)

$$\begin{array}{c} \quad\quad\quad\quad\quad NH_2 \\ \quad\quad\quad\quad\quad | \\ HO-C_6H_4-CH_2-C-COOH \\ \quad\quad\quad\quad\quad | \\ \quad\quad\quad\quad\quad H \end{array}$$

Sulphur Containing Amino Acids

10. Cysteine (amino thiolpropionic acid) and Cystine (or di-cysteine)

```
                    NH2
                    |
 H   S—CH2—C—COOH
                    |
                    H

                    NH2
                    |
 H   S—CH2—C—COOH
                    |
                    H
```

Cysteine is oxidized to cystine on protein hydrolysis but both are believed to be individually found in proteins.

11. Methionine

```
                                   NH2
                                   |
CH3—S—CH2—CH2—C—COOH
                                   |
                                   H
```

Heterocyclic Amino Acids

12. Tryptophan (α amino β indolepropionic acid)

```
                            NH2
                            |
 (indole ring)—CH2—C—COOH
        NH                  |
                            H
```

13. Proline (α pyrrolidine carboxylic acid)

```
CH2—CH2
|       |
CH2   CH—COOH
   \  /
    NH
```

14. Hydroxyproline

$$
\begin{array}{lll}
HO{-}CH & {-} & CH_2 \\
\quad | & & | \\
\quad CH_2 & & CH{-}COOH \\
\quad \diagdown & & \diagup \\
 & NH &
\end{array}
$$

Acidic Amino Acids

15. Aspartic acid (α amino succinic acid)

$$
\begin{array}{c}
\qquad\qquad\qquad\quad NH_2 \\
\qquad\qquad\qquad\quad | \\
HOOC{-}CH_2{-}C{-}COOH \\
\qquad\qquad\qquad\quad | \\
\qquad\qquad\qquad\quad H
\end{array}
$$

The amide asparagine is also present in proteins

$$
\begin{array}{c}
\qquad\qquad\qquad\quad NH_2 \\
\qquad\qquad\qquad\quad | \\
H_2NOC{-}CH_2{-}C{-}COOH \\
\qquad\qquad\qquad\quad | \\
\qquad\qquad\qquad\quad H
\end{array}
$$

16. Glutamic acid (α amino glutaric acid)

$$
\begin{array}{c}
\qquad\qquad\qquad\qquad\qquad NH_2 \\
\qquad\qquad\qquad\qquad\qquad | \\
HOOC{-}CH_2{-}CH_2{-}C{-}COOH \\
\qquad\qquad\qquad\qquad\qquad | \\
\qquad\qquad\qquad\qquad\qquad H
\end{array}
$$

and the amide glutamine

$$
\begin{array}{c}
\qquad\qquad\qquad\qquad\qquad NH_2 \\
\qquad\qquad\qquad\qquad\qquad | \\
H_2NOC{-}CH_2{-}CH_2{-}C{-}COOH \\
\qquad\qquad\qquad\qquad\qquad | \\
\qquad\qquad\qquad\qquad\qquad H
\end{array}
$$

Basic Amino Acids

17. Arginine (α amino γ guanido-valeric acid)

$$
\begin{array}{ccccccc}
 & & & & & NH_2 & \\
 & & & & & | & \\
NH_2{-} & C & {-}NH{-}CH_2{-} & CH_2{-} & CH_2{-} & C & {-}COOH \\
 & \| & & & & | & \\
 & NH & & & & H &
\end{array}
$$

18. Lysine (1:5 diamino-caproic acid)

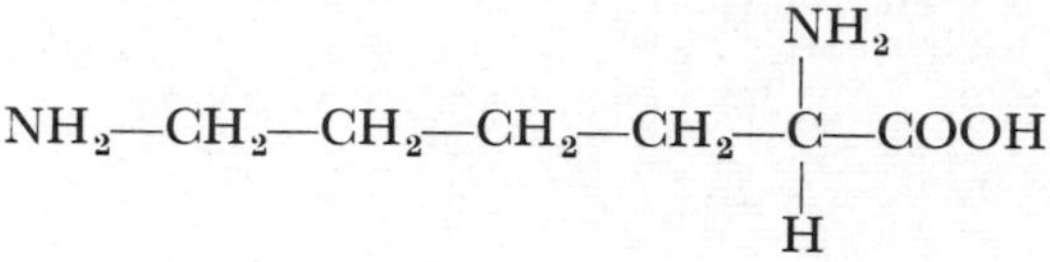

$$\begin{array}{ccccccccccccc} & & & & & & & & & & NH_2 & & \\ & & & & & & & & & & | & & \\ NH_2 & — & CH_2 & — & CH_2 & — & CH_2 & — & CH_2 & — & C & — & COOH \\ & & & & & & & & & & | & & \\ & & & & & & & & & & H & & \end{array}$$

19. Histidine (α amino iminazolyl-propionic acid)

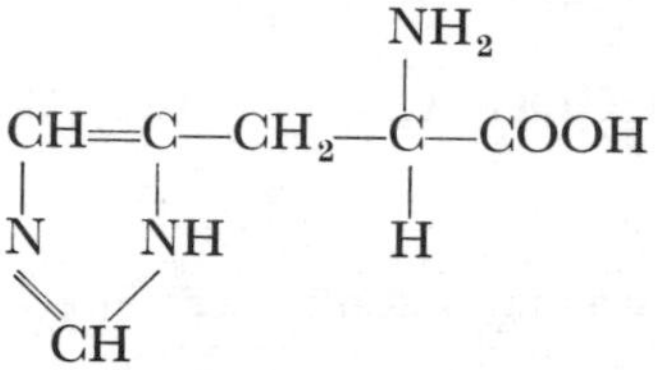

$$\begin{array}{ccccccccc} & & & & & & NH_2 & & \\ & & & & & & | & & \\ CH & = & C & — & CH_2 & — & C & — & COOH \\ | & & | & & & & | & & \\ N & & NH & & & & H & & \\ & \backslash\backslash & / & & & & & & \\ & CH & & & & & & & \end{array}$$

These are the normal amino acids found in protein structure. There are others occurring in nature but not necessarily associated with proteins.

In the protein molecule the amino acids are linked together by the peptide bond. This is the result of the reaction of the carboxyl group of one amino acid with the amino group of another

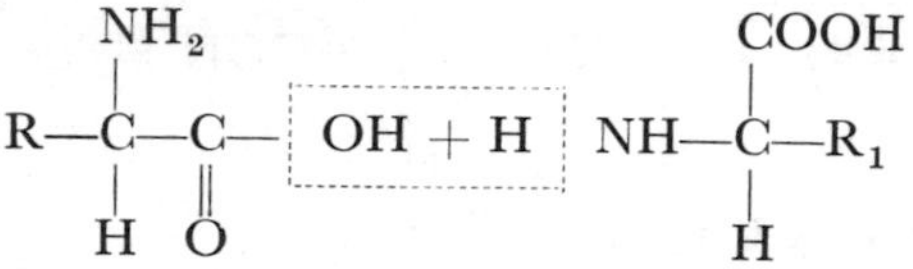

$$\begin{array}{cccccccccccc} & & NH_2 & & & & & & & COOH & & \\ & & | & & & & & & & | & & \\ R & — & C & — & C & — & \boxed{OH + H} & NH & — & C & — & R_1 \\ & & | & & \| & & & & & | & & \\ & & H & & O & & & & & H & & \end{array}$$

giving

$$\begin{array}{ccccccccccc} & & NH_2 & & & & & & COOH & & \\ & & | & & & & & & | & & \\ R & — & C & — & C & — & NH & — & C & — & R_1 \\ & & | & & \| & & & & | & & \\ & & H & & O & & & & H & & \end{array}$$

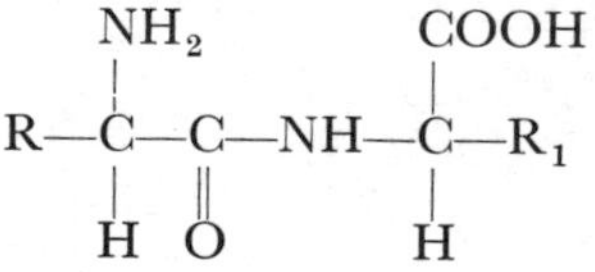

This compound is a dipeptide and if more amino acids are attached at the amino and carboxyl group a chain will be formed, a polypeptide, which is in fact the backbone structure of all proteins. These compounds are obtained by the partial hydrolysis of proteins and larger fragments are called proteoses and peptones.

An example of a peptide found in blood and tissues is glutathione

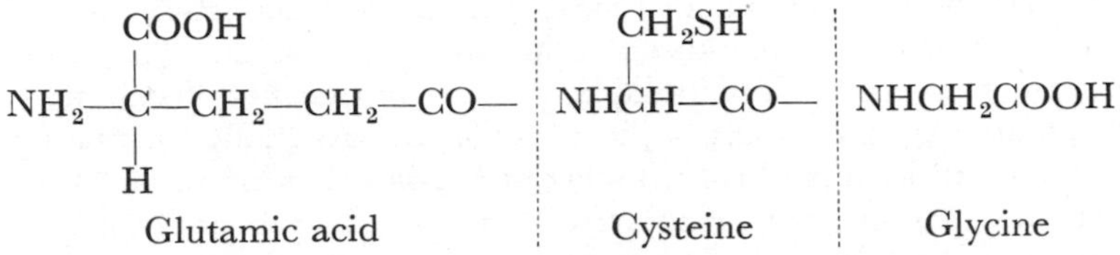

$$
\begin{array}{l|l|l}
\quad\;\; COOH & \quad CH_2SH & \\
\quad\;\;\; | & \quad\;\; | & \\
NH_2\text{—}C\text{—}CH_2\text{—}CH_2\text{—}CO\text{—} & NHCH\text{—}CO\text{—} & NHCH_2COOH \\
\quad\;\;\; | & & \\
\quad\;\; H & & \\
\text{Glutamic acid} & \text{Cysteine} & \text{Glycine}
\end{array}
$$

Proteins then are large molecules made up of amino acids. Some have non-amino groups attached to the chain as well and this gives the classification, simple proteins and conjugated proteins. These are subdivided as follows:

Simple Proteins (amino acid structure only)

Albumins, globulins, glutelins, prolamines, albuminoids

Conjugated Proteins (simple proteins combined with other groups)

Nucleoproteins	plus nucleic acid
Glucoproteins	plus carbohydrate
Phosphoproteins	plus phosphoric acid (Nucleoproteins also contain phosphorus)
Chromoproteins	plus a group containing a metal such as haem, chlorophyll.
Lipoproteins	plus fats

One further group may be included called the derived proteins. These are large protein fragments derived by either partial hydrolysis or the removal of the prosthetic group from the conjugated proteins, or the denaturation of the protein.

Another classification is possible based on the structure of the protein, two general types being known. First the fibrous proteins that make up muscle tissue and tendons. In these the polypeptide chain is believed to exist in a spiral form and is possible of extension. The second type is the globular proteins such as the albumins. These are usually water soluble but very little as yet is known of their structure, which is thought to be layers of peptide chains linked together by disulphide bonds between cysteine units.

General Reactions of Proteins

Proteins, having a large molecular weight, form colloidal solutions. They are unable to pass through some membranes such as parchment and collodion. This is an advantage in some purification methods as low molecular impurity can be dialysed out.

Proteins are precipitated in several ways:

1. By salts of heavy metals such as lead acetate and silver nitrate, also by magnesium, ammonium and sodium sulphates. The latter two salts are commonly used in protein purification techniques. Fractionation of a mixture is possible by gradually increasing salt concentration, and centrifuging precipitates as they are formed. Generally the protein is unaltered and will readily redissolve.

2. By acids. A drop of strong mineral acid will give a heavy precipitate but with addition of more acid this dissolves and hydrolysis takes place. Other acids react differently, some forming insoluble salts. Phosphotungstic acid, tannic acid and picric acid are examples. Two very common precipitants are trichloracetic acid and sulphosalicylic acid, the latter being used clinically for the detection of protein in urine.

3. By solvents miscible with water such as alcohols and acetone. These may tend to denature proteins but this is minimized by performing the precipitation at 0°C.

4. By heat. This depends on the pH of the solution to some extent causing coagulation, denaturing the protein and making it insoluble.

Proteins and amino acids behave both as acids and bases because they have carboxyl and amino groups together. That is they are amphoteric substances. Protein solutions may also be regarded as buffers as ionization occurs, and the pH of minimum ionization where the concentration of acidic ions equal the basic ions, is called the isoelectric point. It is at this point that proteins are most readily coagulated by heat.

Biuret test

The colour reactions given by a protein depend to a large extent on the amino acids in its make-up, but one test is given by all proteins as it depends on the presence of two peptide links. This is the biuret reaction.

Biuret is the substance formed when urea is heated.

$$2CO\,(NH_2)_2 \longrightarrow NH_2\text{—}CO\text{—}NH\text{—}CO\text{—}NH_2$$

To a few cm^3 of a dilute protein solution add an equal volume of 10 per cent NaOH. Mix, then add dropwise 0·5 per cent copper sulphate solution until a purple colour develops. Do not confuse this with the Cu colour given by copper hydroxide precipitate that will appear if too much copper is added. The colour will vary with size of protein molecule. Peptides give a pink colour.

Ninhydrin reaction

Ninhydrin (Triketohydrindene hydrate) will give a colour with proteins and its products of hydrolysis. Dissolve a few milligrammes of ninhydrin in about 10 cm^3 of water. Prepare a fresh solution daily as it will not keep. Add two drops to every cm^3 of protein solution, and boil. A blue or purple colour will develop. During the reaction both ammonia and CO_2 are liberated.

The tests that follow depend on the presence of certain amino acids.

Xanthoproteic reaction

This test will be positive if amino acids containing a benzene ring are present such as tyrosine, phenylalanine and tryptophan.

To 3–5 cm^3 of protein solution add 1 cm^3 of concentrated nitric acid. Mix and boil. A yellow colour will develop. Cool and carefully add excess ammonia. The yellow colour will gradually turn to orange.

Millon's reaction

This requires the presence of a hydroxybenzene radical and is therefore specific for tyrosin in amino acid mixtures.

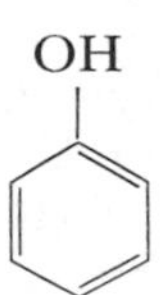

Millons reagent is a digest of mercury with nitric acid. Modifications of this reagent are now preferred as they are less subject to interference by high salt concentrations.

To 1–5 cm^3 protein solution add 1 cm^3 of a 15 per cent solution of mercuric sulphate in 15 per cent v/v sulphuric acid. Place in a boiling water bath for 10 min, cool, and add 1 cm^3 of a 1 per cent sodium nitrite solution. A red colour will develop if positive.

Hopkins-Cole Reaction

This is a test for the indole nucleus and will therefore give a positive if tryptophan is present.

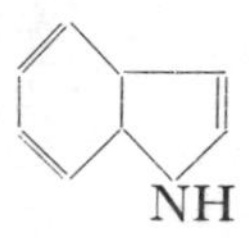

To 1–2 cm^3 of protein solution add an equal volume of glyoxylic acid solution (Hopkins-Cole Reagent). Carefully run in concentrated sulphuric acid so that it layers underneath. A purple-violet ring will develop at the interface.

Sulphur test

The presence of cystine, cysteine and methionine can be detected by converting the sulphur in their molecule into a sulphide, and testing for this with lead acetate solution.

Add an equal volume of 10 M NaOH to a protein solution and boil for a few minutes. Cool and add a few drops of lead acetate solution, a black or brown colour indicates sulphur.

Quantitative Methods of Protein and Amino Acid Analysis

There are three common methods of estimation of proteins. The first two are absorptiometric methods based on the Biuret and ninhydrin colours, and the third less accurate method a direct measurement of u.v. absorption.

Quantitative Biuret Method (M. Dittebrandt)*

This method is sensitive, and maximum protein content used is 0·5 mg.

Prepare 1 cm^3 quantities of protein, standards and blank. Add 1·5 cm^3 biuret reagent to each tube. Incubate in a water bath at 37°C for 30 min. Cool the tubes and compare the colours at 555 nm in a spectrophotometer.

There are many variations of this method and it is readily scaled up. If an absorptiometer is used an Ilford 625 spectrum filter is suitable.

* *Am. J. clin. Path.* **18** (1948) 439.

Biuret reagent

Dissolve 9 g sodium potassium tartrate in 400 cm^3 of 0·2 M NaOH. Add 3 g ground hydrated copper sulphate, followed by 5 g potassium iodide. Make up to 1 dm^3 with 0·2 M CO_2 free NaOH.

Quantitative Ninhydrin Method

Dispense 0·1 cm^3 of protein sample, standards and blank into small glass-stoppered centrifuge tubes. Add 0·5 cm^3 ninhydrin reagent, mix, stopper and immerse in boiling water for 20 min. Cool, and add 2 cm^3 of 50 per cent *n*-propanol in water, mix by inversion. If any deposit is formed, centrifuge the tubes. Pour off into cuvettes and compare the absorbances at 570 nm.

Reagents

Ninhydrin reagent: Dissolve 0·4 g ninhydrin in 12·5 cm^3 peroxide free methyl cellosolve. Mix with 25 cm^3 of 0·2 M, pH 5 citrate buffer containing 40 mg Stannous chloride ($SnCl_2 \cdot 2H_2O$).

Citrate buffer: 2·101 g citric acid monohydrate plus 20 cm^3 M NaOH, made up to 50 cm^3 with water. Store in a refrigerator at 2°C. When required for use dilute 1 : 1 with water giving pH 5.

Methyl cellosolve must be redistilled before use.

Many workers have measured quantitatively the ammonia that is formed in this reaction. Van Slyke and his associates have estimated the carbon dioxide liberated using a manometric method.

The amino acid analysis of proteins is of great importance and three methods are more generally employed. The first is the use of ion exchange columns. This method has been fully developed by Moore and Stein and completely automatic apparatuses are now on the market making it possible to perform an analysis in 2 h.

The second method is the use of high voltage electrophoresis or paper chromotography. The amino acids are located with ninhydrin and the colour extracted quantitatively into solutions and these read in an absorptiometer. The results may be obtained within 24 h.

The third method is by microbiological assay. It has been found that certain organisms require amino acids for growth. To determine the amount of an amino acid in a protein hydrolysate a growth medium is prepared that is lacking in this particular constituent. A known quantity of the hydrolysate is added, the medium sterilized and then inoculated with the micro-organism. After a suitable incubation period growth is either measured turbidometrically, or the acid produced by the bacterium is titrated volumetrically. Readings are compared with standards and blanks prepared simultaneously.

Finally there is one method which determines the quantity of whole protein present. This depends on light absorption at 280 nm due to the presence of tyrosine, phenylalanine or tryptophan in the molecule. The other amino acids do not exhibit absorption above 250 nm. If a protein has one or more of these amino acids its concentration may be measured providing no interfering substances are present. At 205 nm absorption is due to the peptide bonds.

ANALYTICAL METHODS FOR BIOLOGICAL MATERIAL

Considerable research has been done on perfecting methods of analysis on physiological material, especially those of clinical importance. As this book is not specifically aimed at medical technology but is more broadly biological, the methods that follow will include one that is well-tried and at least one other which may find application in analysis of other tissues and species. In this respect values of normal constituents of the blood listed will contain figures for species other than man.

Physiological Fluids

Collection

When whole blood is required for analysis it must be mixed with an anti-coagulant immediately on withdrawal. This is usually dried on to the walls of the container. An anti-coagulant is chosen that will not interfere with the analysis. The two most commonly used are potassium oxalate and heparin using 2 mg/cm^3 and 0·2 mg/cm^3 respectively. Other anti-coagulants are sodium citrate, sodium oxalate and sodium fluoride, the latter also acting as a preservative, and sodium ethylene-diamine-tetra-acetate (EDTA).

In man, blood is usually drawn from a vein in the arm, made prominent by the use of a tourniquet and clenching and unclenching of the fist. The area is swabbed with alcohol and allowed to dry. The needle is laid along the vein with the bevel uppermost and using a thumb to prevent the vein moving the needle is inserted. As soon as the blood flows, the tourniquet is removed and when the required amount has been collected, the puncture covered with clean cotton-wool and the needle removed. Small quantities are obtained by puncturing the finger, the back of the thumb just below the nail or the lobe of an ear. 0·2 cm^3 samples are easily obtained by this method. A blood pipette must be used, this is calibrated to contain the volume. A glass mouthpiece is connected to it by rubber tubing. Blood is drawn to the mark and the pipette wiped on the outside with tissue. The contents are then blown into isotonic saline solution, the pipette being washed by filling and emptying with the solution. Isotonic sodium sulphate is commonly used as this interferes less with the analyses. It is made up by dissolving 13·2 g of anhydrous sodium sulphate in 1 dm^3 of distilled water.

When serum is required blood is immediately placed into a clean screw-capped bottle. The bottle is placed in a stand, or if there is a large number, in a wire basket, set at an angle of 30/45° from the horizontal, and left at room temperature until the clot has formed. The bottles are placed in a refrigerator at +2–4°C overnight, when the serum may be poured off. If separation proves difficult, the bottle can be centrifuged. It is as well to select bottles for centrifugation that have first been given a test run at 2000 $\times$ *g* and then used at 1500 $\times$ *g* this being sufficient to sediment the clot and erythrocytes.

Blood clotting is a complicated process and as yet not fully understood. The clot is made up of erythrocytes trapped in a fine matrix of a protein material, fibrin. The formation of fibrin is caused by a series of reactions. Tissues contain a substance called thromboplastin which is liberated at the site of injury. Thromboplastin with calcium ions converts prothrombin that is circulating in the blood into the enzyme thrombin. This catalyses the production of fibrin from its precurser fibrinogen. Thrombin is very active and very little is required. Other factors are involved such as platelets but their role is not yet fully understood. Prevention of clotting is accomplished by interfering or inhibiting this chain of events. Oxalate and citrate remove calcium ions, and heparin inhibits the reaction between thrombin and fibrinogen.

Coagulation time is determined by placing 1 ml of venous blood into a small 7·5 $\times$ 1 cm test-tube and placing it in a bath at 37°C. Remove the tube at intervals and tilt it to see if the clot has formed. When it is possible to invert the tube, note the time it has taken to form. In normal blood this will be between 3–5 min. Several methods have been proposed the normals varying slightly.

Fibrin

Fibrin may be prepared by taking freshly drawn blood in a beaker and whipping it with glass rods or twigs. Blood cells and clots are washed off with water and the fibrin preserved in glycerol. Fibrin may be stained with carmine. The stain is prepared by dissolving 0·5 g of carmine with 10 drops 0·880 ammonia and making up to 100 cm^3 with water. Fibrin is steeped in the dye until it is fully stained (1–2 days), then washed with dilute acetic acid and preserved in glycerol.

Composition of Blood

Blood consists of plasma, which is the fluid portion, cells and platelets, the solids taking up 45 per cent of its volume. Plasma has solid content of 9 per cent and 85 per cent of this is protein.

Blood plays many roles, it distributes oxygen, breaks down products of digestion and carries away waste products. It helps to maintain water balance, body temperature, combats disease and conveys hormones to their site of action. The analysis of its content aids the biochemist in assessing the efficiency of these functions.

Plasma proteins—These are albumin, the globulins and fibrinogen. Albumin and the globulins are most easily analysed by electrophoretic means as described in Chapter 7. Separation in bulk can be achieved by salt fractionation as described on page 450, fibrinogen being least soluble, the globulins more soluble, and albumin most soluble. The proteins are formed mainly in the liver. The other substances in solution are salts, lipids, carbohydrates, hormones, vitamins, products of intermediary metabolism such as pyruvic acid, and non-protein nitrogenous substances such as urea.

The Formed Elements, or Blood Cells

Red blood cells contain 35 per cent solids, almost all of which is the protein haemoglobin. This is a conjugated protein of molecular weight 68 000, the protein portion is globin and the prosthetic group is an iron containing compound called haem. These may be separated by adding acid. Haemoglobin can be released from the erythrocytes by haemolysis or laking. This is brought about by making the extracellular fluid hypotonic, that is, lowering its osmotic pressure, which bursts the cells. The protein is very soluble in water.

Haemoglobin has the ability to combine with oxygen and the compound formed is called oxyhaemoglobin (HbO_2). The oxygen may be released by physical means, such as partially evacuating a flask containing HbO_2. This indicates that haemoglobin is not oxidized in the usual chemical sense.

Several derivatives may be obtained from haemoglobin. Carboxyhaemoglobin is formed by passing carbon monoxide through a haemoglobin solution. Haemoglobin has a very high affinity for carbon monoxide, 250 times that of oxygen.

Haemoglobin can be oxidized easily giving methaemoglobin by adding a few drops of a fresh solution of potassium ferricyanide to a dilute solution of laked blood. The red colour is displaced by a brown one.

Haemochromagens are denatured haemoglobin but with the haem portion still attached.

The prosthetic group, haem, is an iron porphyrin, a substance composed of four pyrrole rings linked together by CH groups. Haem is easily oxidized to haematin and this is converted by the action of dilute hydrochloric acid to haemin.

```
                         CH2
                         ||
                 H3C     CH
                  |      |
                  C======C
                  |      |
 H3C    CH-------C      C====CH    CH3
  |     ||         \  /        |     |
  C-----C           N          C-----C
  ||     \          :         //     ||
  ||      N-----FeCl....N            ||
  ||     /          |         \      ||
  C-----C           |          C-----C
  |     ||          N          ||    |
 H2C    CH-------C     C-------CH    CH
  |              ||    ||            ||
 H2C             C-----C             CH2
  |              |     |
 COOH           H2C    CH3
                 |
                H2C
                 |
                COOH
```

This substance is easily crystallized and this fact is made use of as a test for blood.

Teichmann's Test

A drop of blood is evaporated to dryness on a microscope slide. Allow the slide to cool, then add one drop of a 0·1 per cent solution of sodium chloride in glacial acetic acid. Place a cover slip on top and heat very gently until the colour turns brown. When cool examine under a low-power microscope for crystals of haemin (Rhombic).

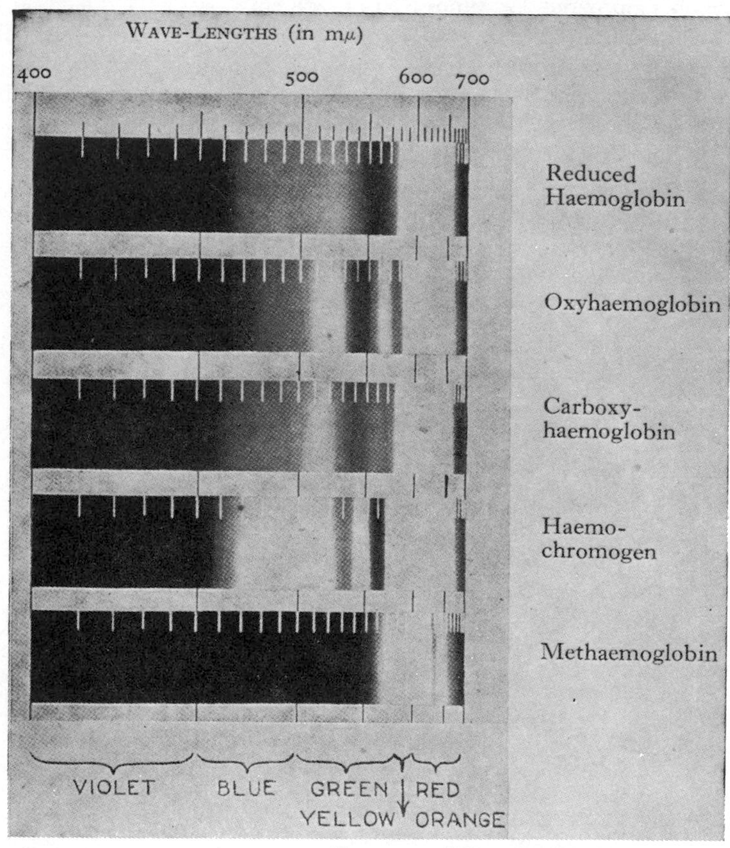

[By courtesy of Messrs. J. and A. Churchill, Ltd]

Figure 11.1. Absorption spectra of haemoglobin and its derivatives

First alkaline haematin is formed by adding a few drops of strong sodium hydroxide solution to 5 cm³ of 1 : 100 dilution of blood and heating until the solution turns yellow. A reducing agent such as sodium dithionate $Na_2S_2O_4$ is then added, and the solution gently shaken until it turns bright pink.

All these compounds give distinctive absorption spectra and may be examined spectroscopically. The spectroscope is a simple optical instrument; the specimen in a test-tube is placed against the slit, thus allowing only light to enter that has passed through the solution. The image of the slit is defracted through a series of prisms and is brought to focus at the eye-piece. Blood is examined by diluting 1:10 with water and removing cell debris by centrifugation. The filtrate is examined with the spectroscope and if the absorption bands are too broad further dilutions can be made. *Figure 11.1* shows several of the blood pigment absorption bands.

White Blood Cells or Leucocytes

These cells are approximately twice the diameter of the erythrocytes and not as numerous, usually there are about 3 for every 1000 red cells. Unlike the erythrocytes the leucocytes have nuclei, these vary in size and shape with the type of cell of which there are five. The ratio of these types to each other and their overall number varies with disease and injury, therefore their enumeration is useful diagnostically. White cells are involved in the combat of disease, pus being almost all derived from leucocytes.

Platelets

These are the last of the formed elements and are not strictly cells as they do not contain nuclei. They contain a clotting factor that is released during haemorrhage. On the disruption of these particles a physiologically active chemical 5-hydroxy tryptamine is released. This is believed to be part of the defence system of the body causing a decrease in blood flow.

Lymph

This body fluid has its own circulation system and is in contact with blood at the lymph nodes. Its composition is similar to plasma containing proteins, organic and inorganic metabolic products and leucocytes. It is in intimate contact with the tissues and is concerned with their nutrition and removal of waste products (catabolism).

Cerebro Spinal Fluid

This is a clear, colourless liquid to be found in the spinal canal. It is derived from blood and is similar to plasma, but has no pigment and very little protein content. Its analysis is important in diagnosing disorders of the nervous system.

Urine

The kidneys excrete between 1 and 2 dm^3 of urine every 24 h. Its yellow colour is due to the pigment urochrome. As the kidneys clear waste products from the blood, its content gives an indication of the functioning of the body and detoxication system. For this purpose several qualitative tests for abnormal constituents are useful.

Normal Constituents	
Nitrogeneous waste 30–35 g/24 h	*Salts 12–14 g/24 h*
Urea (30 g/24 h)	Chloride, Phosphate, Sulphate — salts of — sodium, potassium, calcium
Creatinine	
Uric acid	
Amino acid	
Ammonia	
	Abnormal constituents
Other trace materials	Protein
Indican	Bile
Oxalic acid	Sugars
Purines	Blood
Phenols	Ketone bodies

The protein is usually albumin and is tested for by centrifuging a few cm^3 of urine to remove suspended deposits and then adding 1 or 2 drops of 20 per cent salicylsulphonic acid. A white precipitate indicates protein.

The most common sugar found is glucose and occasionally lactose which is excreted by pregnant and nursing women. Before testing for sugars any protein present must be removed, this being done by the following coagulation procedure. Acidify the urine with a few drops of 2 M acetic acid, boil, then cool under a cold tap and centrifuge. Pour off the supernatant and divide it into several tubes for testing. Carry out Benedict's Test as described on page 440. If this is positive and lactose is suspected, prepare the osazone as described on page 441 to determine the type of sugar present.

The more modern method is to run a chromatogram of the urine in parallel with markers. If the sugar content is low, the sample of urine is first desalted (page 298) so that a larger volume can be safely applied to the paper. The chromatogram is developed overnight in iso-propanol 100 cm^3/water 40 cm^3 solvent mixture by the ascending method. After drying the chromatogram is dipped in an aniline-diphenylamine reagent which is made as follows:

Dissolve 1 g diphenylamine in acetone, add 1 cm^3 aniline and make up to 100 cm^3 with acetone. Add 10 cm^3 of 85 per cent phosphoric acid.

After dipping in this reagent dry the paper at 100°C for a short while until the coloured spots appear. Do not prolong the heating or the paper will char. The only disadvantage in this method is the time delay. Quicker results can be obtained with the thin layer chromatographic technique.

If sugar is present there may also be ketone bodies, these are detected by Rothera's Test. Saturate the urine sample with ammonium sulphate, add 2 crystals of sodium nitroprusside and shake. Place the tube in a stand and carefully layer on 0·880 ammonia solution, if positive a violet colour develops at the interface.

Bile is indicated if a froth persists after shaking a tube of urine. It is positively identified by the nitric acid test. For this urine is allowed to pass through a filter paper on which a large proportion of the bile pigments will remain absorbed. Open up the paper and allow a drop of concentrated nitric acid to fall on to it. If bile pigments are present green and blue coloured rings appear round the edges of the drop.

Blood may be present if the salicylsulphonic acid test is positive. The deposit after centrifugation can be examined microscopically for blood cells, and if suspected the deposit is shaken with 1 cm^3 of distilled water, boiled and cooled. Add equal volumes of 0·4 per cent *o*-tolidine in glacial acetic acid and 10 volumes hydrogen peroxide. A blue colour indicates the presence of blood.
N.B. *o*-tolidine is carcinogenic.

Quantitative Analytical Procedures

It is necessary in many assays of the constituents of blood and tissues to provide a protein-free filtrate. Many protein precipitating methods will be found in the literature, a list of a number of them is given in 'Micro-analysis in Medical Biochemistry' by Wootton. Not all are suitable for every procedure so the method of the original author will be given in each case. For simplification the principle and sequence of operations is given on the left-hand page which is divided into four columns: (1) The reagent (2) Its volume (3) The vessel used and its treatment (4) Time of Treatment. The method is followed by a short discussion.

Although a substance may be found in various tissues or fluids the methods for its determination are grouped together. In many cases the method only requires simple adaptation for analysis, for instance, in blood, urine or sap and thus avoids repetition. This is opposite to the usual convention found in books of physiological chemistry, because the field of biological chemistry is greater, requiring analysis of a wide variety of materials from, for example, animals, plants, marine life, sewage, fungi, etc.

ESTIMATION OF REDUCING SUGARS IN BIOLOGICAL MATERIAL

Iodimetric Method of Somogyi (1945)

Principle

Sugar reduces the copper in an alkaline copper solution, a fixed amount of iodine is liberated in the solution and oxidizes this reduced copper. The excess iodine is titrated with standard thiosulphate solution.

Method

Sample A	5 cm^3	In 20 × 2·5 cm test-tube	
Copper reagent B	5 cm^3	Mix with a stream of N_2, close tube with glass bulb or funnel	
		Immerse in boiling water	10 min
		Cool to 30°C in cold water	3 min
Pot. iodide C	2 cm^3	From burette *without* mixing	
M Sulphuric acid	1·5 cm^3	From fast flowing burette *with* mixing	
		Leave	5 min
		Shake, then titrate with	
Thiosulphate D		Use starch and phenol red solution for end point	

Prepare test, blanks and standards using the above procedure. This method is suitable for glucose concentrations of 0·03–3 mg. Subtract the test and standard titrations from the blanks and plot a graph of the standard titres against their concentrations.

Solutions

Sample A	This should be made just pink to phenolphthalein. Many salts will interfere giving different values for reducing sugars. If the sample is blood this is laked with water 1:8 and deproteinized using 1 volume 10 per cent zinc sulphate $7H_2O$ followed after mixing, with 1 volume of 0·5 M sodium hydroxide. Allow to stand a few minutes then separate the protein.

Copper Reagent B

Rochelle salt	40 g
Disodium hydrogen phosphate $Na_2HPO_4 \cdot 12H_2O$	71 g
M sodium hydroxide	100 cm^3
	Dissolve in 500 cm^3 water
10% cupric sulphate $CuSO_4 \cdot 5H_2O$	80 cm^3
0·892 per cent potassium iodate	100 cm^3
Anhydrous sodium sulphate	180 g

Dissolve these in this order, make up 1000 cm^3 with water. Stand 3 days to settle, filter through sintered glass filter.

Potassium Iodide 25% C	Use iodate free reagent and make alkaline with a little sodium hydroxide solution.
Sodium thiosulphate D	Standard 0·1 M is diluted 20 times with a solution of 5 cm^3 M NaOH in 1000 cm^3 of CO_2 free distilled water.
Phenol red indicator	100 mg dissolved in 28 cm^3 0·01 M NaOH and diluted to 250 cm^3 with water.
Starch solution	1 g of soluble starch is dissolved in 100 cm^3 of hot distilled water.

ESTIMATION OF GLUCOSE IN BIOLOGICAL MATERIAL

Enzymatic Method of Saifer & Gerstenfeld, *J. Lab. Clin Med.* **51** (1958) 448

Principle

Glucose oxidase oxidizes glucose to gluconic acid and hydrogen peroxide, the latter is broken down by peroxidase to water and oxygen, which in turn oxidizes *o*-dianisidine to a yellow dye, the colour produced is proportional to the glucose. The procedure is suitable for blood, urine and C.S.F. and other protein-free filtrates.

Method

Oxidase reagent	8 cm^3	In 15 × 1·5 cm test-tube placed in water bath at 37 °C. Equilibrate	4–5 min
Sample	1 cm^3	Mix. Incubate exactly. Remove from bath and add	30 min
M Hydrochloric acid	0·5 cm^3	Mix, and stand	5 min

Read in an absorptiometer or spectrophotometer at 400 nm against a reagent blank made by substituting distilled water for the sample. Compare with a series of standards.

For C.S.F. use weaker standards and for urine the sample may have to be diluted to bring the reading within the range of the standards.

Reagents

Glucose oxidase	This must be bought commercially and is supplied as a dry powder in a vial. Dilute this according to the maker's instructions and use on the day of preparation.
Glucose standard	Stock solution 1 per cent glucose in 0·25 per cent Benzoic Acid. Standard solution Dilute the stock solution 1:10 with 0·25 per cent Benzoic acid. For C.S.F. dilute 1:20 or 1:40.
Sample for analysis	Dilute 0·5 cm^3 plasma, urine or C.S.F. to 5 cm^3 with water, add 2·5 cm^3 0·5 M sodium hydroxide and mix. Add 2·5 cm^3 10 per cent zinc sulphate. Centrifuge coagulated protein and pipette supernatant for assay. The anti-coagulant for blood may be oxalate, heparin or E.D.T.A. The zinc sulphate and sodium hydroxide must neutralize each other and must be adjusted by titration, using phenolphthalein as an indicator.

SERUM OR PLASMA ALKALINE PHOSPHATASE

Of the phosphatase enzymes alkaline phosphatase, optimum pH 9 is the most prevalent in blood. It hydrolyses phosphate esters liberating inorganic phosphorus and is the basis of several methods of estimation, each using a different substrate. The enzyme or enzymes exhibit a different activity to each substrate and therefore every method has its own unit and normal values. The unit is the number of milligrammes of one of the liberated components per 100 cm^3 of blood in 1 h at 37°C.

Modified Bodansky Method of Hawk, Oser & Summerson

Principle:

After incubation of sodium glycerophosphate with serum the liberated phosphate is estimated using the technique of Fiske and Subbarow (see Chapter 6).

Method

Substrate	9 cm^3	In 15 cm^3 centrifuge tube standing in a rack immersed in a bath at 37°C. Equilibrate	5 min
		Add	
Serum	1 cm^3	Stopper the tube and leave Transfer rack with tubes to a bath at 0°C	1 h 5 min
		Add	
30% T.C.A.	2 cm^3	Mix and stand Centrifuge	5 min
Filtrate	8 cm^3	In tube graduated at 10 cm^3 Add	
Molybdate	1 cm^3	Mix. Add	
Amino napthol			
Sulphonic acid	0·4 cm^3 to	Mix and add	
Water	10 cm^3	Mix and stand	5 min

Read in a spectrophotometer at 700 nm

Control: Add serum (1 cm^3) *after* the 30 per cent T.C.A. then proceed as before.

Standards: 8 cm^3 of the standard solution is used instead of the filtrate from the 30 per cent T.C.A.

Blank: 8 cm^3 of 5 per cent T.C.A. is used instead of the standard.

Calculation

$$\frac{\text{A sample}}{\text{A standard}} \times \text{conc. of standard} = \text{mg phosphate per } \tfrac{2}{3}\ \text{cm}^3 \text{ serum}$$

Multiply the above by $\frac{3}{2} \times 100$ = mg phosphate liberated per 100 cm^3 serum = units Bodansky

The control gives the serum phosphate content and if significant must be subtracted from the unknown. It may be used instead of the blank in a spectrophotometer provided its absorption is not too great.

Solutions

Substrate

Sodium-glycerophosphate	0·5 g
Sodium diethyl barbiturate	0·424 g

Water to 100 cm^3

Store under petroleum ether b.p. 20–40°C, in a refrigerator at 2°C.

Molybdate solution

Ammonium molybdate 25 g in 200 cm^3 water

Transfer quantitatively into a 1 dm^3 volumetric flask containing 300 cm^3 of 5 M sulphuric acid. Dilute to mark with water.

Amino-naphthol sulphonic acid solution

1-amino-2-naphthol-4-sulphonic acid 0·5 g in

15 per cent sodium bisulphite solution 195 cm^3 and add

20 per cent sodium sulphite 5 cm^3

Add more of the sodium sulphite if solution is not complete.

When stored in the cold in a dark bottle the solution is stable for one month.

Standard phosphate solution stock

Potassium dihydrogen orthophosphate 0·351 g

Make up to 1 dm^3 with water volumetrically

Standard

Stock solution 6·25 cm^3

30 per cent T.C.A. 16·7 cm^3

Make up to 100 ml with water volumetrically

This solution contains 0·5 mg phosphorus/100 cm^3

8 cm^3 contains 0·04 mg phosphorus.

Method of D. J. Bell *Biochem. J.* **75** (1960) 224

Principle:

After incubation of the serum with 4-nitrophenyl phosphate the liberated yellow 4-nitrophenyl is measured directly in a spectrophotometer at its absorption peak of 397 nm.

Method

Buffer	1 cm^3	in 15 × 1·75 cm test-tube and add	
Plasma	0·05 cm^3	Mix and add	
Substrate	0·1 cm^3	Mix and place in a bath at 37°C	60 min
		Dilute with	
0·05 M NaOH	3–25 cm^3	Depending on concentration	

A blank is made by adding the substrate last after dilution with the 0·05 M sodium hydroxide.

For use with a 1 cm cuvette a series of standards of 0·25–10·0 g/cm^3 are made by diluting the stock standard with 0·05 M sodium hydroxide. Over this range Beer's law is obeyed. The optical density is measured at 397 nm. Results are expressed as milligrammes nitrophenyl liberated by 100 cm^3 plasma in 60 min at 37°C.

Calculation

The gram per cm^3 obtained from standard curve multiplied by the volume of digest multiplied by 2000 equals milligrams nitrophenyl per 100 cm^3.

Reagents

Buffer	0·1 M sodium carbonate	7 parts
	0·1 M sodium bicarbonate	3 parts
Substrate	0·066 M disodium 4-nitrophenyl phosphate. This reagent must be purified by recrystallization and washing three times with water saturated butyl alcohol and then three times with ether to remove free nitrophenyl. Dissolved ether is then removed by bubbling moist air through the solution.	
Stock standard	200 g/cm^3 4-nitrophenol in water. *Caution:* 4-nitrophenol is a scheduled poison.	

ESTIMATION OF UREA

Urea can be estimated in three ways:

1. By precipitating the urea in combination with xanthydrol and estimating the nitrogen content using Kjeldahl's method.
2. By measuring the nitrogen given off by the reaction of sodium hypobromite with urea.
3. The conversion of urea by the action of the enzyme urease and the subsequent estimation of the ammonia.

The second method is the least accurate but is the most direct. The third method is the most popular and can be used with any known method of ammonia assay. Those used are the macro and micro distillations of ammonia into standard acid, the transfer of alkali liberated ammonia in a gas stream to standard acid, or finally the measurement of the colour produced with Nessler's reagent. Nessler's reagent can be used after the aeration and distillation method or directly into a protein free filtrate. This last method is preferred as only the simplest apparatus is required. A micro diffusion technique involving Conway units has been devised.

Method of Archer & Robb *Q. Jl. Med.* **18** (1925) 274

Principle

The sample is incubated with urease and the ammonium carbonate formed reacted with Nessler's reagent. The brown-yellow colour produced is proportional to the urea in the sample.

Method			
Water	2 cm^3	in 15 cm^3 centrifuge tube and add	
Sample	0·2 cm^3	from a blood pipette. Mix and add	
Urease	0·2 cm^3	Mix. Incubate in bath at 55 °C.	15 min
		Precipitate the protein with	
Tungstate sol.	0·3 cm^3	Mix and add	
0·33 M H_2SO_4	0·3 cm^3	Mix and add	
Water	5·0 cm^3	Mix and stand	5 min
		Centrifuge	
Supernatant	5·0 cm^3	In a test-tube, dilute with	
Water	5·0 cm^3	and add	
Nessler's reagent	2 cm^3	Mix	
		Read within 10 min in a spectrophotometer at 500 nm or if a Dubosq type colorimeter is employed use an Ilford blue filter (622).	
		Compare with urea standards or ammonium sulphate standards.	

A reagent blank is made by using water instead of the sample in the above procedure. This method is suitable for blood or protein-containing solutions.

Reagents

Urease:

The enzyme is available in three forms:

1. As soya bean flour 20 mg of which can be used/tube.
2. Absorbed on to paper strips.
3. Tablets. Grind one urease tablet in 5 cm^3 of 30 per cent ethanol and use 0·2 cm^3 of this suspension/tube.

Tungstate solution:

10 per cent sodium tungstate $Na_2WO_4 \cdot 2H_2O$ in water.

Standard solution:

Stock solution 0·22 per cent ammonium sulphate.

Dilute the stock solution 20 times. 1 cm^3 of this diluted to 10 cm^3 with water and treated with 2 cm^3 Nessler's reagent gives a colour equivalent to 40 mg urea/100 cm^3 of sample.

Nessler's reagent

Several formulae are available for this reagent but it is important that for this method the solution of Folin & Wu J.B.C. **38** (1919) 81 be used. This reagent is usually purchased but is made up by the following method:

Potassium mercuric iodide

Potassium iodide	100 g in a 500 cm^3 flask. Add
Iodine	100 g. Add
Water	100 cm^3 and dissolve. Add
Mercury	150 g. Shake the flask until iodine dissolves, cooling under a running water tap.

Decant into a 2 dm^3 measure, wash the remaining mercury several times and introduce washings to the measure. Make up to 2 dm^3 with water.

Final mixture

10% sodium hydroxide	3500 cm^3 in a 5 cm^3 vessel, add
pot. mercuric iodide	750 cm^3, add
distilled water	750 cm^3 Mix.

This solution can be used diluted 1 to 5 or 10 with the ammonia solution.

CHAPTER 12

ENZYMES

Life requires energy for growth, maintenance and reproduction. This energy is made available to animals and plants by the ingestion of food or the absorption of nutrients. The living cell is the working unit of life and is able to utilize these substances, breaking them down and converting them to the compounds it requires for its continued existence and for the particular function the cell has in the order of life. These changes involve numerous chemical reactions and energy transformations and are greatly accelerated by catalysts, substances that although involved in the reaction, emerge unchanged.

It was thought a century ago that these catalysts occurred and could act only in the cell. Towards the end of the nineteenth century a 'ferment' was extracted from yeast cells which oxidized glucose to alcohol and carbon dioxide. The substance responsible was called zymase and similar ferments were called en-zymes, from the Greek, 'in yeast'. There are many different enzymes and one cell may contain a thousand or more. All enzymes are proteins and have the same general reactions but are usually very sensitive to changes in their environment. A change in temperature, pH, or attempts at extraction from associated molecules may quickly result in irreversible inactivation and coagulation. They are soluble or form colloidal solutions. Many enzymes have been isolated and crystallized, urease being one of the first. Urease has the ability to split urea into ammonia and carbon dioxide.

$$O{=}C\begin{matrix}\diagup NH_2 \\ \diagdown NH_2\end{matrix} + H_2O = 2NH_3 + CO_2$$

Urease is extracted from Jack beans and is used in methods of estimating urea concentration in urine and blood. This enzyme is very stable in contrast to the majority which are not, and this is the reason why so few have been crystallized although many have been brought to a high degree of purity.

In enzyme nomenclature the compound attacked is the *substrate*, and the enzyme is named after the substrate with the suffix *-ase* added. In the above reaction urea is the substrate, urease is the enzyme. The enzymes are classified in groups also ending in -ase according to what they do. Urease belongs to a group that attacks amides called amidases. Enzymes that were first discovered are still called by the name originally given them, e.g. trypsin, rennin and papain.

MECHANISM OF ENZYME ACTION

The most striking characteristic of enzymes is their specificity. They will only react with one compound, a closely related group of compounds or a particular linkage in a molecule. Urease is specific for urea and will not react with any similar compound. Carbohydrases only act on carbohydrates and not on lipids or proteins, even then different enzymes are required to attack individual sugars such as glucose, fructose and maltose. Dipeptidase only attacks dipeptides, peptidases react with compounds containing a peptide link. Glycosidases attack the link between two carbohydrates. Sucrase splits sucrose to glucose and fructose. Maltase only attacks maltose and not cellubiose although both are made up of glucose units, the linkage being different.

Mode of Action

It was propounded by Emil Fischer that a substrate fits on to or into an enzyme, rather like a key in a lock, before a reaction takes place. The spatial arrangement of the atoms of the substrate molecule is related to a similar arrangement on the enzyme.

Factors Affecting Enzyme Action

Temperature—At low temperatures, e.g. 0°C, enzyme activity is slight but the enzyme itself is not damaged. As the temperature rises so the activity increases, and this rate of increase is about doubled for every 10°C rise in temperature. This 10°C change is called the Q_{10} or temperature coefficient. Eventually a point is reached where activity ceases to increase and then falls rapidly. This occurs because of the protein nature of enzymes, denaturation being caused by these conditions. All enzymes are denatured by boiling except a few, which under certain pH conditions withstand boiling a short time, e.g. ribonuclease. In some cases a proportion of activity may return on cooling. The point of maximum activity is called the optimum temperature. This is usually about 40°C for animal enzymes and 50–60°C for plant enzymes.

Hydrogen Ion Concentration

As with temperature, each enzyme shows greatest activity at an optimum pH and slight changes in either direction show considerable decrease in activity. This optimum pH is usually that in which the enzyme is sited in nature. For salivary amylase it is about neutral but for pepsin of gastric juice it is pH2. This optimum depends on several factors, the type of buffer may affect it and the degree of ionization of the protein or protein substrate combination.

Inhibition and Activation

The activity of an enzyme may be reduced or halted by 'poisons' or inhibitors. These are chemical agents and they fall into two groups, competitive and non-competitive inhibitors.

Substances having a similar molecule to a substrate may be able to fit on to the enzyme at the active site. This excludes the substrate and therefore no reaction occurs. The inhibitor may then be released leaving the enzyme unharmed. This is competitive inhibition and is observed when activity fluctuates with the ratio of substrate to inhibitor. If substrate concentration is increased the activity rises. Non-competitive inhibition is often irreversible, the inhibitor inactivating the enzyme by combining with it. The activity in this case is controlled by the inhibitor, the raising of the substrate concentration not affecting it. Salts of heavy metals are examples of this type of inhibitor.

The study of enzyme action and inhibition is important as many antibiotics and drugs owe their activity to inhibitive properties. It is believed that the drug sulphanilamide competes for the site where an enzyme utilizes P.A.B. (para-amino-benzoic acid).

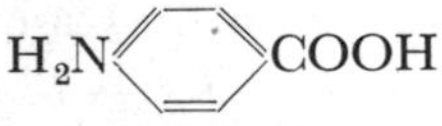

$H_2N-C_6H_4-COOH$ P.A.B.

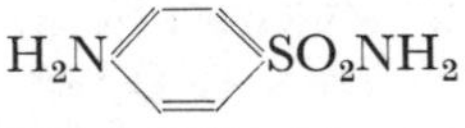

$H_2N-C_6H_4-SO_2NH_2$ Sulphanilamide

P.A.B. is essential for the growth of bacteria.

Specific inhibitors are prepared by nature and are called anti-enzymes, e.g. the walls of the intestine are protected from attack by the proteinases pepsin and trypsin by the production of anti-enzymes in the mucosa. An intestinal parasite, the round worm ascaris, survives the presence of these proteinases as it also produces an inhibitor. It is interesting to note that ascaris is quickly digested by the plant proteinase papain indicating the specificity of the anti-enzyme it produces.

Many enzymes require the presence of certain chemical agents before they become active. Inorganic salts are needed for some, e.g. ptyalin, salivary amylase, needs the chloride ion, alkaline phosphatase requires magnesium. Other enzymes have a specific requirement for one or more organic compounds such as the vitamins. These compounds are called *co-enzymes* and are the prosthetic group which when combined with enzyme protein called the *apo-enzyme*, make a whole or active enzyme, the *holo-enzyme*. Vitamin B6 pyridoxal phosphate is known to be required by many decarboxylases and deaminases and to be involved in transamination and sulphur transfer.

Cocarboxylase (vitamin B1 phosphate), combines with carboxylase and decarboxylates pyruvic acid to acetaldehyde and carbon dioxide.

$$\begin{matrix} CH_3 \\ | \\ CO \\ | \\ COOH \end{matrix} \longrightarrow \begin{matrix} CH_3 \\ | \\ CHO \end{matrix} + CO_2$$

Many co-enzymes were first given a number or letter before their structure was accurately known, e.g. co-enzyme 1 is diphosphopyridine nucleotide, co-enzyme 2 is triphosphopyridine nucleotide, co-R is biotin or vitamin H.

These co-enzymes may in some cases be detached from the apo-enzyme by dialysis, the co-enzyme, being a much smaller molecule, passes through the permeable membrane.

Some enzymes occur in an inactive form called *pro-enzymes* or *zymogens*. These are activitated by *kinases* or by traces of their active form. The best known examples are the pancreatic pro-enzymes trypsinogen and chymotrypsinogen. When these are secreted into the intestine they come in contact with the enzyme enterokinase which converts trypsinogen to trypsin which in turn converts more trypsinogen to trypsin and also chymotrypsinogen to chymotrypsin. The proteinase pepsin of the stomach is secreted as pepsinogen and is activated by hydrochloric acid. Active pepsin will then activate its own precursor. This is called auto-catalysis.

Other substances may activate an enzyme although they are not essential components for the reaction. HCN and H_2S are strong inhibitors of oxidases but enhance the activity of papain. Some enzymes are activated by the presence of metal ions, e.g. phosphatases (Mg) but others are inhibited by them, e.g. urease (Ag).

Some activators are not true activators as they enhance the activity of the enzyme by removing or binding traces of an inhibitor. This type was detected when after purification of the enzyme the action of the activator decreased or disappeared.

Effects of Substrate, Enzyme Concentration and Products of Reaction

When studying enzyme reaction *in vitro* many factors must be borne in mind. For control conditions the enzyme must have the right environment at its optimum temperature and pH. The next factor to consider is the rate of reaction and the effect of enzyme and substrate concentrations on it. To study the rate of reaction the velocity must be controlled and this is usually directly

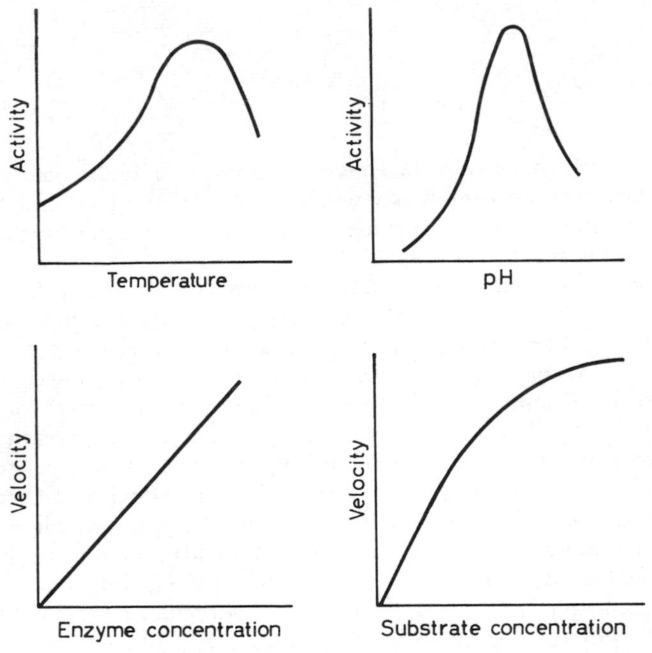

Figure 12.1. Graphs showing factors influencing enzyme activity

proportional to the enzyme concentration provided there is sufficient substrate present. Increasing the substrate whilst keeping the enzyme concentration constant will also increase the velocity, until a saturation point is reached where the enzyme is at its peak performance (*Figure 12.1*). On further increase the velocity will be maintained for a time and then decrease. Sucrase will behave in this manner, the velocity increasing until the sucrose concentration reaches 5 per cent, above this point it decreases.

The rate of reaction may also be affected by the end products inhibiting the enzyme.

Adaptive Enzymes

Some organisms are able to manufacture enzymes only when their substrate is present. The organism Pseudomonas saccharophila if grown in the presence of sucrose will produce the enzyme sucrose phosphorylase. If it is grown in the presence of certain other carbohydrates instead, such as starch, glucose or maltose, this enzyme is not produced in detectable amounts.

Adaptive enzymes are most easily observed in bacteria but have been demonstrated in other organisms. Although an adaptive enzyme may be produced due to the replacement of a normal constituent in the media, the organism does not necessarily lose its ability to utilize the original constituent. For example, the bacterium Streptoccocus lactis, when separated from the medium after growth in glucose broth, will not ferment galactose, only glucose, but if grown in galactose broth the separated cells will ferment galactose and glucose. This means that the enzyme utilizing galactose is adaptive, that is, it is only produced when grown in the presence of galactose.

Kinetics of Enzyme Action

It will now be realized that enzyme reactions are complicated processes and the study of these processes, kinetics, is essential for their understanding and practical utilization. The normal equations used by physical chemists have to be modified because of the many factors involved and usually the enzyme is not obtainable in a pure state.

Michaelis and Menton in 1913 put forward a theory involving the following dissociation:

$$E + S \rightleftharpoons ES$$
$$ES \rightleftharpoons E + \text{Products}$$

This assumes the formation of an intermediate complex ES. The reaction may be written:

$$E + S \underset{K_2}{\overset{K_1}{\rightleftharpoons}} ES \overset{K_3}{\rightarrow} E + P$$

The reverse reaction of the last stage may be ignored. This assumes that in an initial reaction P is negligible. By applying the law of mass action Michaelis and later Briggs and Haldane derived the following formula:

$$v = \frac{V[S]}{K_m + [S]} \quad \text{or} \quad v = \frac{V}{1 + \frac{K_m}{[S]}}$$

where: v = initial velocity
V = maximum velocity
[S] = substrate concentration
K_m = Michaelis constant

Figure 12.2a is a plot of the velocity v against substrate concentration [S]. K_m is the substrate concentration at half the maximum velocity V. There is an obvious difficulty in finding V and several ways have been postulated for plotting results. The most common is that of Lineweaver and Burke who used a transposed version of the Michaelis equation:

$$\frac{1}{v} = \frac{1}{V} + \frac{K_m}{V} \cdot \frac{1}{[S]}$$

The plot of $1/v$ against $1/[S]$ immediately shows V and K_m *Figure 12.2b*. This type of plot is especially useful for the study of enzyme inhibitors. By finding the rate (velocity) of reaction with and without various concentrations of inhibitors, an indication of the type of inhibition is seen (*Figure 12.2c* and *d*). It will be seen that for competitive inhibition V remains constant and for non-competitive inhibition K_m is unchanging. A detailed derivation of the formulae will be found in volumes mentioned in the bibliography, especially *Enzymes* by Dixon and Webb, and *Catalysis and Inhibition of Chemical Reactions* by P. G. Ashmore.

The necessary data for plotting these graphs may be obtained using the following techniques.

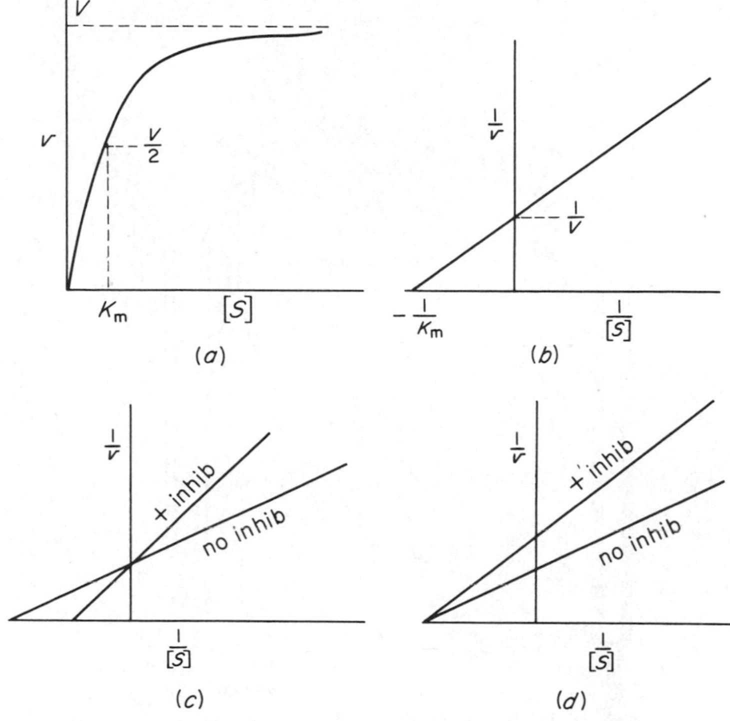

Figure 12.2. Enzyme kinetics

MEASUREMENT OF ENZYME ACTIVITY

The activity of an enzyme is expressed as a measure of the work it does. Therefore it is necessary either to measure the quantity of the products formed or the disappearance of the substrate. This measurement is done in two ways, chemically involving gravimetric, volumetric or absorptiometric analysis, or by manometry, that is measuring a gas evolved or utilized.

The unit of activity varies with the method employed. In fluids the activity may be expressed as weight of end product formed per volume per unit time. For example, Bodansky's alkaline phosphatase unit is the milligrams of inorganic phosphate liberated per 100 cm^3 of blood per hour. The activity may also be the time required for a known weight of enzyme preparation to catalyse

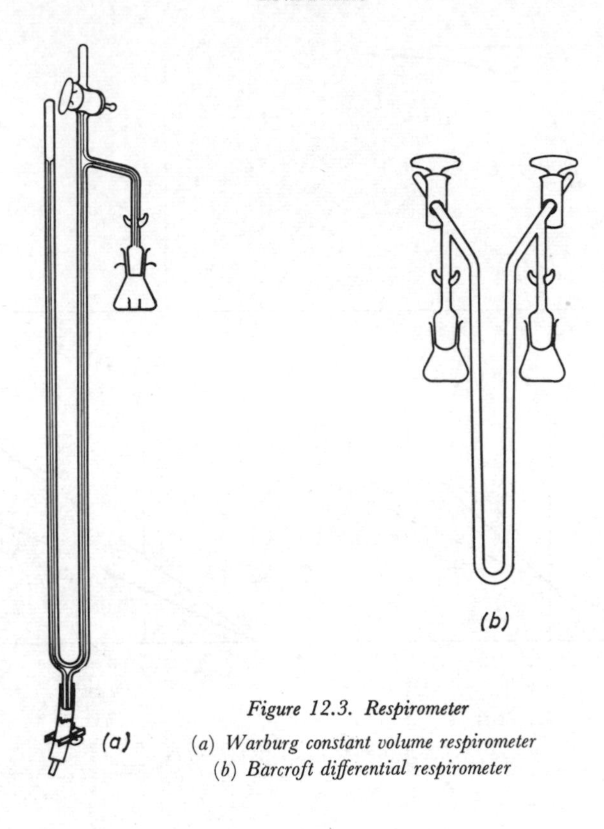

Figure 12.3. Respirometer
(*a*) *Warburg constant volume respirometer*
(*b*) *Barcroft differential respirometer*

the conversion of a known amount of substrate. Manometrically the activity is expressed as a rate of gas evolved or taken up and this is given the symbol 'Q'. The recognized convention is as follows:
Q is the mm³ of gas taken up or given off per milligram dry weight of enzyme preparation per hour.

The experimental conditions are indicated as follows: $Q_{O_2}^{37°C}$ is the mm^3 of O_2 taken up at 37°C per milligram dry weight of enzyme per hour.

In some instances it is of value to compare activity with the phosphorus or nitrogen content and this is indicated after the Q, e.g. $Q_{CO_2}^{28°C}$ (N) = mm^3 CO_2 evolved at 28°C/mg nitrogen/h.

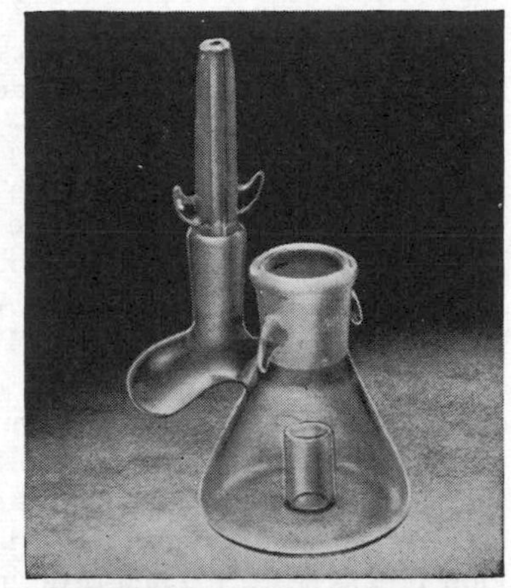

Figure 12.4. Warburg flasks

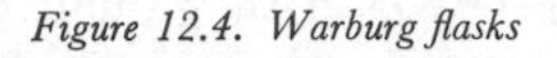

[By courtesy of Hoslab Ltd]

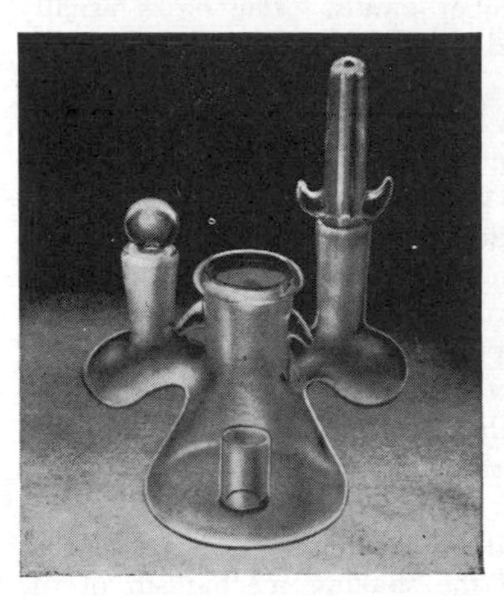

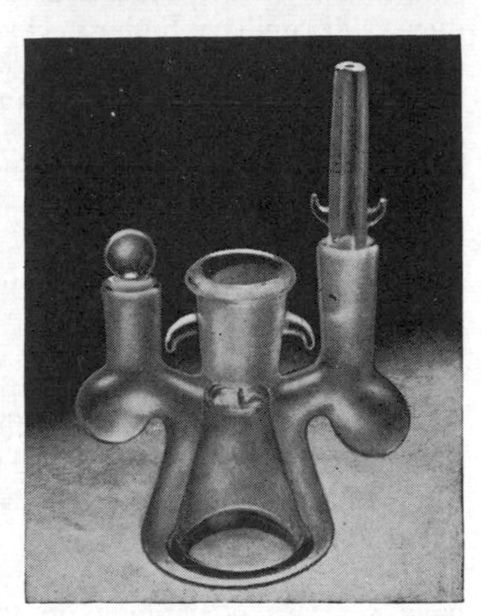

Manometric Methods of Enzyme Activity Measurement

Changes in quantities of a gas are most easily measured with accuracy by keeping the volume constant in a reaction vessel, and measuring changes in pressure by means of a manometer. The calculations are simplified by ensuring that there is only one variable, the pressure, making possible the use of a constant.

Two types of respirometer, that is, reaction flasks or flask with manometer, are in general use. These are the Warburg constant volume respirometer (*Figure 12.3a*) and the Barcroft differential respirometer (*Figure 12.3b*). These consist of either a flask or flasks attached to a manometer by side arms. This allows the flasks to be immersed in a constant temperature bath.

The manometers are capillary U-tubes of 1–2 mm bore. The Warburg manometer has a side arm at the bottom of the U to which is attached a piece of rubber tubing closed at its other end and fitted with a screw clip. Adjustment of the clip varies the level of the manometer fluid. The manometers are graduated in millimetres and centimetres and may be in two forms reading either from 0–300 mm or having a centre zero and graduated from this in both directions to 150 mm. The limb with the side arm has a three-way stopcock, one exit being through the plug itself. The connection to the flask is by standard ground glass joint usually C.10. When ordering replacements be sure to obtain them with capillary tubing. Hooks are provided so that the flask can be secured by springs or rubber bands. The flasks usually have a volume of 15–30 cm^3 and are of varied designs. Those commonly used have one or two side arms (*Figure 12.4*) and a centre well which allow different reactants to be kept separate until needed. The addition of liquid from the side arm is performed by tipping the flask carefully. It should be possible to do this, and to be able to wash out the side arm with fluid from the flask, without running any fluid into the other side arms or the centre well. On receiving new flasks it is wise to experiment with each of them using water to see that tipping without mixing is possible.

The Barcroft respirometer differs in that a flask is attached to both limbs of the manometer and that there is no means of adjusting the manometer fluid. Both halves of the respirometer should be identical.

The respirometers are mounted on wood or metal backs that have a means of clipping on to the shaking mechanism of the temperature regulated bath.

The advantage of the Barcroft respirometer is that conditions in one side of the manometer are the same as in the other and variations in temperature are compensated for automatically. The disadvantage is the difficult calculation of a suitable constant. The Warburg respirometer is open to the atmosphere and subject to changes of barometric pressure. This disadvantage is overcome by the use of another respirometer containing only water in the flask, called a thermobarometer. This registers changes in temperature and atmospheric pressure only, the reading of which may then be taken into account in the calculations. The Warburg respirometer is the more popular instrument and its calibration and use will be described, similar methods being used for the Barcroft respirometer. For a full account of manometric methods the reader is referred to works on the subject mentioned in the appendix.

Calibration

When using the Warburg apparatus the fluid in the closed limb is set to a reference point and the level in the open one is noted. Any gas changes in the flask will bring about a change in pressure, this is measured by adjusting the fluid in the closed limb to the reference point and again reading the level in the open limb. The generated pressure is the difference in readings of the open limb. This is converted to a volume by the use of the universal gas equation the derived formula being $x = hK$, where x is the volume of gas in mm³, h is the difference in readings of the open limb, K is a constant.

$$\text{Where } K = \frac{V_g \dfrac{273}{T} + V_f\alpha}{P_o}$$

Where V_f = Volume of fluid in flask in mm³.

V_g = Volume of flask plus manometer to reference point less V_f

T = Temperature of flask in degrees absolute

α = Solubility of the gas being taken up or evolved in reaction

P_o = Standard pressure of one atmosphere in manometer fluid (Krebs, Brodies) = 10 000 mm

This constant must be calculated for every flask and manometer combination. The calibration has been simplified by Umbreit so that for a given set of conditions the constant may be calculated from the volume of the flask and manometer combination using Table 12.1.

TABLE 12.1 Data for the Calibration of Warburg flasks

Temperature	*s*	ΔK_{O_2}	ΔK_{CO_2}
20	$0\cdot0932 \times 10^{-3}$	$-0\cdot090$	$-0\cdot005$
25	$0\cdot0916$	$-0\cdot089$	$-0\cdot016$
28	$0\cdot0907$	$-0\cdot088$	$-0\cdot020$
30	$0\cdot0901$	$-0\cdot087$	$-0\cdot023$
33	$0\cdot0892$	$-0\cdot087$	$-0\cdot026$
35	$0\cdot0886$	$-0\cdot086$	$-0\cdot029$
37	$0\cdot0881$	$-0\cdot086$	$-0\cdot031$
40	$0\cdot0872$	$-0\cdot085$	$-0\cdot034$

The constant for the flask with no fluid in it is

$$\frac{V\dfrac{273}{T}}{P_o} = K_e$$

Where V is the volume of flask and manometer or $V_g + V_f$. K_e can be calculated for any chosen working temperature by experimentally determining V and calculating $(273/T)/P_o$, this portion of the formula, a constant s will be found calculated in Table 12.1.

Therefore $K_e = V \times s$.

To calculate the final K the volume of fluid to be used in the flask must be known. Another constant ΔK which is dependent on t and α may be calculated from the formula

$$\Delta K = \frac{-\,1{,}000\,\dfrac{237}{T} + 1{,}000\alpha}{P_o}$$

This has been tabulated for various temperatures and gases that are usually employed (Table 12.1). For each cm³ of fluid in the flask the amount indicated in the table is added to the K_e giving K. This is best illustrated by a few examples.

A respirometer has a V of 20·5 cm³ and it is desired to use it for measurement of oxygen uptake at 30°C with a flask fluid volume of 3 cm³.

$$K_e = V \times s = 20{,}500 \times 0\cdot0901 \times 10 = 1\cdot83$$
$$K = K_e + (3 \times -0\cdot087)$$
$$= 1\cdot83 + (-0\cdot261)$$
$$= 1\cdot83 - 0\cdot260$$
$$= 1\cdot57$$

2

If the K was desired for the same conditions but with 2 cm³ it can be obtained as above or simply adjust by adding 0·087.

$$K\ (2\ \text{cm}^3) = 1{\cdot}83 + (2 \times -0{\cdot}087) = 1{\cdot}66$$

$$\text{or } K\ (2\ \text{cm}^3) = K\ (3\ \text{cm}^3) + 0{\cdot}087 = 1{\cdot}57 + 0{\cdot}087 = 1{\cdot}66$$

Calibration of Respirometer

Several methods will be found in the literature, most are concerned with first obtaining volume V. This is most accurately done

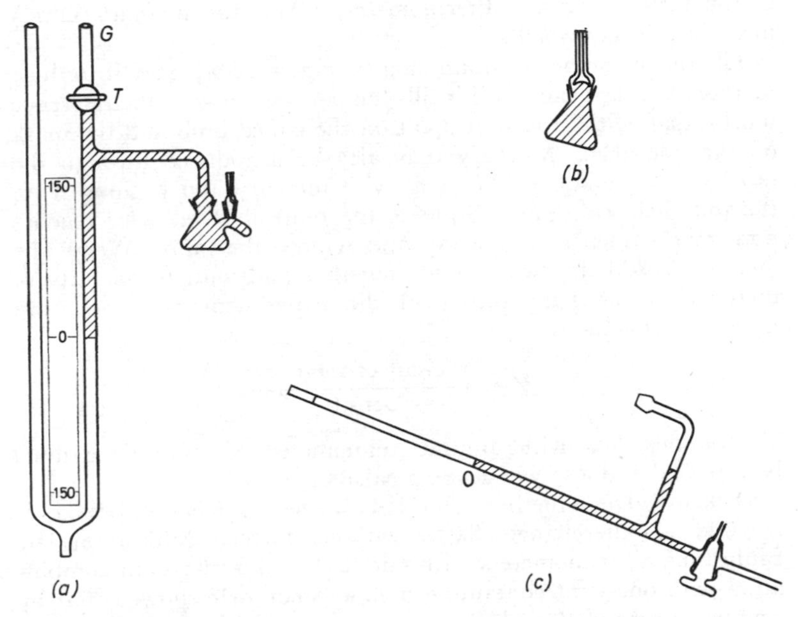

Figure 12.5. The calibrated volume of the Warburg respirometer (see text)

by filling the volume with mercury, weighing the mercury and calculating V from the density. The volume required to be calibrated is the shaded area in *Figure 12.5a*.

The mercury used must be clean, preferably redistilled. The filling of the flasks or manometers should be performed over a large tray, and a balance for weighing to 0·1 g either in the tray or beside it. All flasks and manometers should be cleaned with cleaning fluid (Chapter 1), rinsed well, thoroughly dried, and kept free of dust. Clamp the manometer on a retort stand. Fill the flask with mercury and remove trapped air by touching the bubbles with a thin glass rod until they rise. Hold the flask by the joint with the thumb and forefinger to minimize heating, examine the underneath, for air bubbles, then seat it on the manometer. To prevent mercury spilling out, the cone should enter the socket at an angle (*Figure 12.5b*). The mercury thread should rise to a position in the side arm, adjust the volume if necessary. Mark the mercury level with glass ink or grease pencil. Carefully withdraw the flask, retaining all the mercury, and take the temperature of the mercury with a thermometer. Pour the mercury into a tared beaker and weigh.

Tilt the manometer at an angle (*Figure 12.5c*) and introduce mercury via the cone, until it fills the portion between the reference marks, that is, the 150 mm point on the closed limb and the mark on the side arm. Mercury may also be introduced through the tap by filling a stiff rubber teat with mercury and pushing it on the tube below the tap. Squeeze the bulb until mercury reaches both marks, then close the tap and remove the bulb. Weigh the mercury. Add the two weights together and with the density of mercury at the temperature of the experiment obtained from tables, the volume is

$$V = \frac{\text{Weight of mercury}}{\text{Density}}$$

To use this flask with another manometer this procedure must be repeated and also whenever breakages occur.

Dickens* gives a method which minimizes this work and makes it possible to interchange flasks without further calibration, by calibrating all manometers with one flask and all flasks in combination with one joint containing a short stem. He showed that by simple arithmetic V could be calculated quickly for any flask and manometer.

If v = volume of flask in combination with standard stopper
m = volume of manometer in combination with standard flask
c = volume of standard flask with standard stopper

$$V = v + m - c$$

* *Biochem. J.* **48** (1951) 385

Calibration is again done with mercury as already described, c need be found once only.

The following calculations are given to clarify the process:

Weight of beaker	25·5 g
Weight of beaker + mercury from standard flask + standard stopper	180·8 g
Weight of mercury	155·3 g

If density of mercury at 20°C = 13·54

$$\text{Volume} = \frac{\text{Mass}}{\text{Density}} = \frac{155{\cdot}3}{13{\cdot}54} = 11{\cdot}48 \text{ cm}^3 = c$$

Flask A1

Weight of beaker + mercury from flask A1 + standard stopper	362·4 g
	25·5 g
Weight of mercury	336·9 g

$$\text{Therefore volume of flask} = \frac{336{\cdot}9}{13{\cdot}54} = 24{\cdot}85 \text{ cm}^3 = v$$

Flask A2

Weight of beaker + mercury from flask A2 + standard stopper	294·6 g
	25·5 g
Weight of mercury	269·1 g

$$\text{Therefore volume of flask} = \frac{269{\cdot}1}{13{\cdot}54} = 19{\cdot}85 \text{ cm}^3 = v$$

Manometer 8

Weight of beaker + mercury from standard flask on manometer 8 to side arm mark	188·0 g
	25·5 g
Weight of mercury	162·5 g
Weight of beaker + mercury from manometer between reference points	27·35 g
	25·5 g
Weight of mercury	1·85 g

Total weight of mercury = 163·5 + 1·85 = 165·35

$$\text{Therefore volume} = \frac{165{\cdot}35}{13{\cdot}54} = 12{\cdot}2 \text{ cm}^3 = m$$

Manometer 9

Weight of beaker + mercury from standard flask on manometer 9 to side arm mark	170·7 g
	25·5 g
Weight of mercury	145·2 g
Weight of beaker + mercury from manometer between reference points	26·9 g
	25·5 g
Weight of mercury	1·4 g

Total weight of mercury = 145·2 + 1·4 = 146·6 g

$$\text{Therefore volume} = \frac{146{\cdot}6}{13{\cdot}54} = 10{\cdot}8 \text{ cm}^3 = m$$

It is worth tabulating v and $m - c$ at this point. Some workers record $m - c$ on the backs of manometers.

Flask	v	*Manometer*	$m - c$
A1	24·85	8	12·2 − 11·48 = +0·72
A2	19·85	9	10·8 − 11·48 = −0·68

The minus quantity occurs due to variations in thickness of the glass in the manometer standard joint.

Manometer	*Flask*	V $(v+m-c)$	$K_e^{28°C}$ $V \times 0{\cdot}0907 \times 10^{-3}$
8	A1	24·85 + 0·72 = 25·57	25 570 × 0·0907 × 10⁻³ = 2·33
9	A2	19·85 + (−0·68) = 19·17	19 170 × 0·0907 × 10⁻³ = 1·74

Let us assume that the respirometers are required for measurements of CO_2 output at 28°C with a fluid volume in the flask of 4 cm³. The constants are:

Manometer	*Flask*	$K_e^{28°C}$	$K_{CO_2}^{28°C}$ (3cm³)
8	A1	2·33	2·33 + (3 × −0·02) = 2·27
9	A2	1·74	1·74 + (3 × −0·02) = 1·68

It has been the author's practice to tabulate the constants for one set of manometers with two sets of flasks and place them back to back between sheets of 3 mm perspex bolted together. This protects them and they can be safely stored by the Warburg apparatus (*Figure 12.6*).

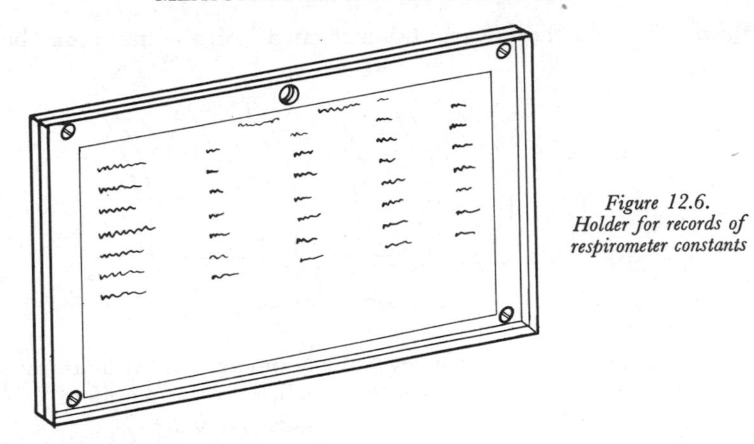

Figure 12.6. Holder for records of respirometer constants

Use of The Respirometers

Records of working experiments are best recorded on a printed form, these are then placed in a loose-leaf folder: They are designed to suit one's own needs as personal tastes or needs differ widely. Paper divided into 1 cm squares on one side can be adapted to most situations.

The clean flasks should be lined up and contents added by pipette. Addition to sidearms and centre well is facilitated by the use of Warburg pipettes (*Figure 12.7*) these have a fine drawn

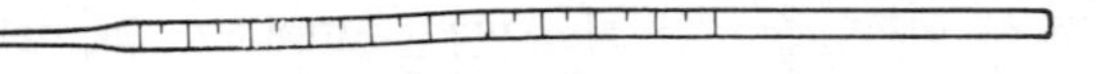

Figure 12.7. The Warburg pipette

tip and are calibrated to the shoulder. Keep the ground faces of the stopper and standard joint dry. Anhydrous lanolin is used as a lubricant and is smeared sparingly on the stoppers and the cone on the manometer. The stopper of one sidearm on each flask has a capillary with an exit in the ground portion of the joint and acts as a tap (*Figure 12.5*). This is used for flushing the flask with gases. The flasks are attached to the manometers and secured by the springs or bands. *Figure 12.5* Tap (*T*) should be open to the atmosphere. The respirometers are then connected to the bath, the bath liquid (water) must cover the flask leaving only capillary tubing exposed. If the atmosphere in the flask is to be changed, rubber tubing from a gassing manifold is connected at (*G*)

(*Figure 12.8*) and the sidearm tap is opened. All the flasks may be

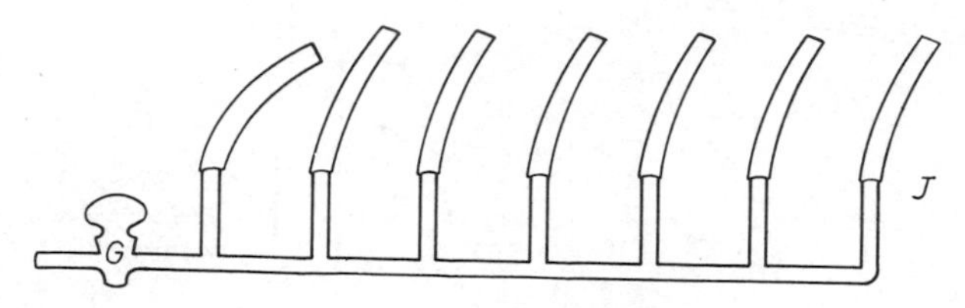

Figure 12.8. Gassing manifold

gassed together, a difference in level of manometer fluid will indicate that gas is flowing. An open tube attached to the manifold and dipping into the bath water acts as a safety valve.

The flasks are shaken and gassed for at least 10 min for temperature equilibration, shaking is then stopped and the tap (T) is turned on each manometer to direct the gas into the air, and the stoppers on the flask sidearms are quickly closed. The rubber gassing tubes are disconnected and all joints are tightened by 'grinding' them in until they resist the turning motion. Adjust the fluid in the open limb to give maximum movement and then readings may be taken. The closed limb is set to zero and the flasks shaken. Readings are taken for the first few minutes to check that there are no leaks, then the experiment may begin. At zero time the substrate is tipped into the flask, and shaking commenced. At regular time intervals shaking is stopped, the manometer fluid brought to zero in the closed limb, and the level in the open limb is recorded.

A thermobarometer must always be included and its readings subtracted or added, depending on type of gas measurement, from the readings of the other respirometers. Its use is illustrated by the following examples. Table 12.2 is an experiment measuring carbon dioxide output. The thermobarometer changes reflect variations common to all of the respirometers. If a rise in the thermobarometer is ignored, a higher yield of carbon dioxide would be indicated, therefore the difference in thermobarometer readings is subtracted. A fall in the thermobarometer would have to be added to the readings of the others. With oxygen uptake the reverse situation holds, that is, if the thermobarometer rises its net rise must be added to the others. (Table 12.3.)

Table 12.2. Measurement of Carbon Dioxide

Manometer	*1*				*2*				*3*					
Time	*Reading*	*Difference*	*Corrected Difference*	*Total*	*Reading*	*Difference*	*Corrected difference*	*Total*	*Reading*	*Difference*	*Corrected difference*	*Total*	*Thermo-barometer*	*Difference*
0	3	0	0	0	6	0	0	0	19	0	0	0	11	0
5	7	4	2	2	10	4	2	2	26	7	5	5	13	2
10	9	2	2	4	12	2	2	4	31	5	5	10	13	0
15	9	0	3	7	12	0	3	7	33	2	5	15	10	3

Table 12.3. Measurement of Oxygen Uptake

Mano-meter	*8*				*9*				*10*					
Time	*Reading*	*Difference*	*Corrected difference*	*Total*	*Reading*	*Difference*	*Corrected difference*	*Total*	*Reading*	*Difference*	*Corrected difference*	*Total*	*Thermo-barometer*	*Difference*
0	113	0	0	0	156	0	0	0	89	0	0	0	11	0
5	101	−12	−14	−14	121	−35	−37	−37	72	−17	−19	−19	13	2
10	88	13	−12	−26	−90	−31	−30	−67	54	−18	−17	−36	12	−1

With practice two columns may be eliminated, the difference and correction being performed mentally, and only the total recorded.

The volume x of gas taken up is calculated as follows:

$$x = hK$$

For manometer 8 above which has a constant of 1·50 the total uptake over 10 min $= 1{\cdot}5 \times -26 = -39$ of oxygen. If 5 mg dry weight of enzyme was used then

$$Q_{O_2} = \frac{-39 \times 60}{5 \times 10} = -47$$

Manometric methods are useful in demonstrating and measuring tissue respiration. Samples of animal and plant tissue, thinly sliced, are immersed in buffered nutrient solutions and the oxygen uptake measured at constant temperature. Plant tissues such as potato, carrot and leaves will be found to respire over very long periods and are able to be used some time after sampling. Animal tissues, however, have to be fresh and delays before measurement must be kept to a minimum.

Measurement of Respiration of Liver Slices

Warburg flasks are first prepared in readiness, these are put in numerical order and the experimental conditions written on a *pro forma* or in a notebook. The Warburg bath is turned on and set at 37°C and 3 cm^3 of buffered Ringer solution is placed in each flask. Buffered Ringer Solution: 9 g sodium chloride, 0·23 g potassium chloride, 0·24 g calcium chloride, are dissolved in water and made to 1 dm^3. The calcium chloride is best added as 10 cm^3 of a 0·22 M solution made from hydrated calcium chloride. This is made up in excess and the solution adjusted to 0·22 M after the calcium has been assayed.

Add 10 cm^3 of M/15 Sorensen phosphate buffer (pH 7·2) to every cm^3 of Ringer solution.

A rat or guinea pig is killed by a blow on the base of the skull and the jugular vein severed using scissors. When the blood has drained, open the abdominal cavity and remove the liver. Wash it with Ringer solution, remove a lobe and cut slices using one of the methods described in Chapter 10. Weigh portions of the slices on tared watch glasses or preferably siliconed or water repellent paper. Transfer samples to the flasks using a damped spatula. Place 0·2 cm^3 of 10 per cent potassium hydroxide into the centre well using a Warburg pipette, followed by a coil of filter paper. This must project into the flask the object being to give evolved carbon dioxide a greater chance of being absorbed.

Place the flasks on the manometers and the manometers on the bath. Connect the manometers to the gassing manifold and flush for 5 min with oxygen, at the same time shaking the flasks. After this time, stop the shaking mechanism, turn the 3-way taps to direct the oxygen into the atmosphere, remove connecting tubing and re-seat the flasks on the manometers. As oxygen uptake is to be measured, adjust the manometer fluid in the open limb so that its level is higher than in the other one. This is already the case if gassing is efficient. If not, quickly open the tap, lower the manometer fluid as far as possible, close the tap and raise the fluid again. Provided there are no leaks in the system the level in the open limb will rise above the level in the other limb. Set the flask shaking and readings may commence.

An example of a record of an experiment is as follows: (Table 12.4).

TABLE 12.4

Flask contents	1	2	T.B.
Buffered Ringer	2·8 cm³	2·8 cm³	3 cm³
Wt. of wet liver slices	220 mg	250 mg	0
Centre well 10 % KOH	0·2 cm³	0·2 cm³	0

Time	*Reading*	*Difference*	*Corrected difference*	*Total*	*Reading*	*Difference*	*Corrected difference*	*Total*	*Reading*	*Difference*
0	160	0	0	0	180	0	0	0	22	0
10	125	−35	−35	−35	141	−39	−30	−39	22	0
20	86	−39	−37	−72	98	−43	−41	−80	20	−2
30	55	−31	−31	−103	61	−37	−37	−117	20	0

If 100 mg of wet tissue has a dry weight of 19·5 mg and the constants are (1) 1·92, (2) 1·93, the $Q_{O_2}^{37°C}$ in oxygen may be calculated as follows:

Manometer 1

$$x = hK = -\ 103 \times 1{\cdot}92 = -\ 198 \text{ mm}^3/30 \text{ min}$$

$$O_2 \text{ uptake/h} = -198 \times 2 = -396$$

$$\text{Dry weight of tissue used} = 19{\cdot}5 \times 2{\cdot}2 = 42{\cdot}9 \text{ mg}$$

$$Q_{O_2} = \frac{-396}{42{\cdot}9} = -\ 9{\cdot}24$$

Manometer 2

$$x = hK = -117 \times 1{\cdot}93 = -226 \text{ mm}^3/30 \text{ min}$$

$$O_2 \text{ uptake/h} = -226 \times 2 = -452$$

$$\text{Dry weight of tissue used} = 19{\cdot}5 \times 2{\cdot}5 = 48{\cdot}7 \text{ mg}$$

$$Q_{O_2} = \frac{-452}{48{\cdot}7} = -9{\cdot}3$$

On completion of experiments all manometer taps are opened to the atmosphere and the respirometers removed from the bath. Grease should be cleaned from the joints, using an alcohol swab if necessary. The flasks are rinsed with water and cleaned by boiling in liquid soap, and finally rinsed several times with distilled water.

Manometer fluid may be 'topped up' by injecting into the reservoir through the rubber tubing, using a syringe and needle.

Manometer Fluids: *Krebs'*

Sodium bromide anhydrous	44 g
Triton X 100	0·3 g
Evans blue	0·3 g
Water to 1 dm^3	

Brodie's

Sodium chloride	46 g
Sodium glycocholate or	
Tauroglycocholate	10 g
Evans blue	0·2 g
Water to 1 dm^3	

Several other manometric methods have been recorded usually microtechniques. The most intriguing are the capillary ultra microrespirometer (E. Kirk*) and The Cartesian Diver (Holter.†)

In the capillary method it is possible to measure the respiration of a single cell. This is placed in a very fine capillary that is belled out at the lower end. After insertion of the tissue the lower end is closed and alkali is introduced into the capillary. This absorbs carbon dioxide. As the cell takes up oxygen the alkali level drops. This movement is observed and measured through a microscope.

The Cartesian Diver is a miniature long-necked flask of about 10 mm^3 volume. A cell suspension is introduced and the neck is sealed first with alkali and then oil. The diver is then immersed in a high density salt solution contained in a closed system. The pressure is adjusted so that the diver is suspended below the surface of the solution and is stationary. As oxygen is taken up by the cells its volume is replaced by the salt solution, this alters its density and the diver sinks. The diver's position is kept constant by pressure adjustment, the pressure change is a measure of the oxygen uptake.

Other methods of enzyme analysis involve the chemical assay of the utilization of the substrate or the products of reaction involving well-known analytical methods. For example, glucose oxidase activity may be followed by the assay of glucose, urease by the assay of ammonia, monoamine oxidase may be followed both by its oxygen uptake manometrically and the production of ammonia by the Conway microdiffusion technique.

---

* *Quantitative Ultramicroanalysis.* 1950. Wiley

† 'The Cartesian Diver'. *C.r. Trav. Lab. Carlsburg* **24** (1943) 399

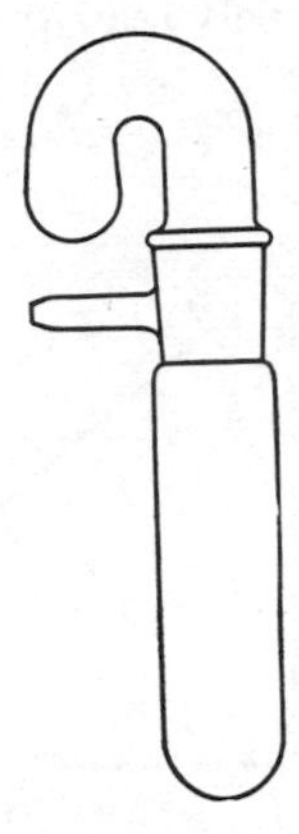

Figure 12.9. The Thunberg tube

One other useful technique is that employed by Thunberg. It gives a method of studying biological oxidation. The process of oxidation involves a chain of enzymes performing oxidation and reduction. A substance becomes oxidized as hydrogen is given up, catalysed by dehydrogenases. If no oxygen is present this hydrogen will reduce methylene blue, the rate of reduction being a measure of activity. This may be followed in a constant temperature cell housing of an absorptiometer or spectrophotometer.

The cell or tube used is capable of being evacuated (*Figure 12.9*). Tissue is placed in the hollow stopper and the substrate with buffer and methylene blue is placed into the tube. After evacuation, using a water pump, the stopper is turned and the tube allowed to reach 37°C. The contents are then mixed and the absorbance recorded with time.

Experiment: In the tube place 0·1 cm^3 of 0·1 M sodium succinate.

5 cm^3 $\frac{M}{15}$ phosphate buffer pH 7

0·2 cm^3 1:2000 methylene blue, and

in the stopper 1 cm^3 of bacterial suspension

or 1 cm^3 10–20 per cent tissue homogenate

Two other tubes are prepared, one without the tissue, the stopper containing buffer only, and the other without the methylene blue. The latter tube is a reagent blank, and the former will show the density at the beginning of the experiment. Place the tubes in a water bath at 37°C for 5 min and evacuate them with a water pump. Close the stoppers and mix the contents. At 5 min intervals zero a colorimeter with the blank and read the optical density of the test solution.

Draw a graph of the rate of oxidation (*Figure 12.10*).

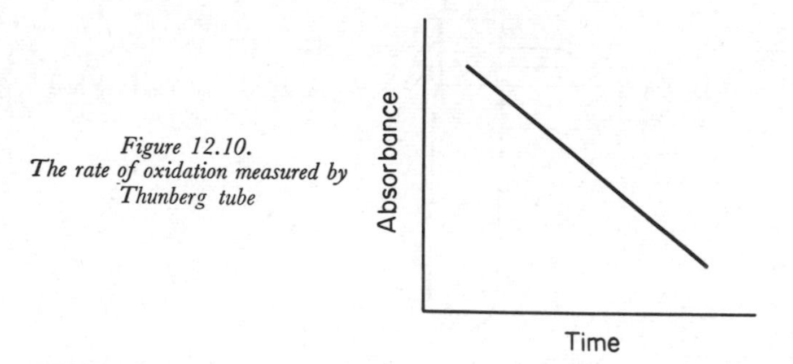

Figure 12.10.
The rate of oxidation measured by Thunberg tube

Classification of Enzymes

Carbohydrates and lipids are classified with respect to their chemical structure. This is not convenient with enzymes as they are all proteins of high molecular weight and as yet unknown structure, except a number of the prosthetic groups or co-enzymes. Enzymes therefore, are classified by the reaction they catalyse, and as the knowledge of, and as the number of enzymes is continually increasing, so the classification undergoes modification. A number of varying lists will be found in the literature. It is possible to say that enzymes bring about three changes, hydrolysis, oxidation and addition. These headings are too broad and the following is a list of the better known sub-groups with a few examples.

Proteolytic Enzymes

These may be divided into proteinases and peptidases.

Proteinases—Pepsin found in gastric juice, trypsin and chymotrypsin in pancreatic juice, and papain have already been mentioned.

Rennin catalyses the clotting of milk and is found in calves' stomachs and other ruminants.

Ficin is a plant proteinase found in the sap of the fig tree and other plants.

These enzymes split proteins into large segments doing so at specific linkages. Pepsin has been shown to attack the linkage between aromatic amino acids.

Peptidases—Carboxypeptidase of the pancreas splits polypeptides and is activated by trypsin.

Dipeptidases are found in the intestine, plant tissues and bacteria. It is believed that there is a specific enzyme for each different dipeptide (two amino acids linked together).

Carbohydrases

These enzymes catalyse the hydrolysis of carbohydrates. Sucrase, maltase, lactase and amylase have already been mentioned and occur in intestinal juice, yeasts and other micro-organisms.

Cellulase will break down cellulose and occurs in many micro-organisms, protozoa and insects, and has been extracted from snails.

Chitinase attacks the chitin in invertebrates and is found in micro-organisms and may also be extracted from snails.

Esterases

These enzymes hydrolyse esters to the alcohol and acid. Lipase is found in the intestine and splits fats and is activated by the presence of bile.

Cholinesterase is found in blood and splits acetyl choline into choline and acetic acid. A highly specific acetylcholinesterase is in nerve fibres and nerve endings.

$$(CH_3)_3{\equiv}\overset{\overset{\displaystyle OH}{|}}{N}{-}CH_2{-}CH_2{-}O{-}\overset{\overset{\displaystyle O}{|}}{C}{-}CH_3 + H_2O$$

$$\longrightarrow (CH_3)_3{\equiv}\overset{\overset{\displaystyle OH}{|}}{N}{-}CH_2CH_2OH + CH_3COOH$$

Phosphatases are found widely spread in animal and plant tissues, and split phosphoric acid esters into the alcohol and phosphoric acid or phosphate. Two phosphatases found in blood are called alkaline and acid phosphatases because of their optimum pH which is pH9 and pH5 respectively. The splitting of glycerophosphoric acid is an example of their action.

$$\begin{array}{l} CH_2OH \\ | \\ CHOH \quad \overset{\displaystyle OH}{|} \\ | \\ CH_2{-}O{-}\underset{\underset{\displaystyle O}{\|}}{P}{-}OH \end{array} + H_2O \longrightarrow \begin{array}{l} CH_2OH \\ | \\ CHOH \\ | \\ CH_2OH \end{array} + H_3PO_4$$

Other specific phosphatases are glucose-6-phosphatase and fructose-1:6-diphosphatase which are part of the mechanism for liberating glucose.

Sulphatases are widely distributed in animal and plant tissues and hydrolyse sulphate esters into the alcohol and the acid or sulphate.

$$\underset{\text{K phenyl sulphate}}{C_6H_5O{\cdot}SO_3K} + H_2O \longrightarrow \underset{\text{Phenol}}{C_6H_5OH} + \underset{\text{K bisulphate}}{KHSO_4}$$

Enzymes hydrolysing amino-compounds giving or transferring ammonia

Amidases—Urease as already mentioned breaks down urea to carbon dioxide and ammonia. Other examples are asparaginase and glutamase and aspartase.

$$\underset{\text{Asparagine}}{\begin{array}{l} CONH_2 \\ | \\ CH_2 \\ | \\ CHNH_2 \\ | \\ COOH \end{array}} + H_2O \longrightarrow \underset{\text{Aspartic acid}}{\begin{array}{l} COOH \\ | \\ CH_2 \\ | \\ CHNH_2 \\ | \\ COOH \end{array}} + NH_3$$

Arginase splits arginine to ornithine and urea.

Deaminases—These remove ammonia from purines and pyrimidines, e.g. adenase and guanase.

$$\begin{array}{l} \quad\;\; HN{-}C{=}O \\ \quad\;\;\;\, | \quad\;\; | \\ NH_2{-}C \quad C{-}NH \\ \quad\;\;\; \| \quad\;\, \| \quad\;\; \rangle CH \\ \quad\;\;\, N{-}C{-}N \end{array} + H_2O \longrightarrow \begin{array}{l} \quad\; HN{-}C{=}O \\ \quad\;\;\; | \quad\;\; | \\ O{=}C \quad C{-}NH \\ \quad\;\;\; | \quad\;\;\, \| \quad\;\; \rangle CH \\ \quad\; HN{-}C{-}N \end{array} + NH_3$$

Carboxylases

These enzymes remove or fix carbon dioxide.

Carboxylase itself is more accurately pyruvic acid decarboxylase (page 473) and requires cocarboxylase. It occurs in plants and micro-organisms and is usually obtained from brewers' yeast.

Aminoacid decarboxylases are found in animals, plants and micro-organisms and have the following general reaction:

$$R{\cdot}CH{\cdot}NH_2{\cdot}COOH \rightarrow CO_2 + R{\cdot}CH_2{\cdot}NH_2$$

Some examples are L-histidine decarboxylase which gives histamine, L-glutamic acid decarboxylase which gives aminobutyric acid, and L-lysine decarboxylase which gives cadaverine.

Enzymes catalysing oxidation and reduction

Energy is made available in living organisms by the oxidation of nutrients. This is performed gradually by a host of enzymes. A compound is said to be oxidized if hydrogen is removed so this group includes dehydrogenases.

Oxidases are metal proteins and are inhibited by traces of H_2S and cyanide.

Glucose oxidase converts glucose in the cyclic form in the presence of water and oxygen to gluconic acid and hydrogen peroxide. This is one of the most specific enzymes as it will not affect any other natural sugar. Because of this property it is used in a method for estimating glucose in biological material (Chapter 11, page 463).

Catalase and peroxidase break down hydrogen peroxide to water and oxygen.

Cytochrome oxidase is an iron enzyme giving cytochrome-c and water from reduced cytochrome-c and oxygen.

Ascorbic acid oxidase is a copper enzyme oxidizing ascorbic acid to dehydroascorbic acid.

Amino acid oxidases deaminate amino acids.

$$R{-}\overset{\displaystyle H}{\underset{\displaystyle NH_2}{C}}{-}COOH \longrightarrow R{-}\overset{\displaystyle O}{\overset{\|}{C}}{-}COOH + NH_3$$

Amine oxidases remove ammonia from amines or diamines.

$$R{\cdot}CH_2{\cdot}NH_2 + H_2O + O_2 \longrightarrow R{\cdot}CHO + H_2O_2 + NH_3$$

Dehydrogenases catalyse the production of oxidizing substances. The presence of a hydrogen acceptor is required, this is the co-enzyme.

Lactic dehydrogenase gives pyruvic acid.

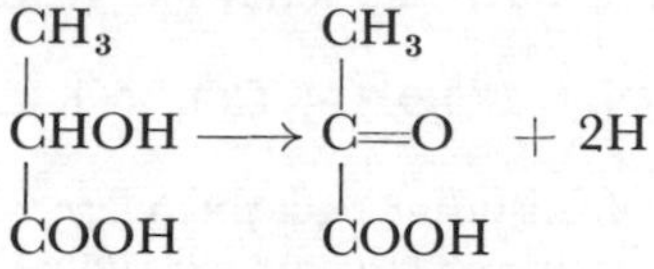

$$
\begin{array}{l}
CH_3 \\
| \\
CHOH \\
| \\
COOH
\end{array}
\longrightarrow
\begin{array}{l}
CH_3 \\
| \\
C{=}O \\
| \\
COOH
\end{array}
+ 2H
$$

Malic dehydrogenase gives oxaloacetic acid from malic acid.

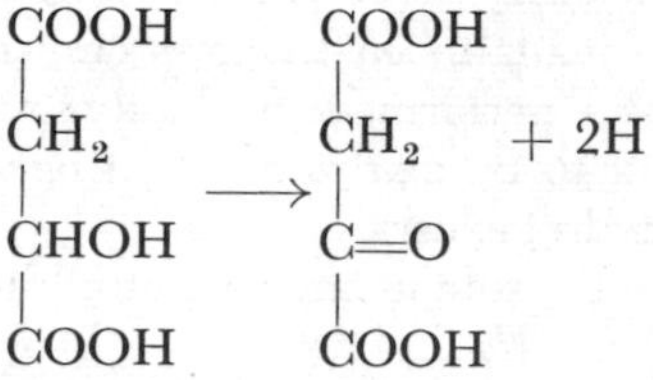

$$
\begin{array}{l}
COOH \\
| \\
CH_2 \\
| \\
CHOH \\
| \\
COOH
\end{array}
\longrightarrow
\begin{array}{l}
COOH \\
| \\
CH_2 \\
| \\
C{=}O \\
| \\
COOH
\end{array}
+ 2H
$$

Adding and Transferring Enzymes

These enzymes are able to add a group or molecule to a compound or transfer a group from one molecule to another. These reactions are usually reversible. Some of the groups involved are H_2O, NH_3, CO_2, SH, Acetyl, Carbamyl and more complex groups.

Fumarase adds water to fumaric acid to form malic acid.

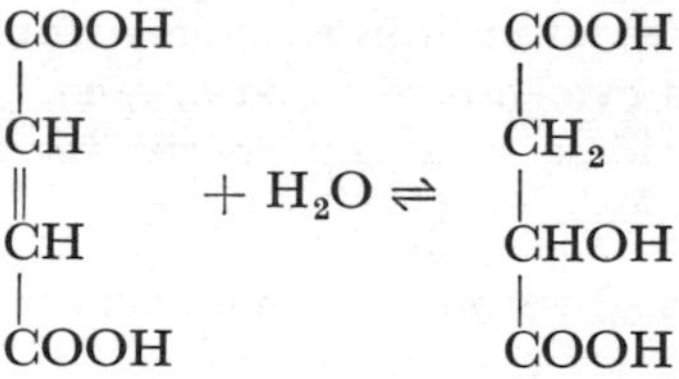

$$
\begin{array}{l}
COOH \\
| \\
CH \\
\| \\
CH \\
| \\
COOH
\end{array}
+ H_2O \rightleftharpoons
\begin{array}{l}
COOH \\
| \\
CH_2 \\
| \\
CHOH \\
| \\
COOH
\end{array}
$$

malic acid

A large group of enzymes in this class are the transaminases which affect the transfer of an amino group from one amino acid to another, or to an acid thereby forming an amino acid, e.g.

Aspartic acid + α ketoglutaric acid $\rightleftharpoons$ oxaloacetic acid + glutamic acid

Glutamic acid + pyruvic acid $\rightleftharpoons$ α ketoglutaric acid + alanine

Another example of a transferring enzyme is α glutamyl transferase.

Glutathione + peptide $\longrightarrow$ cysteinylglycine + α glutamylpeptide.

Phosphorylases

This important group of enzymes has the ability to break down polysaccharides and at the same time add a phosphate group to the sugar residues. Other phosphorylases transfer a phosphate group from one compound to another.

Glycogen phosphorylase catalyses the break-down of glycogen in the presence of phosphate to glucose monophosphate.

Sucrose phosphorylase catalyses the reaction between sucrose and phosphate to give fructose and glucose monophosphate.

There are many other enzymes too numerous to mention of varied action and complexity.

APPENDIX 1

THE LITERATURE OF CHEMISTRY

Every scientific worker should have some knowledge of the literature of his subject and the library is as important as the bench to any scientist who undertakes original work. The technique for using a large chemical library and for literature searching is fairly complicated and it would be inappropriate to describe it here. Those who need to use a library regularly and in depth are urged to read:

R. T. Bottle (editor)	*Use of the Chemical Literature*, 2nd edn., 1969.	(Butterworths)

which gives an extensive survey of the literature together with clear directions for literature searching including useful practical exercises. Other useful volumes are:

G. M. Dyson	*A Short Guide to Chemical Literature*, 1958.	(Longmans Green)
M. G. Mellon	*Chemical Publications, their Nature and Use*, 1958.	(McGraw-Hill)
E. J. Crane, A. M. Patterson and E. B. Marr	*A Guide to the Literature of Chemistry*, 1957.	(Wiley)

A more general and philosophical approach including a discussion of computers and modern storage systems is found in:

B. C. Vickery	*Techniques of Information Retrieval*, 1970.	(Butterworths)

The account which follows is necessarily brief, and covers the field of chemistry from the great encyclopaedias to the abstracting services.

1. *Encyclopaedias and Large Reference Works*

There are several encylopaedias of technical chemistry which cover most of the important topics of pure chemistry and which have numerous references to the literature. Perhaps the most useful of these:

Kirk and Othmer	*Encyclopaedia of Chemical Technology*, 2nd edn. Published in 18 vols.	(Interscience)

Thorpe	*Dictionary of Applied Chemistry* in parts 1937–56.	(Longmans Green)
Ullman	*Encyclopädie der Technischen Chemie*, from 1951.	(Urban and Schwarzenberg)

Smaller works giving background information are:

Handbook of Chemistry and Physics, 53rd edn. 1973.	(Chemical Rubber Co.)
Lange's Handbook of Chemistry	(Handbook Publishers, Sandersly, Ohio)

The larger reference works are directed to one of the main branches of the subject.

In inorganic chemistry the most important is

Gmelin's Handbuch der Anorganischen Chemie

which began publication in 1924. This is extremely comprehensive although some of the earlier volumes are now out of date and supplementary volumes are now being published. It is, of course, in German. The English equivalent is:

Mellor's Comprehensive Treatise on Inorganic Chemistry

which was published in 16 volumes between 1922 and 1937. The publication of supplementary volumes began in 1956. An American reference work is:

Comprehensive Inorganic Chemistry

edited by Sneed, Maynard and Brastead. This is rather more modern in approach although a little less detailed than Gmelin or Mellor.

Organic chemistry is plentifully supplied with reference works. The most important of these is:

Beilstein's Handbuch der organischen chemie

This massive work is now in its fourth edition, the first being in 1881–3, the latest being published in 26 volumes and 2 indexes between 1918 and 1937 and covering the literature up to the end of 1909. Three supplements have been published; the first in 15 volumes between 1928 and 1938, overlapping the publication of the fourth edition; the second supplement began publication in 1941 and is complete; the third supplement is still in course of publication and will cover the period 1930–1949. The system of classification in Beilstein is rather complex and is difficult for the beginner to follow. A useful guide is available:

E. H. Huntress	*A Brief Introduction to the Use of Beilstein's Handbuch der organischen chemie*	(Wylie, New York)

Although the reference literature is dominated by Beilstein other useful publications are available. Among the most useful are:

Faraday's	*An Encyclopaedia of Hydrocarbon Compounds*	(Chemindex Ltd.)
Radt, F. (editor)	*Elsevier's Encyclopaedia of Organic Chemistry*	(Elsevier)

This is still in the course of publication

Rodd, E. H. (editor)	*Chemistry of Carbon Compounds*	(Elsevier)

A smaller work which is very useful is:

Heilbron and Bunbury's Dictionary of Organic Compounds	(Eyre and Spottiswood and Spon)

A new edition in five volumes was published in 1965. Annual supplements have followed with a formula index in 1971. This series is much easier to use than the larger publications and is a valuable source of information where an exhaustive search is not necessary.

An Encyclopaedia of organic chemical techniques is:

Technique of Organic Chemistry	(John Wiley–Interscience)

which has been published in volumes since 1945. The second edition began in 1949 with vol. 16 appearing in 1969. Each volume covers a specific technique or a related group of techniques.

Physical chemistry has not been served with massive reference works in the same way as the other branches of the science.

The main general text of this type is:

Partington, J. R.	*An Advanced Treatise on Physical Chemistry*	(Longmans Green)

in five volumes published between 1949 and 1954.

In biochemistry the most exhaustive work in English is:

Florkin, M. and Stotz, E. H. (editors)	*Comprehensive Biochemistry*	(Elsevier)

which has so far reached vol. 28 with more to be issued.

Practical biochemistry is represented by several series

Biochemical Preparations	(Wiley)

Began publishing in 1949 and a new volume has appeared almost every year.

Colowick, S. P. and Kaplan, N. O. (editors)	*Methods in Enzymology*	(Academic Press)

Began in 1955 with vol. 25 appearing in 1972. Despite the specialist title of this work it covers a wide range of biochemical techniques in very great detail and is of importance to all biochemists, not only to enzymologists.

A recently inaugurated series is:

Work, T. S. and Work, E. (editors)	*Laboratory Techniques in Biochemistry and Molecular Biology*	(North Holland)

Volume I appeared in 1969. A second volume has been published and several additions to the series are in preparation. Each volume contains a small number of very comprehensive practical reviews of important techniques by acknowledged experts. The individual sections are available in paperback.

Useful single volume reference books are:

Long, C. (editor)	*The Biochemist's Handbook*, 1961.	(Spon)
Dawson, R. M. C., Elliot, D. C., Elliot, W. H. and Jones, K. M. (editors)	*Data for Biochemical Research*, 1969.	(Oxford)

The former has several useful articles on modern laboratory techniques by experts in their respective fields.

The whole of analytical chemistry is covered in great detail in:

Kolthoff, I. M. and Elving, P. J.	*Treatise on Analytical Chemistry*	(John Wiley–Interscience)

The three 'volumes' began publication in 1959. The first two 'volumes' have each reached up to fifteen separate books so far. The third 'volume' has reached the third 'part' with several more in preparation.

2. *Annual Reviews, Series, etc.*

The large reference works quoted above can never be up-to-date, and, as information increases at an explosive rate they fall further and further behind. The main advances which occur are noted much more rapidly in the regular annual reviews and reports.

The most general of these are:

Annual Reports of the Progress of Chemistry	(Chemical Society)
Annual Reports of the Progress of Applied Chemistry	(Society of Chemical Industry)

Other annual publications are related to specialized fields. Among the more important are:

Annual Review of Physical Chemistry	(Since 1950)
Annual Review of Biochemistry	(Since 1932)
Advances in Protein Chemistry	(Since 1944)
Advances in Carbohydrate Chemistry	(Since 1945)
Advances in Enzymology	(Since 1941)
Physiological Reviews	(Since 1920)

Besides these there are several regular publications dealing entirely with preparative chemistry:

Organic Syntheses
Inorganic Syntheses

3. *Journals*

As science expands the number of journals expands and every year sees an extension of the publications devoted to smaller and smaller areas of science. The journals are listed in:

World List of Scientific Periodicals (Butterworths)

which is itself a massive tome.

It would be impossible to give anything like a comprehensive list of useful journals but a few which may be of particular use to readers of this book must be mentioned.

There are few journals dealing mainly with apparatus and techniques. In this country the most important are:

Laboratory Practice	(United Trade Press)
Journal of Science Technology	(Institute of Science Technology)
Scientific Instrument Review	(Physical Society)

These should be studied regularly by those whose main interest is equipment and techniques.

A useful monthly publication is :

Laboratory Equipment Digest (Gerard Mann Ltd.)

Although this exists mainly to publicize new apparatus, and is given free to most of its 'subscribers' it describes much of the new apparatus available and often contains useful articles on particular pieces of equipment. It is much more than just an advertising sheet.

An important German publication is:

Glas und Instrumenten Technik (*G.I.T.*) (Bondway Publishing Co.)

which publishes excellent articles on apparatus.

It is not proposed to deal with the other important journals here. Every laboratory worker will soon find (or be directed to) those most appropriate for his special interest.

4. *Abstract Services*

No one person could possibly read all the journals containing information in his subject. This has led to the rise of abstracting services which summarize the material and present it regularly in greatly condensed form. After reading the abstracts the scientist can then refer to those articles and papers which interest him most.

In chemistry the most important is:

Chemical Abstracts (American Chemical Society)

in which over 8000 journals are abstracted.

For earlier work:

British Abstracts.

Which has also been titled

British Chemical Abstracts
and *British Chemical and Physiological Abstracts*

may be found useful. This ceased publication in 1953.

The cumulative indexes of the abstracting journals above, which are published regularly every 10 years are ideal for literature searching. Other abstracting journals are more specialized covering the literature on particular topics.

An extension of the abstracting system is the recent publication of journals which list the contents of a selection of other journals. The scientist may then read those articles whose titles appear appropriate to his interests. These are useful because the contents lists can be published much more quickly than the abstracts and it is possible to keep up with the very latest publications in a particular field.

Among these is:

Current Contents (Institute for Scientific Information)

which covers over 1000 journals and publishes their contents often only a few days after publication of the articles.

5. *Other Useful Publications*

Among all the literature one form is often forgotten or omitted. Especially useful to those interested in scientific techniques are the publications of the laboratory supply houses. The leaflets, house journals, and catalogues of these organizations are a mine of information on the latest methods and should not be ignored or used merely with the order book. The catalogues especially merit study as they often contain valuable information about the form of apparatus from which an insight into the selection of the best apparatus for a particular job may be cultivated.

APPENDIX 2

BIBLIOGRAPHY

General Practical Chemistry

1. Atkins, H. J. B. (editor) — *Tools of Biological Research*, 1960 (series of 3 volumes). — Blackwell

2. Ewing, G. M. — *Instrumental Methods of Chemical Analysis*, 3rd edn., 1969. — McGraw-Hill

3. Holness, H. — *Small Scale Organic Preparations*, 1959. — Pitman

4. Holness, H. — *Small Scale Inorganic Preparations*, 1960 — Pitman

5. Elvidge, J. A. and Sammes, P. G. — *A Course in Modern Techniques of Organic Chemistry*, 2nd edn., 1966. — Butterworths

6. Milton, R. F. and Waters, W. A. — *Methods of Quantitative Micro-Analysis*, 2nd edn., 1955. — Edward Arnold

7. Sixma, F. L. J. and Wynberg, H. — *A Manual of Physical Methods in Organic Chemistry*, 1964. — Wiley (New York)

8. Strouts, C. R. N., Gillfillan, J. H. and Wilson, H. N. — *Chemical Analysis. The Working Tools*, 1962. — Oxford

9. Swift, E. H. and Butler, E. A. — *Quantitative Measurements and Chemical Equilibria*, 1972. — W. H. Freeman

10. Vogel, A. I. — *A Textbook of Macro and Semimicro Qualitative Inorganic Analysis*, 1964. — Longmans Green

11. Vogel, A. I. — *A Textbook of Practical Organic Chemistry*, 3rd edn., 1956. — Longmans Green

12. Vogel, A. I. — *A Textbook of Quantitative Inorganic Analysis*, 3rd edn., 1961. — Longmans Green

Author	Title	Publisher
13. Willard, H. H., Merritt, L. L. and Dean, J. A.	*Instrumental Methods of Analysis*, 4th edn., 1965.	Van Nostrand (New York)
14. —	*The Instrument Manual*, 4th edn., 1970.	United Trade Press

Chapter 1

Author	Title	Publisher
Edwards, J. A.	*Laboratory Management and Techniques*, 1960.	Butterworths
Elliot, A. and Home Dickinson, J.	*Laboratory Instruments, their Design and Application*, 2nd edn., 1959.	Chapman & Hall
Guy, K.	*Laboratory Administration and Organization*, 2nd edn., 1973.	Butterworths
Lewis, H. F.	*Laboratory Planning*, 1962.	Rheinhold
Shand, E. B.	*Glass Engineering Handbook* 2nd edn., 1959.	McGraw-Hill (New York)
—	*British Catalogue of Plastics*.	National Trade Press
—	*Booklets on Perspex, PTFE, PVC, Polypropylene.*	I.C.I.
—	*Leaflets on Pyrex Laboratory Glass.*	James A. Jobling and Co.

Also Nos. 2, 6, 8, 11, 12 and 14, above.

Chapter 2

Author	Title	Publisher
Coulston, E. A. and Herrington, E.F.G.	*Laboratory Distillation Practice*, 1958.	Newnes
Davies, E. H.	*Durran's Solvents*, 1971.	Chapman & Hall
Krell, E.	*Handbook of Laboratory Distillation*, 1963.	Elsevier
Weisz, H.	*Microanalysis by the Ring-Oven Technique*, 1970.	Pergamon
—	*Whatman Technical Bulletins.*	H. Reeve Angel and Co.
—	*Filtrations in the Chemical Laboratory.*	Carl Schleicher and Schüll

Also Nos. 1, 6, 8, 11, 12, above

BIBLIOGRAPHY

Chapter 3

Author	Title	Publisher
Bates, R. G.	*Determination of pH, Theory and Practice*, 1964.	Chapman and Hall
Bell, R. P.	*Acids and Bases*, 1965.	Methuen
Browning, D. R.	*Electrometric Methods*, 1969.	McGraw-Hill
Dole, M.	*The Glass Electrode*, 1941.	Chapman & Hall
Dombrow	*Instrumental Methods in Chemistry Vol. 1. Electrometric Methods*, 1967.	Pitman
Levison, L. L.	*An Introduction to Electroanalysis*, 1964.	Butterworths
Tomicek, O.	*Chemical Indicators*, 1951.	Butterworths
Wilson, A.	*pH meters*, 1970.	Kogan Page
—	*pH Values and their Determination.*	B. D. H. Limited
—	*Buffer Substances, Buffer Solutions, Buffer Titrisols.*	E. Merck Ltd.

Also Nos. 2, 7, 8, 12, 13, 14 above.

Chapter 4

Author	Title	Publisher
Feigl, F.	*Spot Tests in Inorganic Analysis*, 5th edn., 1958.	Elsevier
Johnson, W. C. (editor)	*Organic Reagents for Metals and Certain Radicals*, 1964.	Hopkins and Williams Ltd.
Macdonald, A. M. G.	*Elementary Titrimetric Analysis*, 1960.	Butterworths
—	*The B.D.H. Spot Test Outfit Handbook*, 1965.	B. D. H. Limited

Also Nos. 3, 9, 10, 12 above.

Chapter 5

Author	Title	Publisher
Cheronis, N. D., Entrikin, J. B. and Hodnett, E. M.	*Semimicro Qualitative Organic Analysis*, 1965.	Wiley
Dixon, J. P.	*Modern Methods in Organic Microanalysis*, 1968.	Van Nostrand
Feigl, F.	*Spot Tests in Organic Analysis*, 7th edn., 1966.	Elsevier

Mann, F. G. and Saunders, E. C.	*Practical Organic Chemistry*, 4th edn., 1960.	Longmans Green

Also Nos. 3, 5, 6, 7, 8, 12, 13 above.

Chapter 6

Bergmans, J.	*Seeing Colours*.	Philips
Calder, A. B.	*Photometric Methods of Analysis*, 1969.	Hilger
Carter, H. and Donker, M.	*Photoelectric Devices in Theory and Practice*, 1963.	Cleaver-Hume
Cook, B. and Jones, K.	*A Programmed Introduction to Infra-red Spectroscopy*, 1972.	Heyden
Cross, A. D.	*Introduction to Practical Infra-Red Spectroscopy*, 2nd edn., 1964.	Butterworths
Dombrow, M.	*Instrumental Methods in Chemistry, Vol. 2, Optical Methods*, 1967.	Pitman
Dvorak, J., Rubeska, I. and Rezak, Z.	*Flame Photometry, Laboratory Practice*, 1971.	
Flett, M. St. C.	*Physical Aids to the Organic Chemist*, 1962.	Elsevier
Ive, G. A. G.	*Photoelectric Handbook*, 1955.	Newnes

Also Nos. 2, 5, 7, 12, 13, 14 above.

Chapter 7

Ambrose, D.	*Gas Chromatography*, 2nd edn., 1971.	Butterworths
Bloemendal, H.	*Zone Electrophoresis on Blocks and Columns*, 1963.	Elsevier
Jones, R. A.	*An Introduction to Gas-Liquid Chromatography*, 1970.	Academic Press
Lederer, E. and Lederer, M.	*Chromatography*, 1957.	Elsevier
Morris, C. J. O. R. and Morris, P.	*Separation Methods in Biochemistry*, 1964.	Pitman
Simpson, C.	*Gas Chromatography*, 1970.	Kogan Page
Smith, I. (editor)	*Chromatographic and Electrophoretic Techniques* (2 volumes), 1969.	Heinemann

Stahl, K.	*Thin Layer Chromatography*, 1969.	Allen and Unwin
Zweig, G., and Whittaker, J. R.	*Paper Chromatography and Electrophoresis* (2 vols.), Vol. I, 1967, Vol. II, 1971.	Academic Press
—	*Ion Exchange Resins.*	B. D. H. Limited
—	*The Dowex Handbook.*	Dow Chemical Co.
—	*Leaflets on Ion Exchange.*	Permutit Ltd.
—	*Leaflets on Sephadex and Ion Exchange Sephadex.*	Pharmacia
—	*Liquid Chromatography Handbook.*	Phoenix Precision Instrument Co.
—	*Leaflets on Whatman 'Chromedia'.*	Reeve Angel Ltd.

Chapter 8

Not many books are available dealing with automation in the laboratory. Most publications in this field are concerned with the specialized topic of analysis in clinical chemistry. There are regular articles on aspects of the subject in the appropriate journals, especially *Analytical Chemistry* and *Laboratory Practice.* A series of papers presented at symposia organized by Technicon Inc. since 1959 are published by *Mediad Inc.* as *Automation in Analytical Chemistry.* Most of the papers are concerned exclusively with the use of the Auto Analyser.

Broughton, P. M. G. *et al.*	*Automatic Dispensing Pipettes: an assessment of 35 commercial instruments* (1967).	Assoc. of Clinical Biochemists
Robinson, R.	*Clinical Chemistry and Automation* (1971).	Griffin
Squirrell, D. C. M.	*Automatic Methods in Volumetric Analysis* (1964).	Hilger
—	*Annals of the New York Academy of Sciences.* Vol. 87. Art. 2. 609–951 (July 1960).	
Whitehead, T. P. (editor)	*Automation and Data Processing in Pathology* (1969).	B.M.A.

Descriptions of instruments and recorders can be found in

Dummer, G. W. Mackenzie-Robertson, J.	*Medical Electronic Equipment*, 1969–70, 1970.	Pergamon

Vol. II *Monitoring, Recording and Computing Equipment.*
Vol. III *Laboratory Systems and Equipment* (Analytical).
Vol. IV *Laboratory Systems and Equipment* (General).

Chapter 9

Author	Title	Publisher
Spinks, W. S.	*Vacuum Technology*, 1963.	Chapman & Hall
Turnbull, A. H.	*Vacuum Technique for Beginners*, 1952.	HMSO–AERE
Heyrovsky, J. and Zuman, P.	*Practical Polarography*, 1968.	Academic Press
Kolthoff, I. M. and Lingane, J. J.	*Polarography*, vols. 1 and 2, 1952.	Wiley
Aronoff, S.	*Techniques of Radio-biochemistry*, 1965.	Iowa State College Press
Faires, R. A. and Parks, B. H.	*Radioisotope Laboratory Techniques*, 3rd edn., 1973.	Butterworths
Holm, N. W. and Berry, R. J.	*Manual on Radiation Dosimetry*, 1970.	Dekker
Rakovic, M.	*Activation Analysis*, 1970.	Iliffe
—	*Ionising Radiations. Precautions for Industrial Users*, 1961, reprint 1963	Min. of Lab./ HMSO
—	*Code of Practice Against Radiation Hazards.*	Imperial College (London)
Benyon, J. H.	*Mass Spectrometry and its Applications to Organic Chemistry*, Reprinted 1967.	Elsevier
Hill, H. C.	*Introduction to Mass Spectrometry*, 1971.	Heyden

Also Nos. 1, 2, 7, 8 above.

Chapter 10

There are few books published on the topics discussed in this chapter. The most fruitful sources of information are the catalogues and leaflets published by the larger laboratory suppliers. Some topics are mentioned in The Instrument Manual (No. 14).

Chapter 11

Author	Title	Publisher
Cowgill, R. W. and Pardee, A. B.	*Experiments in Biochemical Research Techniques*, 1957.	Chapman & Hall

Eastoe, J. E. and Courts, A.	*Practical Analytical Methods for Connective Tissue Protein*, 1963.	Spon
Henry, R. J.	*Clinical Chemistry*, 1964.	Harper & Row
Wootton, I. D. P.	*Micro-Analysis in Medical Biochemistry*, 5th edn., 1973.	Churchill
Mills, G. T., Munro, H. N., Leaf, G. and Davidson, J. N.	*Practical Biochemistry*, 1958.	John Smith and Son
Varley, H.	*Practical Clinical Biochemistry*, 4th edn., 1967.	Heinemann

Chapter 12

Dixon, M.	*Manometric Methods, as Applied to Measurement of Cell Respiration and other Processes*, 3rd edn., 1951.	Cambridge
Dixon, M. and Webb, E. C.	*Enzymes*, 2nd edn., 1964.	Longmans
Umbreit, W. W., Burris, R. H. and Stauffer, J. F.	*Manometric Techniques.*	Burgess (Minneapolis)

Books on Animal Techniques

Porter, G. and Lane-Petter, W. (editors)	*Notes for Breeders of Common Laboratory Animals*, 1962.	Academic Press (New York)
Lane-Petter, W.	*Provision of Laboratory Animals for Research: A Practical Guide*, 1961.	Elsevier
Worden, A. N. and Lane-Petter, W. (editors)	*The UFAW Handbook on the Care and Management of Laboratory Animals*, 2nd edn., 1957.	Universities Federation for Animal Welfare
Short, D. J. and Woodnott, D. P.	*The A.T.A. Manual of Laboratory Animal Practice and Techniques*, 1963.	Crosby Lockwood

INDEX